Gels: Synthesis, Characterization and Applications in High Performance Chemistry (2nd Edition)

Gels: Synthesis, Characterization and Applications in High Performance Chemistry (2nd Edition)

Guest Editors

Viorel-Puiu Paun
Maria-Alexandra Paun

MDPI

Basel • Beijing • Wuhan • Barcelona • Belgrade • Novi Sad • Cluj • Manchester

Guest Editors
Viorel-Puiu Paun
Department of Physics
University Politehnica
of Bucharest
Bucharest
Romania

Maria-Alexandra Paun
Access to Market and
Cybersecurity
Federal Office of
Communications (OFCOM)
Bienne
Switzerland

Editorial Office
MDPI AG
Grosspeteranlage 5
4052 Basel, Switzerland

This is a reprint of the Special Issue, published open access by the journal *Gels* (ISSN 2310-2861), freely accessible at: https://www.mdpi.com/journal/gels/special_issues/ZNWSXGPQ9I.

For citation purposes, cite each article independently as indicated on the article page online and as indicated below:

Lastname, A.A.; Lastname, B.B. Article Title. *Journal Name* **Year**, *Volume Number*, Page Range.

ISBN 978-3-7258-5429-5 (Hbk)
ISBN 978-3-7258-5430-1 (PDF)
https://doi.org/10.3390/books978-3-7258-5430-1

Contents

About the Editors

Viorel-Puiu Paun

Viorel-Puiu Paun was a full university professor of Physics at Politehnica University of Bucharest, Faculty of Applied Sciences, Physics Department, and a distinguished member of the Academy of Romanian Scientists. A person of great character, he taught several younger generations of the most accomplished students the complex subject of physics, and he did so with great academic passion and personal joy. He has conducted innovative research in several important fields and published extensively through great devotion. The main areas of his scientific interest were nonlinear dynamics and chaos theory, as well as their applications in different physicochemical systems (nanostructures and composites), biological systems, and human organs. Viorel-Puiu Paun published more than 150 papers in national and international journals, 115 ISI-indexed journal papers, 50 communications in national and international meetings (more than 30 on invitation), 577 citations without self-citations (data from 2024), respectively, and 20 books, chapters in books, and monographs. His Hirsch factor value is currently 24. He was a guest/visiting professor at nine prestigious universities in Europe, including the University of Cambridge, UK (2015); Ecole Polytechnique Fédérale de Lausanne (EPFL), Switzerland (1996-1997, 2014); University of Florence, Italy (2011); ENSICA Toulouse, France (2007); and the School of Physical Sciences, Kent University, Canterbury, UK (2004, 2006). Prof. Viorel-Puiu Paun unexpectedly passed away in April 2024, and his profound scientific legacy has left the scientific world with a multitude of innovative ideas and exquisite scientific papers, books, Special Issues, and many other materials to support learning in physics, mathematics, chemistry, and other subjects. Through his accomplished scientific work, he is immortalized in the academic world of science as one of the greatest.

Maria-Alexandra Paun

Born on Pi Day, the same day as Albert Einstein, Maria-Alexandra Paun is currently a Telecommunications Engineer at the Swiss Federal Office of Communication (OFCOM), working in radio equipment conformity and access to market, in addition to being a project leader in the field of drones. She received, with highest honours, an Engineer Diploma in 2008 from Politehnica University of Bucharest, FILS, Romania, with a major in Computer Science, graduating top of class. The Master Project was conducted at the Swiss Federal Institute of Technology (EPFL, Switzerland). During the 2008-2009 period, she was awarded a postgraduate Marie Curie research fellowship at the University of Kent at Canterbury, UK, where she worked in the field of Optical Coherence Tomography (OCT), developing, amongst other things, important computer algorithms for wavefront aberration correction that were used in various non-invasive tissue imaging endeavours. Between 2009 and 2013, she was an Assistant Ph.D. student at EPFL. She received her Ph.D. degree in June 2013 through her studies on the subject of Hall effect sensors. During the 2013-2015 period, she was a Visiting Researcher at the University of Cambridge, UK, with two postdoctoral fellowships from the Swiss National Science Foundation (SNSF). Between 2016 and 2019, she was a Scientific Researcher at EPFL, Switzerland, where she focused on the development of electronics for medical devices (MedTech).

She has published more than 80 ISI-indexed scientific journal papers, and her Hirsch (h-index) is currently 15. An IEEE Senior Member, she is very committed to volunteer work within IEEE, and she was Chair of the IEEE R8 (Europe, Middle East, and Africa) Women in Engineering (WIE) Committee (2021-2022) as well as Chair of the IEEE Women in Engineering (WIE) Affinity Group in Switzerland (2011-2024). She has held many executive positions in IEEE Switzerland ExCom and was Chair of the IEEE Switzerland Section for two mandates from 2017 to 2021.

gels

Editorial

Gels: Synthesis, Characterization and Applications in High Performance Chemistry (2nd Edition)

Viorel-Puiu Paun [1,2,†] and Maria-Alexandra Paun [3,*]

1 Department of Physics, Faculty of Applied Sciences, University Politehnica of Bucharest, 060042 Bucharest, Romania; viorel_paun2006@yahoo.com
2 Academy of Romanian Scientists, 050094 Bucharest, Romania
3 Federal Office of Communications (OFCOM), 2501 Bienne, Switzerland
* Correspondence: maria_paun2003@yahoo.com
† We dedicate this to the memory of the editor, Prof. Viorel-Puiu Paun, who passed away during this special issue period.

This Editorial of the Special Issue proposed and managed by Prof. Viorel-Puiu Paun, Guest Editor, has been partially prepared by his daughter, Dr. Maria-Alexandra Paun, in addition to the valuable notes of her late father. The significant work, started by her father in 2023, was carried on by Dr. Maria-Alexandra Paun, Guest Editor, after the passing of Prof. Viorel-Puiu Paun in April 2024 and until the closing of this Special Issue in October 2024. Most of the papers published in this Special Issue (11 out of 15) had Prof. Viorel-Puiu Paun as Academic Editor. In memoriam of their loving father and husband, the deeply and profoundly saddened family has prepared the following obituary lines to highlight the excellent scientific and humane qualities of Prof. Viorel-Puiu Paun.

The Romanian and international research community have lost a prominent personality in the field of physics and mathematics. Prof. Viorel Puiu Paun, an outstanding university professor from the Politehnica University of Bucharest and a member of the Romanian Academy of Scientists, has passed on to the eternal. He was a remarkable father to his two children, Maria-Alexandra and Vladimir-Alexandru and husband to his wife, Jenica. On behalf of the deeply saddened family in mourning, together with my mother, Mrs. Jenica Paun, and my brother, Dr. Vladimir-Alexandru Paun, we would like to express our eternal gratitude for the most remarkable person he was, the most loving, caring, devoted, and supportive father one could ever have. We love him dearly, and he will always be with us in our memories and in our lives forever!

He dedicated his whole life first to his family and secondly to education, science, and knowledge. He was a full professor in the Physics Department of the Politehnica University of Bucharest, Faculty of Automation and Computers and Faculty of Applied Sciences, where he taught physics at the highest level and guided students in their doctoral theses.

An internationally distinguished personality and prolific scientist, renowned, appreciated, and recognized by people from all over the world, recipient of many awards and distinctions, with an impressive intellect, many challenging subjects were very easy for him as he had a level of knowledge and understanding of science that was beyond any reality; he was in an outstanding and privileged realm of the most impressive minds that ever existed in this world. His extensive education, with two master's degrees in both mathematics and physics, two PhDs in physics and biomedical engineering, and other important training, strongly coupled with a vivacious interest in a multitude of different fields and an agile scientific curiosity, made Prof. Viorel-Puiu Paun an excellent scientist, an accomplished interdisciplinary personality, and an admired academician. In addition, he

https://doi.org/10.3390/gels11050351

had a personal charm that made him very endearing and respected throughout the world, in academic environments of the highest levels.

His most important contributions include more than 200 scientific publications, with more than 150 articles published in reputable ISI-indexed international journals with high impact factors, and a significant number of more than 20 books and course materials. For his published research activity, he received numerous citations and held a high h-index of 24. He had a natural gift and passion for teaching, performing research, and writing scientific papers, which he carried with him until the very last day of his life. He was the first person to teach the Romanian community about the importance of publishing in ISI-indexed journals, as he led several workshops showing people how to publish and share/interpret the metrics associated with publications.

The expression of his utmost devotion to teaching and research was the fact that he continued to teach and publish until the very last moment. He was very active and remained so with numerous scientific publications and Special Issues in progress.

His publishing activity can be found on the following platforms:

(1) ResearchGate: Viorel-Puiu PAUN | Professor (Full) | PhD | University Politehnica of Bucharest, Bucharest | UPB | Department of Physics | Research profile (http://www.researchgate.net/profile/Viorel-Puiu-Paun, accessed on 1 May 2025).
(2) Google Scholar: Viorel-Puiu Paun-Google Academic (https://scholar.google.com/citations?user=59QL-XAAAAAJ&hl=ro, accessed on 1 May 2025).
(3) Web of Science: Clarivate (http://www.webofscience.com/wos/, accessed on 1 May 2025).

God rest him in peace. We will always love him, and we will never forget him. Eternal gratitude and love from his family.

With eternal love,

Jenica, Maria-Alexandra, and Vladimir-Alexandru Paun

The present Editorial showcases the published contents of the Special Issue titled "Gels: Synthesis, Characterization, and Applications in High-Performance Chemistry (2nd Edition)", proposed, written, and managed by Prof. Viorel-Puiu Paun, Guest Editor (and continued by Dr. Maria-Alexandra Paun, after the passing away of Prof. Viorel-Puiu Paun), in 2023–2024, which provided an innovative continuation of the first Special Issue, titled "Gels: Synthesis, Characterization and Applications in High Performance Chemistry", also prepared, written, and managed by Prof. Viorel-Puiu Paun, Guest Editor, published in 2022–2023, which was a great success, with 16 distinct published papers plus an Editorial, signed by Prof. Viorel-Puiu Paun, and published in 2023 as a book, with more than 46,000 views at the time of the preparation of this document.

Organogels, hydrogels, and ionic gels are investigated both theoretically and experimentally. Detailed research is focused on both their synthesis and their applications in high-performance chemistry and its important branches. All the gels mentioned above are characterized from structural and supramolecular points of view via FTIR, NMR, X-ray diffraction, and POM. On this occasion, the proposed keywords, respectively, organogels, hydrogels, ionic gels, chitosan, and fractal analysis, which were considered extremely significant and valuable, were addressed in great technical detail in this innovative scientific study essay. Consequently, all these reference themes, as well as those inherently associated with them, were approached and elegantly synthesized in the Editorial.

Articles that appear in this Special Issue focus on one or more of the topics listed above. At this moment in time, with all scientific works having been published in the journal under the auspices of the Special Issue proposed by the Guest Editor, Prof. Viorel-Puiu, who later became the Editor of this volume, together with Dr. Maria-Alexandra Paun, we can state our reflection that the central topic pertains to the refined analysis of hydrogels, with an emphasis on chitosan. The employment of xerogels for different applications was also investigated through a plethora of scientific means, such as gravimetrical measurements, SEM imaging, and fractal analysis of SEM pictures. The fractal analysis of such gels was pioneered and thoroughly examined by Prof. Viorel-Puiu Paun in this Special Issue and other of his past publications. Images were scrutinized and reviewed by cleverly calculating the fractal dimension and the lacunarity as a quantitative measure of the homogeneity of the material and its texture through their topological analysis in some of the published papers. Moreover, highly interest-worthy aspects using different types of gels in reinforcement, flood prevention, ecological restoration, combustion enhancement, corrosion protection, acoustic propagation, electrochemical storage, synthesis of nanoparticles, gas sensing, ozone detection, the evaluation of coatings, the development of concrete, and obtaining magnetic nanoparticles were considered in the scientific papers included in this Special Issue.

The works edited and included in this Special Issue are the exact ones that appeared in *Gels* in the Special Issue with the same name; more precisely, it includes 15 distinct published papers in 2023–2024, out of which two are reviews, with more than 31,000 views at the time of the preparation of this document, plus an Editorial signed by the Editors of this volume. The papers are presented succinctly in continuation.

The first work presented is titled "Acoustic Fractional Propagation in Terms of Porous Xerogel and Fractal Parameters" by the authors Maria-Alexandra Paun, Vladimir-Alexandru Paun, and Viorel-Puiu Paun. This article portrays solid xerogel-type materials, based on chitosan, TEGylated phenothiazine, and TEG (tri-ethylene glycol), dotted with a large number of pores, that are effectively represented in their constitutive structure. They were assumed to be fractal geometrical entities and adjudged as such. The acoustic fractional propagation equation in a fractal porous medium was successfully applied and solved with the help of Bessel functions. In addition, the fractal character was demonstrated by the produced fractal analysis and has been proven on the evaluated scanning electron microscopy (SEM) pictures of porous xerogel compounds. The fractal parameters (more precisely, the fractal dimension), the lacunarity, and the Hurst index were calculated with great accuracy [1].

The second work presented is titled "Fractal Analysis of Four Xerogels Based on TEGylated Phenothiazine and Chitosan" by the authors Maria-Alexandra Paun, Mihai-Virgil Nichita, Vladimir-Alexandru Paun, and Viorel-Puiu Paun. The present article describes novel massive materials (in the solid phase) based on TEGylated phenothiazine and chitosan that possess great capability to recover mercury ions from constituent aqueous solutions. These were produced by chitosan hydrogelation accompanied by the formyl subsidiary item of TEGylated phenothiazine, attended by lyophilization. The delineation and structure description of the obtained material or supramolecular assembly were realized by FTIR (Fourier transform infrared) spectroscopy, X-ray diffraction, and POM (polarized light optical microscopy). The morphology of their texture was kept under observation by SEM (scanning electron microscopy). The obtained SEM images were evaluated by fractal analysis. The fractal parameters of interest were calculated, including the fractal dimension and lacunarity [2].

The third work presented is titled "Study of the Possibility of Using Sol–Gel Technology to Obtain Magnetic Nanoparticles Based on Transition Metal Ferrites" by the authors Nina Shabelskaya, Sergey Sulima, Elena Sulima, Oleg Medennikov, Marina Kulikova, Tatyana Kolesnikova, and Svetlana Sushkova. The article presents results for the magnetic nanoparticles sol–gel method synthesis of cobalt (II) ferrite and the organic–inorganic composite materials based on it. The obtained materials were characterized using X-ray phase analysis, scanning and transmission electron microscopy, and the Scherrer and Brunauer–

Emmett–Teller (BET) methods. A composite materials formation mechanism was proposed, which includes a gelation stage where transition element cation chelate complexes react with citric acid and subsequently decompose under heating. The fundamental possibility of obtaining an organic–inorganic composite material based on cobalt (II) ferrite and an organic carrier using the presented method has been proven. Composite material formation was established to lead to a significant (5–9 times) increase in the sample surface area. Materials with a developed surface were formed; the surface area measured by the BET method was 83–143 m^2/g. The resulting composite materials had sufficient magnetic properties to be mobile in a magnetic field. Consequently, wide possibilities for polyfunctional materials synthesis open up for various applications in medicine [3].

The fourth work presented is titled "Evaluation of Low-Toxic Hybrid Sol-Gel Coatings with Organic pH-Sensitive Inhibitors for Corrosion Protection of AA2024 Aluminium Alloy" by the authors Eva Jaldo Serrano, Jesús López-Sánchez, Federico García-Galván, Aida Serrano, Óscar Rodríguez de la Fuente, Violeta Barranco, Juan Carlos Galván, and Noemí Carmona. Today's environmental needs require reductions in the weight of vehicles, thus reducing fuel consumption and associated emissions. For this reason, the use of light alloys is being studied, which, due to their reactivity, must be protected before use. In this work, the effectiveness of a hybrid sol–gel coating doped with various organic, environmentally friendly corrosion inhibitors applied to an AA2024 lightweight aluminum alloy was evaluated. Some of the inhibitors tested included pH indicators, acting as both corrosion inhibitors and optical sensors for the surface of the alloy. Samples are subjected to a corrosion test in a simulated saline environment and characterized before and after the test. The experimental results regarding their best inhibitor performance for their potential application in the transport industry were evaluated [4].

The fifth work presented is titled "Chemiresistors with In_2O_3 Nanostructured Sensitive Films Used for Ozone Detection at Room Temperature" by the authors Mariana Chelu, Paul Chesler, Cristian Hornoiu, Mihai Anastasescu, Jose Maria Calderon-Moreno, Daiana Mitrea, Costin Brasoveanu, Carmen Moldovan, and Mariuca Gartner. The detection of greenhouse gases is essential because harmful gases in the air diffuse rapidly over large areas in a relatively short period of time, causing air pollution that contributes to climate change with catastrophic consequences over time. Among the materials with favorable morphologies for gas detection (nanofibers, nanorods, and nanosheets), large specific surfaces, high sensitivity, and low production costs, nanostructured porous films of In_2O_3 obtained by the sol–gel method, deposited on alumina transducers, with gold (Au) interdigitated electrodes (IDE) and platinum (Pt) heating circuits, were selected. Sensitive films contained 10 deposited layers, involving intermediate and final thermal treatments to stabilize the sensitive film. The fabricated sensor was characterized using AFM, SEM, EDX, and XRD. The film morphology was complex, containing fibrillar formations and some quasi-spherical conglomerates. The deposited sensitive films were rough, thus favoring gas adsorption. Ozone sensing tests were performed at different temperatures. The highest response of the ozone sensor was recorded at room temperature, considered to be the working temperature for this specific sensor [5].

The sixth work (a review) is titled "An Elucidative Review of the Nanomaterial Effect on the Durability and Calcium-Silicate-Hydrate (C-S-H) Gel Development of Concrete" by the authors Farqad Yousuf Al-saffar, Leong Sing Wong, and Suvash Chandra Paul. Concrete as a building material is susceptible to degradation by environmental threats such as thermal diffusion, acid and sulphate infiltration, and chloride penetration. Hence, the inclusion of nanomaterials in concrete has a positive effect in terms of promoting its mechanical strength and durability performance, as well as resulting in energy savings due to reduced cement consumption in concrete production. This review article discusses the novel advances in research regarding C-S-H gel promotion and concrete durability improvement using nanomaterials. In summary, this review deals with topics relevant to the influence of nanomaterials on concrete's resistance to heat, acid, sulphate, chlorides, and wear deterioration, as well as the impact on concrete microstructure and chemical bonding. The significance of this review is a critical discussion on the cementation mechanism of nanoparticles in enhancing durability properties owing to their nanofiller effect, pozzolanic reactivity, and nucleation effect. The utilization of nanoparticles enhanced the hydrolysis of cement, leading to a rise in the production of C-S-H gel. Consequently, this improvement in concrete microstructure led to a reduction in the number of capillary pores and pore connectivity, thereby improving the concrete's water resistance. Microstructural and chemical evidence obtained using SEM and XRD indicated that nanomaterials facilitated the formation of cement gel either by reacting pozzolanically with portlandite to generate more C-S-H gel or by functioning as nucleation sites. Due to an increased rate of C-S-H gel formation, concrete enhanced with nanoparticles exhibited greater durability against heat damage, external attack by acids and sulphates, chloride diffusion, and surface abrasion. The durability improvement following nanomaterial incorporation into concrete can be summarized as enhanced residual mechanical strength, reduced concrete mass loss, reduced coefficients for thermal and chloride diffusion, improved performance against sulphates and acid attack, and increased surface resistance to abrasion [6].

The seventh work presented is titled "Layered Sol–Gel Deposition of a Sn, Ti, Zn, and Pr Mixed Oxide Thin Film with Electrical Properties for Gas Sensing" by the authors Izabella Dascalu, Cristian Hornoiu, Jose Maria Calderon Moreno, Petre Osiceanu, and Simona Somacescu. This article presents a layered mixed oxide thin film composed of Sn, Ti, Zn, and Pr obtained by sol–gel deposition for gas sensing applications. The film was characterized by X-ray diffraction (XRD), X-ray photoelectron spectroscopy (XPS), Raman spectroscopy, UV–Vis spectroscopy, scanning electron microscopy with energy dispersive X-ray spectroscopy (SEM-EDX), and electrochemical impedance spectroscopy (EIS). X-ray diffraction results showed the presence of a single crystalline phase with a cassiterite-like structure. Raman spectroscopy revealed characteristic bands of oxygen-deficient SnO_2-based nanocrystallites. The band gap energy calculated from UV–Vis spectroscopy was Eg = 3.83 eV. The XPS proved their presence on the surface of all elements introduced by the inorganic precursors, as well as their oxidation states. Thus, Sn^{4+}, Ti^{4+}, Zn^{2+}, and Pr^{3+} were detected on the surface. Moreover, by XPS, we highlighted the presence of OH groups and water adsorbed on the surface. SEM showed the five-layer morphology of the film after five successive depositions. Electrochemical properties were determined by EIS-impedance spectroscopy. The selectivity for gas sensing was also investigated for methane, propane, and formaldehyde, and the gas sensing mechanism was explained. The results indicated that the mixed oxide thin film exhibited high sensitivity and selectivity towards specific gases [7].

The eighth work presented is titled "A SEM-EDX Study on the Structure of Phenyl Phosphinic Hybrids Containing Boron and Zirconium" by the authors Petru Merghes, Narcis Varan, Gheorghe Ilia, Iosif Hulka, and Vasile Simulescu. The SEM-EDX method was used to investigate the structure and morphology of organic–inorganic hybrids containing zirconium, boron, and phosphorus compounds synthesized by the sol–gel method. We started by using, for the first time together, zirconyl chloride hexa-hydrate ($ZrOCl_2 \cdot 6H_2O$), phenyl phosphinic acid, and triethyl borate as precursors and reagents at different molar ratios. The obtained hybrids showed a very high thermal stability and were not soluble in water or in organic solvents. As a consequence, such hybrid solid materials are suitable for applications at high temperatures. The obtained hybrids had complex 3D structures and formed organic–inorganic networks containing Zr-O-Zr, Zr-O-P, and Zr-O-B bridges.

Such organic–inorganic networks are also expected to form supramolecular structures and to have many potential applications in different fields of great interest, such as catalysis, medicine, agriculture, energy storage, fuel cells, sensors, electrochemical devices, and supramolecular chemistry [8].

The ninth work presented is titled "Synthesis of Highly Luminescent Silica-Coated Upconversion Nanoparticles from Lanthanide Oxides or Nitrates Using Co-Precipitation and Sol–Gel Methods" by the authors Ana Iglesias-Mejuto, Alyne Lamy-Mendes, João Pina, Benilde F. O. Costa, Carlos A. García-González, and Luisa Durães. Upconversion nanoparticles (UCNPs) are under consideration for their use as bioimaging probes with enhanced optical performance for real-time follow-up under non-invasive conditions. Photostable and core-shell $NaYF_4{:}Yb^{3+}$, Er^{3+}-SiO_2 UCNPs obtained by a novel and simple co-precipitation method from lanthanide nitrates or oxides were herein synthesized for the first time. The sol–gel Stöber method followed by oven or supercritical gel drying was used to confer biocompatible surface properties to UCNPs by the formation of an ultrathin silica coating. Upconversion (UC) spectra were studied to evaluate the fluorescence of UCNPs upon red/near-infrared (NIR) irradiation. ζ-potential measurements, TEM analyses, XRD patterns, and long-term physicochemical stability were also assessed and confirmed that the UCNPs co-precipitation synthesis is a shape- and phase-controlling approach. The bio- and hemocompatibility of the UCNPs formulation with the highest fluorescence intensity was evaluated with murine fibroblasts and human blood, respectively, and provided excellent results that endorse the efficacy of the silica gel coating. The herein synthesized UCNPs can be regarded as efficient fluorescent probes for bioimaging purposes with the high luminescence, physicochemical stability, and biocompatibility required for biomedical applications [9].

The tenth work presented is titled "Electrochemical Storage Behavior of a High-Capacity Mg-Doped P2-Type $Na_{2/3}Fe_{1-y}Mn_yO_2$ Cathode Material Synthesized by a Sol–Gel Method" by the authors Mobinul Islam, Md. Shahriar Ahmed, Daseul Han, Gazi A. K. M. Rafiqul Bari, and Kyung-Wan Nam. Grid-scale energy storage applications can benefit from rechargeable sodium-ion batteries. As a potential material for making non-cobalt, nickel-free, cost-effective cathodes, Earth-abundant $Na_{2/3}Fe_{1/2}Mn_{1/2}O_2$ is of particular interest. However, Mn^{3+} ions are particularly susceptible to the Jahn–Teller effect, which can lead to an unstable structure and continuous capacity degradation. Modifying the crystal structure by aliovalent doping is considered an effective strategy to alleviate the Jahn–Teller effect. Using a sol–gel synthesis route followed by heat treatment, we succeeded in preparing an Mg-doped $Na_{2/3}Fe_{1-y}Mn_yO_2$ cathode. Its electrochemical properties and charge compensation mechanism were then studied using synchrotron-based X-ray absorption spectroscopy and in situ X-ray diffraction techniques. The results revealed that Mg doping reduced the number of Mn^{3+} Jahn–Teller centers and alleviated high-voltage phase transition. However, Mg doping was unable to suppress the P2–P′2 phase transition at a low-voltage discharge. An initial discharge capacity of about 196 mAh g^{-1} was obtained at a current density of 20 mAh g^{-1}, and 60% of rate capability was maintained at a current density of 200 mAh g^{-1} in a voltage range of 1.5–4.3 V. This study will greatly contribute to the ongoing search for advanced and efficient cathodes from Earth-abundant elements for rechargeable sodium-ion batteries operable at room temperature [10].

The eleventh work presented is titled "Combustion Enhancement of Gel Propellant Containing High Concentration Aluminum Particles Based on Carbon Synergistic Effect" by the authors Jiyuan Chen, Hui Zhao, Weifeng Li, and Haifeng Liu. The addition of aluminum particles to gel propellants can improve combustion performance. However, the agglomeration of aluminum during the combustion process can result in a series of negative effects. In this paper, the aluminum agglomeration inhibition method of gel propellant based on the carbon synergistic effect is proposed. Carbon particles exhibit excellent combustion properties, and the gaseous product CO_2 generated during combustion can mitigate the agglomeration of aluminum. The research demonstrates that incorporating carbon particles into an aluminum-containing gel effectively reduces the incomplete combustion of the aluminum particles and increases the volumetric calorific value of the gel. When the mass fraction of carbon is 5 wt.%, the volume calorific value of the gel reaches its highest. Meanwhile, the rheological experiments show that the addition of carbon particles can improve the shear-thinning properties of the gel, which is beneficial to the atomization and combustion processes of the gel [11].

The twelfth work presented is titled "Eco-Friendly Sol–Gel Coatings with Organic Corrosion Inhibitors for Lightweight AZ61 Alloy" by the authors Jorge Domínguez-Martínez, Jesús López-Sánchez, Federico García-Galván, Aída Serrano, Violeta Barranco, Juan Carlos Galván, Óscar Rodríguez de la Fuente, and Noemí Carmona. The latest advances in technology and materials science have catalyzed a transformative shift towards the adoption of environmentally conscious and lightweight materials across key sectors, such as the aeronautics, biomedical, and automotive industries. Noteworthy among these innovations are the magnesium–aluminum (Mg-Al) alloys employed in aeronautical applications, contributing to the overall reduction in aircraft weight and subsequently diminishing fuel consumption and mitigating atmospheric emissions. The present work delves into a study of the anti-corrosive properties inherent in various sol–gel coatings, leveraging a range of environmentally friendly corrosion inhibitors specifically tailored for samples of the AZ61 alloy. Methodologically, the work involves the synthesis and application of sol–gel coatings on AZ61 alloy containing eco-friendly inhibitors: L-cysteine, *N*-acetyl-cysteine, curcumin, and methylene blue. Subsequently, an accelerated corrosion test in a simulated saline environment was performed. Through microstructural and compositional analyses, the best inhibitor responses were achieved with inhibitors containing S and N heteroatoms and conjugated double bonds in their structure, probably due to the creation of a continuous $MgCl_2$ layer. This research contributes to the ongoing discourse on protective eco-coatings, aligning with the broader paradigm shift towards sustainable and lightweight materials in key industries [12].

The thirteenth work (a review) presented is titled "Green and Low-Cost Modified Pisha Sandstone Geopolymer Gel Materials for Ecological Restoration: A Phase Review" by the authors Changming Li, Yubing Fu, Haifeng Cheng, Yaozong Wang, Dongyang Jia, and Hui Liu. Pisha sandstone (PS) is a special interbedded rock in the middle reaches of the Yellow River that experiences severe weathering and is loose and broken. Due to severe multiple erosion events, the Pisha sandstone region is called "the most severe water loss and soil erosion in the world" and "the ecological cancer of the earth". As a special pozzolanic mineral, PS has the potential to be used as a precursor for the synthesis of green and low-carbon geopolymer gel materials and applied in ecological restoration. This paper aimed to undertake a phase review of the precursors for geopolymer gel materials. The genesis and distribution, physical and chemical characterization, erosion characteristics, and advances in the ecological restoration of PS are all summarized. Furthermore, current advances in the use of PS for the synthesis of geopolymer gel materials in terms of mechanical properties and durability are discussed. The production of Pisha sandstone geopolymer gels through the binder jetting technique and 3D printing techniques is prospected. Meanwhile, the prospects for the resource application of PS in mine rehabilitation and sustainable ecology are discussed. In the future, multifactor-driven comprehensive measures should be further investigated in order to achieve ecological restoration in the Pisha sandstone region and promote high-quality development in the Yellow River Basin [13].

The fourteenth work presented is titled "Development of Environmentally Friendly and Economical Flood-Prevention Stones Based on the Sediments of the Yellow River", by the authors Ying Liu, Hao Xiao, Yongxiang Jia, Yajun Lv, Li Dai, and Chen Yang. The deposition of Yellow River sediment in the middle and lower reaches is a significant factor in the siltation of reservoirs and the occurrence of serious flooding along the river. The efficient and valuable utilization of Yellow River sediment has already become a key research topic in this field. In this study, Yellow River sediment was employed as the primary material, in conjunction with commercially available slag, fly ash, and quicklime as the binder, to develop a novel type of artificial flood-prevention stone. Following a 28-day standard curing procedure, the highest compressive strength of the prepared artificial stone was recorded at 4.29 MPa, with a value exceeding 0.7 MPa under wet conditions. The results demonstrated that the prepared artificial stone met the specifications for artificial flood-prevention stones. The curing mechanism, as evidenced by analyses from SEM and XRD testing, indicated that the alkali excitation process in the binder, which produced C-A-S-H gel, was the key factor in enhancing the compressive strength of the specimens. Notably, an evaluation of the amount of CO_2 emissions and the cost of the artificial stone concluded that the preparation process was both environmentally friendly and cost-effective [14].

The fifteenth and last work presented is titled "Sol–Gel Derived Alumina Particles for the Reinforcement of Copper Films on Brass Substrates", by the authors Samah Sasi Maoloud Mohamed, Marija M. Vuksanović, Dana G. Vasiljević-Radović, Ljiljana Janković Mandić, Radmila M. Jančić Heinneman, Aleksandar D. Marinković, and Ivana O. Mladenović. The aim of this study was to provide tailored alumina particles suitable for reinforcing the metal matrix film. The sol–gel method was chosen to prepare submicron-sized particles and to control crystal structure by calcination. In this study, copper-based metal matrix composite (MMC) films were developed on brass substrates with different electrodeposition times and alumina concentrations. Scanning electron microscopy (FE-SEM), with energy-dispersive spectroscopy (EDS), TEM, and X-ray diffraction (XRD), was used to characterize the reinforcing phase. The MMC Cu-Al_2O_3 films were synthesized electrochemically using the co-electrodeposition method. Microstructural and topographical analyses of pure (alumina-free) Cu films and the Cu films with incorporated Al_2O_3 particles were performed using FE-SEM/EDS and AFM, respectively. Hardness and adhesion resistance were investigated using the Vickers microindentation test and evaluated by applying the Chen–Gao (C-G) mathematical model. The sessile drop method was used for measuring contact angles for water. The microhardness and adhesion of the MMC Cu-Al_2O_3 films were improved when Al_2O_3 was added. The concentration of alumina particles in the electrolyte correlates with an increase in absolute film hardness in the way that 1.0 wt.% of alumina in electrolytes results in a 9.96% increase compared to the pure copper film, and the improvement is maximal in the film obtained from electrolytes containing 3.0 wt.% alumina, giving the film 2.128 GPa, a 134% hardness value similar to that of the pure copper film. The surface roughness of the MMC film increased from 2.8 to 6.9 times compared to the Cu film without particles. The decrease in the water contact angle of Cu films with incorporated alumina particles relative to the pure Cu films was from 84.94° to 58.78° [15].

Finally, we would like to express our deep appreciation and gratitude that the authors of the papers/chapters of this book, fifteen works in total, who undertook the efforts of highlighting the remarkable applications of gels in pharmaceutical uses, industry-related endeavors, ecology, chemistry, and imaging and sensing techniques aiming to facilitate and ameliorate the quality of life for people in our modern society.

Conflicts of Interest: The authors declare no conflicts of interest.

References

1. Paun, M.; Paun, V.; Paun, V. Acoustic Fractional Propagation in Terms of Porous Xerogel and Fractal Parameters. *Gels* **2024**, *10*, 83. [CrossRef] [PubMed]
2. Paun, M.; Nichita, M.; Paun, V.; Paun, V. Fractal Analysis of Four Xerogels Based on TEGylated Phenothiazine and Chitosan. *Gels* **2023**, *9*, 435. [CrossRef] [PubMed]
3. Shabelskaya, N.; Sulima, S.; Sulima, E.; Medennikov, O.; Kulikova, M.; Kolesnikova, T.; Sushkova, S. Study of the Possibility of Using Sol–Gel Technology to Obtain Magnetic Nanoparticles Based on Transition Metal Ferrites. *Gels* **2023**, *9*, 217. [CrossRef] [PubMed]
4. Jaldo Serrano, E.; López-Sánchez, J.; García-Galván, F.; Serrano, A.; Rodríguez de la Fuente, Ó.; Barranco, V.; Galván, J.; Carmona, N. Evaluation of Low-Toxic Hybrid Sol-Gel Coatings with Organic pH-Sensitive Inhibitors for Corrosion Protection of AA2024 Aluminium Alloy. *Gels* **2023**, *9*, 294. [CrossRef] [PubMed]
5. Chelu, M.; Chesler, P.; Hornoiu, C.; Anastasescu, M.; Calderon-Moreno, J.; Mitrea, D.; Brasoveanu, C.; Moldovan, C.; Gartner, M. Chemiresistors with In_2O_3 Nanostructured Sensitive Films Used for Ozone Detection at Room Temperature. *Gels* **2023**, *9*, 355. [CrossRef] [PubMed]
6. Al-saffar, F.; Wong, L.; Paul, S. An Elucidative Review of the Nanomaterial Effect on the Durability and Calcium-Silicate-Hydrate (C-S-H) Gel Development of Concrete. *Gels* **2023**, *9*, 613. [CrossRef] [PubMed]
7. Dascalu, I.; Hornoiu, C.; Calderon Moreno, J.; Osiceanu, P.; Somacescu, S. Layered Sol–Gel Deposition of a Sn, Ti, Zn, and Pr Mixed Oxide Thin Film with Electrical Properties for Gas Sensing. *Gels* **2023**, *9*, 638. [CrossRef] [PubMed]
8. Merghes, P.; Varan, N.; Ilia, G.; Hulka, I.; Simulescu, V. A SEM-EDX Study on the Structure of Phenyl Phosphinic Hybrids Containing Boron and Zirconium. *Gels* **2023**, *9*, 706. [CrossRef] [PubMed]
9. Iglesias-Mejuto, A.; Lamy-Mendes, A.; Pina, J.; Costa, B.; García-González, C.; Durães, L. Synthesis of Highly Luminescent Silica-Coated Upconversion Nanoparticles from Lanthanide Oxides or Nitrates Using Co-Precipitation and Sol–Gel Methods. *Gels* **2024**, *10*, 13. [CrossRef] [PubMed]
10. Islam, M.; Ahmed, M.; Han, D.; Bari, G.; Nam, K. Electrochemical Storage Behavior of a High-Capacity Mg-Doped P2-Type Na2/3Fe1−yMnyO_2 Cathode Material Synthesized by a Sol–Gel Method. *Gels* **2024**, *10*, 24. [CrossRef] [PubMed]
11. Chen, J.; Zhao, H.; Li, W.; Liu, H. Combustion Enhancement of Gel Propellant Containing High Concentration Aluminum Particles Based on Carbon Synergistic Effect. *Gels* **2024**, *10*, 89. [CrossRef] [PubMed]
12. Domínguez-Martínez, J.; López-Sánchez, J.; García-Galván, F.; Serrano, A.; Barranco, V.; Galván, J.; Rodríguez de la Fuente, Ó.; Carmona, N. Eco-Friendly Sol–Gel Coatings with Organic Corrosion Inhibitors for Lightweight AZ61 Alloy. *Gels* **2024**, *10*, 168. [CrossRef] [PubMed]
13. Li, C.; Fu, Y.; Cheng, H.; Wang, Y.; Jia, D.; Liu, H. Green and Low-Cost Modified Pisha Sandstone Geopolymer Gel Materials for Ecological Restoration: A Phase Review. *Gels* **2024**, *10*, 302. [CrossRef] [PubMed]
14. Liu, Y.; Xiao, H.; Jia, Y.; Lv, Y.; Dai, L.; Yang, C. Development of Environmentally Friendly and Economical Flood-Prevention Stones Based on the Sediments of the Yellow River. *Gels* **2024**, *10*, 622. [CrossRef] [PubMed]
15. Mohamed, S.; Vuksanović, M.; Vasiljević-Radović, D.; Janković Mandić, L.; Jančić Heinneman, R.; Marinković, A.; Mladenović, I. Sol–Gel Derived Alumina Particles for the Reinforcement of Copper Films on Brass Substrates. *Gels* **2024**, *10*, 648. [CrossRef] [PubMed]

gels

Article

Acoustic Fractional Propagation in Terms of Porous Xerogel and Fractal Parameters

Maria-Alexandra Paun [1,*,†], Vladimir-Alexandru Paun [2,†] and Viorel-Puiu Paun [3,4,†]

1 Division Radio Monitoring and Equipment, Section Market Access and Conformity, Federal Office of Communications (OFCOM), 2501 Bienne, Switzerland
2 Five Rescue Research Laboratory, 75004 Paris, France; vladimir.alexandru.paun@ieee.org
3 Physics Department, Faculty of Applied Sciences, University Politehnica of Bucharest, 060042 Bucharest, Romania; viorel.paun@physics.pub.ro
4 Academy of Romanian Scientists, 50085 Bucharest, Romania
* Correspondence: maria_paun2003@yahoo.com
† These authors contributed equally to this work.

Abstract: This article portrays solid xerogel-type materials, based on chitosan, TEGylated phenothiazine, and TEG (tri-ethylene glycol), dotted with a large number of pores, that are effectively represented in their constitutive structure. They were assumed to be fractal geometrical entities and adjudged as such. The acoustic fractional propagation equation in a fractal porous media was successfully applied and solved with the help of Bessel functions. In addition, the fractal character was demonstrated by the produced fractal analysis, and it has been proven on the evaluated scanning electron microscopy (SEM) pictures of porous xerogel compounds. The fractal parameters (more precisely, the fractal dimension), the lacunarity, and the Hurst index were calculated with great accuracy.

Keywords: acoustic equation; fractional propagation; porous xerogel; SEM images; fractal parameters

1. Introduction

Porous substances comprise countless products made by human hands as well as genuine materials. In reality, all solid materials are porous to a certain degree, or on a certain measurement scale. However, the only famous genuine materials in preponderance that have a natural porosity high enough to have notable quantities of liquids or gases pass through them are soil/ground and rocks.

Porosity is defined as the ratio between the volume of voids amidst the solid particles of a material (solid substance) in its natural state and the total volume of the material. It is denoted by p and expressed as a decimal fraction or percentage. The current values of the porosity coefficient are generally distributed for some better-known materials such as the porosity of rocks in the natural layer. An example of this is between limestone or shale with a value of 0.10 or 0.01 (a minimum on the scale in terms of the nature of the layer) and clay with a value of 0.45–0.55 (the maximum), within the limits of the respective type of materials. The porosity of a material, being an intrinsic quantity, depends on the (effective) size of the granules, on the arrangement of the granules, on the non-uniformity coefficient, and on the shape of the granules [1–3].

Xerogel is a solid colloidal material obtained by drying a gel (e.g., gelatin or silica gel) at low temperatures (not higher than 100 °C), which is endowed with a consistent final structural shrinkage. Usually, a xerogel maintains a high porosity (which can reach 50%) and a huge specific surface area (150–900 m^2/g), with very small pores (1–10 nm). On the other hand, when the liquid is removed under supercritical conditions, the structure does not tend to shrink and an aerogel is obtained, as a highly porous material (90%) with an extremely low density. We can also say that the sol–gel process is a method used to produce xerogels. By heating a xerogel at high temperatures, sintering of the porous gel occurs and ultimately becomes a dense glass.

The present paper provides information regarding how it has finally been possible to show that gas flow and liquid passage (and different other basic transport properties) also crossed other porous media such as those in the category of xerogels. The current presentation can optionally follow several theoretical paths. However, we will opt for only one of the ways, where the power–law distributions more correctly depicts the pore space and such distributions are congruent with the self-similarity connected with fractal or multi-fractal patterns [1–3]. The so-called fractal environment is an environment that benefits from the self-similarity quality regardless of the scale at which it is represented. However, numerous phenomena of real-world manifest restricted statistical or fractal properties [2]. But, it is not obligatory that these fractal objects have the same organization, no matter what the scale; they can be nearly self-similar, or they can be self-similar only for discrete values of the scale that appertain to a straight-line segment $[r_{min}, r_{max}]$, where r_{min} (r_{max}) is the smallest (the largest) scale value present in the considered geometrical object [3].

In this article, the advanced continuum-type mechanics of porous media with a generally anisotropic, mathematical-product-like fractal geometry are considered. Relying on a fractal derivative, the approach leads to global balance laws in terms of fractal functions and their derivatives. This is based on product measures and then converting them to integer-order differential equations in conventional (Euclidean) space. The theoretical strategy, in the last instance, follows the approach to continuum-type mechanics of heterogeneous porous media of the fractal type. The final objective is to develop a continuum physics-type framework for the diffusion of liquids and gases in elastic–inelastic deformable solid porous materials. More precisely, the reasonable association of this type of behavior with an acoustic fractional propagation equation was successfully achieved.

Another technique for solving fractional differential equations is that based on the introduction and implementation of an optimal auxiliary function method to solve a system of fractional-order Whitham–Broer–Kaup equations; these are a class of nonlinear partial differential equations with broad applications in mathematical physics. This method provides a systematic and efficient approach to finding accurate solutions for complex systems of fractional-order equations [4–6].

The fractal environment can be fashioned as mensurable metric assortments, each having non-integer dimensions [7–11]. The non-integer Hausdorff dimension [7] is commonplace in fractal theory and also a major property. The isotropic fractal material is distinguished by the mass, M, relation of a ball region, V, of fractal material, $M = M(V)$, which is function of M_0, R, and R_0. This is basically the most general connection between them. More precisely:

$$M(V) = M_0\left(\frac{R}{R_0}\right)^D \tag{1}$$

where R_0 characterizes the associated size of the fractal environments [7,8] and D is the mass measure of the fractal environment.

The anisotropy property of fractal materials can be taken into consideration utilizing a power–law formula, for the mass of a parallelepiped section Vp of the studied object, in the present mathematical form:

$$M(V_p) = M_0\left(\frac{L_x}{R_0}\right)^{\alpha_1}\left(\frac{L_y}{R_0}\right)^{\alpha_2}\left(\frac{L_z}{R_0}\right)^{\alpha_3}, \text{with } min\{L_x,\ L_y, L_z\} \gg R_0 \tag{2}$$

where α_i are non-integer values for the mathematical dimension on the coordinate axes X_i and $i = 1,2,3$ are counting pointers. In this context L_x, L_y, and L_z are the three edges meeting at the same vortex. The parameter α_i delineates/relates to the expansion in the mass of this considered environment when one of parallelepiped dimensions is enlarged along one axis, and the other dimensions of the parallelepiped along the other axes are not modified. Then, the expression $D = \alpha_1 + \alpha_2 + \alpha_3$ represents the fractal mass size of the anisotropic fractal environment. Thus, we can now define fractal material as an environment that has a non-integer mass dimension, according to Equations (1) and (2). Though the non-integer size does not completely depict all of the physics aspects of the fractal environment, it is considered a relevant property of fractal material [8].

This paper includes a consistent introduction in Section 1. This introduction presents xerogels and their particular porosity, together with their fractal behavior. Section 2 is titled 2. Results and Discussion. In Section 2.1, fractal parameters including the fractal dimension, the lacunarity concept, and the Hurst exponent are introduced, in order to be able to understand the notions of the particular morphology of the porous xerogels considered in continuation. In Section 2.2, the fractional differential equation of acoustic wave propagation in a fractal porous media is established and solved in relation to the dynamics of gas movement through xerogel porous media. Circulation is primarily due to the presence of pores in the material and its fractal nature. Section 2.3 includes the results and discussion, referring to the results obtained by studying the structure of the SEM-investigated samples. In addition, it presents the calculated values of the associated fractal parameters. Section 3 is dedicated to presenting the conclusions of this article. Finally, Section 4 details the materials and methods, with the xerogel materials and synthesis in Section 4.1, the xerogel characterization in Section 4.2, and the morphology in Section 4.3.

2. Results and Discussion

2.1. Fractal Parameters

In terms of fractal assumption of natural reality, the fractal parameters are the most important theoretical components. Of these, the most widely used are the fractal dimension, the lacunarity, the succolarity concept, and the Hurst exponent, being fluently utilized to permit access to the structural circumstances of porous-material intricate assessment in reciprocation [7–11]. However, we now specify that succolarity is less necessary for evaluating the fractal aspect of the behavior of porous materials. This was the reason for not using it, as it was practically useless on this occasion. The realistic application and correct valuation of the three aforementioned necessary parameters are arduous, due to the complicated definitions and laborious procedures of calculus [9]. In the following section, we introduce the accredited definitions, the substantial significance of the fractal characteristics, and the evaluation procedure based on the box-counting method from images in principle, but on alternative practices too.

2.1.1. Fractal Dimension

A fractal object can be defined taking into account the fact that its fragments have an identical/self-same look with the entirety, that is, they are geometrically similar even if subsequent simple modifications are applied. In addition, if the fractal dimension is made up of a certain number of duplicates, each transformed to a determined scale and eventually slightly remodeled, this will be named a self-similar object. The self-similarity quality is really important and nearly all natural objects are capable of being self-similar. Projective transformations, which are defined as combinations of translating, and rotations of space, together with scaling operations, can be used to express the changes that the entire object passes through to get into the new shape. Hereafter, we will refer to the most familiar method of calculating the fractal dimension as the box-counting method.

To calculate the fractal dimension by the box-counting dimension, the considered figures must be segregated into quadrats (2D) or cubes (3D), with a side/edge length of ε (1D). To completely comprise the fractal geometrical object, only the same type of cassettes are utilized. These boxes, which overlap the measured initial geometrical object, will have a supposed N complete number.

Note: The total length value of the measured curve undergoes an essential conversion according to each ruler of scale used, and the maximum number of boxes differs consistently, as the size of each box undergoes modifications.

The meaning of the box-counting dimension, Db, is the ratio limit of $logN(\varepsilon)$ and $log(1/\varepsilon)$, in the situation in which ε (the length) drives/tends to zero, such as in the expression:

$$Db = \lim_{\varepsilon \to 0} \frac{logN(\varepsilon)}{log(1/\varepsilon)} = -\lim_{\varepsilon \to 0} \frac{logN(\varepsilon)}{log\varepsilon} \quad (3)$$

Conformable to the box (cassettes) numeration technique and viewed as an easily accessible global measurement, the box-counting additive routine [6,9] can be assumed as being precious both for simulated fractals as well as genuine fractals, indifferent of their degree of complexity.

2.1.2. Lacunarity

A good lacunarity definition is the one that determines with certainty, first of all, the gap size, or the pore distribution in the considered corpus. Furthermore, it takes into consideration the volume of real holes and apportions them without physical content matter, in comparison to the total accessible volume [8]. If the geometrical body also has major holes or defects, (in the sense of being lacunar), the more likely its voids are to be larger, due to the included special open formal figures, such as disks (2D) or spheres (3D). Thus, for a fractal object, the more numerous the defects or holes (i.e., presenting more lacuna), the bigger the lacunarity is considered to be. On the other hand, lacunarity evaluated as small demonstrates a superior homogeneity value in terms of the assessed images [7–9].

Equation (4) $\Lambda(\varepsilon)$ is the lacunarity value and ε is the box dimension dataset [7,9].

$$\Lambda(\varepsilon) = \frac{Z^{(2)}}{\left(Z^{(1)}\right)^2} \quad (4)$$

$Z^{(1)}$ and $Z^{(2)}$ represent the first and second moments, respectively:

$$Z^{(1)} = \sum_{\varepsilon} P \bullet Q(P,\varepsilon) = \sum_{\varepsilon} P \bullet \frac{n(P,\varepsilon)}{(M-\varepsilon+1)^2} \quad (5)$$

and

$$Z^{(2)} = \sum_{\varepsilon} P^2 \bullet Q(P,\varepsilon) = \sum_{\varepsilon} P^2 \bullet \frac{n(P,\varepsilon)}{(M-\varepsilon+1)^2} \quad (6)$$

In Equations (5) and (6), M is the map dimension and P is equal to the box mass. The $n(P,\varepsilon)$ is the box number that contains P pixels, and $Q(P,\varepsilon)$ is the statistical probability.

2.1.3. Hurst Exponent Evaluation

The Hurst index (exponent), H, depicts the motion roughness. The smaller the value, the greater the roughness and vice versa. When the H index is equal to 0.5, the movement is declared to be a Brownian process, and when it is much smaller than 0.5, there is a growing incremental negational correlation (very craggy movement). When it is greater than 0.5, a positive correlation of the incrementals occurs (flattened movement). The fractal dimension (D) and the Hurst exponent (H) of the fractal identical object are linearly related by the formula $D = 2 - H$, where $0 \leq H \leq 1$ and $1 \leq D \leq 2$. The Hurst exponent can be immediately calculated with the expression $H = 2 - D$, using the fractal dimension computed with the box-counting method [10,11].

2.1.4. Implications of the Calculated Fractal Parameters on the Properties or Behaviors of the Xerogel-Type Materials

The fractal dimension quantifies the change in the mass of solid as a function of the observation scale. In addition, the fractal dimensions exhibit the evaluated surface roughness degree. It is known that the strength of xerogels is increased by adding additives, and the shrinkage is also restrained in the course of drying. The calculated fractal dimensions and the thermodynamic correlations have almost the same tendencies. Consequently, the increase in strength with the application of additives and the restriction of shrinkage with the drying process are precisely reflected in the fractal dimension values. Moreover, the higher fractal dimension could also be due to a change in the microstructure happening during the aging or drying process, which could increase the connectivity.

A lacunarity intuitional definition refers to the fact that it is a good measure of the distribution of crevasses (gaps). More precisely, it is a reflection of the volume of gaps/holes (the portions without physical material present) when compared to the entire available volume. In the fractality assumption, the lacunarity concept relates to and defines/explains (mathematically) the presence of current holes (that is why it is named after a porous texture). On top of that, we find it to be synonymous with the quantitative radiography of the "correct or real texture" [7,8].

Note: The fractal dimension is a little smaller, but the lacunarity is greater due to the fact that the xerogels used in the current study have more voids/interstices, such that they are able to diffuse the transmitted gases or liquids in larger quantities and faster.

2.2. Fractional Propagation Equation in Fractal Porous Media

The acoustic fractional propagation equation [12–16] in a fractal porous media is given by:

$$\frac{\partial^2 p}{\partial x^2}(x,t) + \left(\frac{1-\alpha}{x}\right)\frac{\partial p}{\partial x}(x,t) - \left(\frac{\Theta^2(\alpha)|x|^{2\alpha-2}}{c^2}\right)\frac{\partial^2 p}{\partial t^2}(x,t) - B\Theta^2(\alpha)|x|^{2\alpha-2}\frac{\partial^{\frac{3}{2}} p(x,t)}{\partial t^{\frac{3}{2}}} = 0 \quad (7)$$

where $p = p(x,t)$ is the acoustic field pressure, and:

$$\Theta(\alpha) = \frac{\pi^{\frac{\alpha}{2}}}{\Gamma\left(\frac{\alpha}{2}\right)} \quad (8)$$

In Equation (8), $(\alpha) = (\alpha_1,\alpha_2,\alpha_{3,})$ is the multi-index and $\Gamma(x)$ is the gamma function [17]. In the fractal porous environment, the wave-front velocity is:

$$c' = \frac{c}{\Theta(\alpha)}|x|^{1-\alpha} = \left(\frac{c}{\pi^{\frac{\alpha}{2}}}\right)\Gamma\left(\frac{\alpha}{2}\right)|L|^{1-\alpha}. \quad (9)$$

which highlights the fact that the wave-front speed in the fractal media is of a self-similar type. In Equation (9), L is correlated to the thickness value of material and c is the wave-front velocity in a non-fractal (without the self-similarity property) porous media [12,13].

In Figure 1a, we present the computed wave-front velocity (c') as a function of the fractal dimension α for various levels of the parameter L, ranging from 0.01 to 1.0. In Figure 1b, the 3D surface plot illustrates the variation in the wave-front velocity (c') as a function of the fractal dimension α and the parameter L. The fractal dimension α is represented along the x-axis, and L is depicted on the y-axis, with both variables ranging from 0.01 to 0.99 to avoid singularities at the extremes.

Each curve is drawn with a color; the first black curve (from bottom to top) is for the $L = 0.01$ value and the last light green curve is for the $L = 1$ thickness value. A horizontal reference line is drawn at 340 m/s, representing the sound speed in air at room temperature, to constitute a threshold of interest. This is possibly related to a physical phenomenon under investigation. This visualization supports our analysis of fractal material behavior, particularly in terms of how the fractal dimension influences the wave propagation speed in such media.

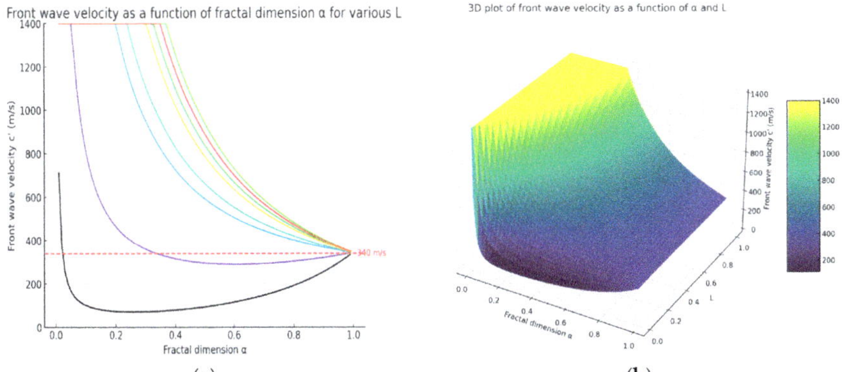

(a) (b)

Figure 1. Wave-front velocity in the fractal media as: (**a**) a fractal dimension function for various L (2D); (**b**) a function of fractal dimension and L (3D).

Note: For very low values of fractal dimension, the wave-front velocity tends to high values, which is due to the mathematical properties of the function $\Gamma(x)$, when x tends to the zero limit. For L values greater than L = 0.3 m, we observe that the wavefront velocity increases rapidly with the fractal dimension, touching supersonic values, and all being located above the dotted horizontal reference line. Considering that, in traditional non-fractal porous materials, the wavefront speed is always lower than the speed of sound in air, we still see here that the wavefront speed can be higher than the speed of sound in air (supersonic speed) for porous fractal materials. This is an interesting novelty for the porous xerogels comportment to take into account.

Solution of Fractional Propagation Equation in Fractal Porous Media

In the current section, we will attempt to resolve propagation Equation (7) in the Laplace domain, utilizing the following initial conditions:

$$p(x,t)|_{t=0} = \frac{\partial p}{\partial t}(x,t)\bigg|_{t=0} = 0, \tag{10}$$

We note $P(x,z)$ and the Laplace transform of $p(x,t)$ is defined by:

$$P(x,z) = \mathcal{L}[p(x,t)] = \int_0^\infty \exp(-zt)p(x,t)dt. \tag{11}$$

The Laplace transform $\mathcal{L}$ of $p(x,t)$, Equation (7) becomes:

$$\frac{1}{W(\alpha,x)}\frac{\partial}{\partial x}\left(\frac{1}{W(\alpha,x)}\frac{\partial P}{\partial x}(x,z)\right) - k^2(z)P(x,z) = 0. \tag{12}$$

where:

$$k^2(z) = \frac{z}{c^2}(z + b'\sqrt{z}), \quad b' = Bc^2\sqrt{\pi}. \tag{13}$$

In Equation (12), $W(\alpha,x)$ is the fractional density of states defined by the measure for integration in non-integer dimensional space, or a fractal space. It is defined by the following expression:

$$W(\alpha_k, x_k) = \frac{\pi^{\alpha_k/2}}{\Gamma(\alpha_k/2)}|x_k|^{\alpha_k - 1} \tag{14}$$

By substituting the expression of $W(\alpha, x)$ in Equation (12) through Equation (14), we acquire:

$$\frac{\partial^2 P(x,z)}{\partial x^2} + \left(\frac{1-\alpha}{x}\right)\frac{\partial P}{\partial x}(x,z) - k^2(z)\Theta^2(\alpha)x^{2\alpha-2}P(x,z) = 0 \tag{15}$$

To determine the solution to Equation (15), we proceed to the function change: $P(x,z) = x^{\nu}y$, where ν is a parameter to be determined [14,15]. Equation (12) becomes:

$$x^2y'' + (2\nu+1-\alpha)xy' + \left[\nu(\nu-1) + (1-\alpha)\nu - k^2\Theta^2x^{2\alpha}\right]y = 0, \tag{16}$$

By taking $\nu = \alpha/2$, Equation (16) is reduced to:

$$x^2y'' + (2\nu+1-\alpha)xy' + \left[\nu(\nu-1) + (1-\alpha)\nu - k^2\Theta^2x^{2\alpha}\right]y = 0, \tag{17}$$

which is transmuted by the variable replacement of $u = k(z)\Theta(\alpha)x^{2\nu}/2\nu$ into:

$$u^2 + uy' - \left(u^2 + \frac{1}{4}\right)y = 0, \tag{18}$$

which is the Bessel differential equation. The general solution is:

$$y(u) = a(z)I_{\frac{1}{2}}(u) + b(z)K_{\frac{1}{2}}(u), \tag{19}$$

with $I_{\frac{1}{2}}$ and $K_{\frac{1}{2}}$ being the modified Bessel functions of index ½.

The solutions of Equation (15) can now be written as:

$$P(x,z) = \sqrt{\frac{2}{\pi}}a(z)\exp\left\{\frac{k(z)\Theta(2\nu)x^{2\nu}}{2\nu}\right\} + \sqrt{\frac{\pi}{2}}a(z)exp\left\{\frac{k(z)\Theta(2\nu)x^{2\nu}}{2\nu}\right\} \tag{20}$$

where $a(z)$ is a new constant [16].

From Equation (17), where the condition $k^2(z)\Theta^2(\alpha) = 1$ is fulfilled and the parameter to be determined is equal to $\nu = 1/2$, we now obtain:

$$x^2y'' + xy' - \left(x^2 + \frac{1}{4}\right)y = 0. \tag{21}$$

Equation (21) is a special case of the zero-order Bessel's equation. For positive values of the argument x, that is, for $x > 0$, the general solution y of the Bessel equation of zero order is a linear combination of zero-order Bessel functions, $J_0(x)$ and $Y_0(x)$:

$$y = c_1J_0(x) + c_2Y_0(x) \tag{22}$$

The graphic representation of the Bessel functions of zero order, $J_0(x)$ and $Y_0(x)$, is shown in Figure 2.

Note: It is easily observed that $\lim_{x\to 0} J_0(x) = 1$ and that $Y_0(x)$ has in $x = 0$ a singularity of a logarithmic type. Thereby, if we are concerned with Bessel's equation solutions of zero order that in origin have finite values, which is frequently the mathematic cover, it is necessary to disclaim Y_0. The graphical represents of the Bessel functions J_0 and Y_0 are introduced in Figure 2. It is extremely interesting to observe from the cited figure that for x high, the functions $J_0(x)$ and $Y_0(x)$ have an oscillatory behavior. Still, such a comportment might be anticipated, even from the original equation [16].

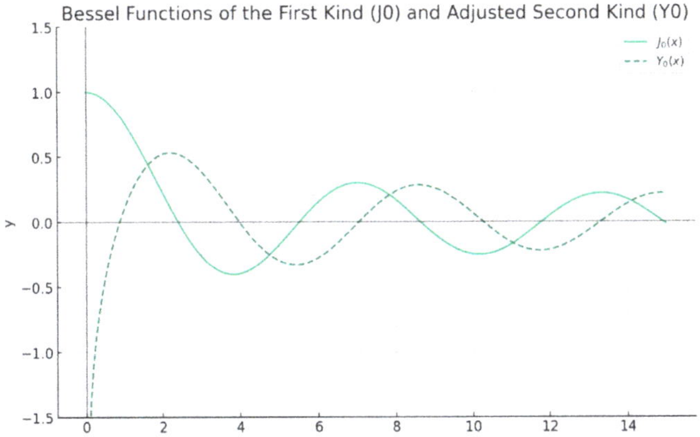

Figure 2. The Bessel functions of zero order.

In the case where the condition k^2 (z) Θ^2 (α) = 1 is fulfilled and the parameter is equal to ν = 1/2, the Equation (21) becomes:

$$x^2y'' + xy' - \left(x^2 + 1/4\right)y = 0. \tag{23}$$

By dividing both members by x^2, Equation (23) is obtained:

$$y'' + \left(\frac{1}{x}\right)y' - \left(1 + \frac{1}{4x^2}\right)y = 0. \tag{24}$$

When x is very large, it is more than reasonable to consider that the terms $(1/x)y'$ and $(\frac{1}{4x^2})$ y are small (practically both tend to zero) and can be neglected. It is therefore logical to admit that Equation (23) can now be approximated with another:

$$y'' - y = 0. \tag{25}$$

This valuable fact is correct to the extent that the Bessel functions are of the oscillatory type; but, this is only true to a certain extent.

For large a x value, the functions J_0 and Y_0 also decay as x increases. Thus, the equation $y'' + y = 0$ does not provide an adequate approximation to the Bessel equation for a large x, and a more delicate analysis is required. For a considerably high x, the functions J_0 and Y_0 also decline as x augments. It is simply shown that the solutions of this differential equation are sin x and cos x. More precisely, the solutions of Bessel's equation for several values of x are identical to the linear combinations of the sin x and cos x functions. Without discussion, it can be shown simply that:

$$J_0(x) \cong \left(\frac{2}{\pi x}\right)^{\frac{1}{2}} \cos\left(x - \frac{\pi}{4}\right) \quad as \quad x \to \infty \tag{26}$$

together with:

$$Y_0(x) \cong \left(\frac{2}{\pi x}\right)^{\frac{1}{2}} \sin\left(x - \frac{\pi}{4}\right) \quad as \quad x \to \infty \tag{27}$$

Note: Let $y_1(x)$ be the solution of differential Equation (21), with:

$$y_1(x) = a_0\left[1 + \sum_{m=1}^{\infty} \frac{(-1)^m x^{2m}}{2^{2m}\left(m!\right)^2}\right], \quad x > 0. \tag{28}$$

Figure 3 represents a polynomial approximation to the $J_0(x)$ function by means of the function $y_1(x)$, which is the solution of differential Equation (28).

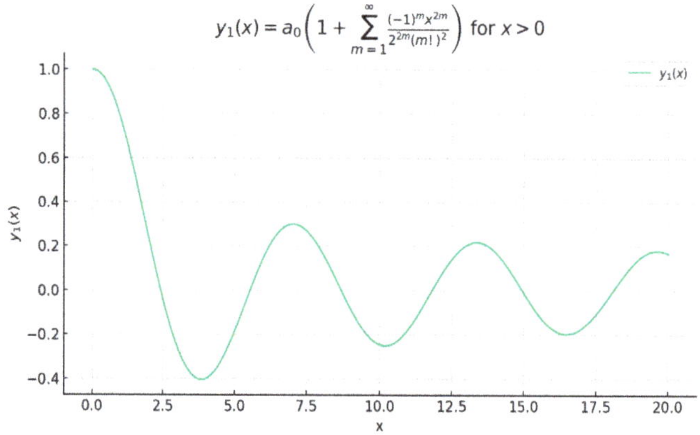

Figure 3. The polynomial solution $y_1(x)$ of differential Equation (28).

It is easy to show that Equation (26) is convergent no matter what the argument x is and that J_0 is analytic in $x = 0$. In Figure 4, one can see the graphs of the function $y = J_0(x)$ along with some of the partial sums of the series under discussion.

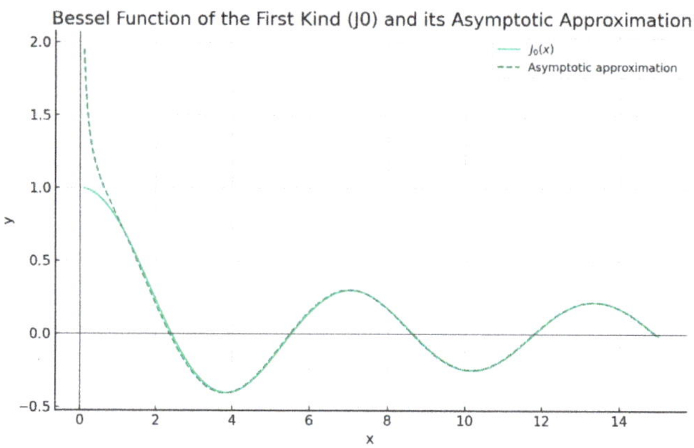

Figure 4. Graphic representation of $J_0(x)$. Asymptotic approximation.

Figure 4 represents the solution of the Bessel differential Equation (25).

It can be said that these asymptotic approximations, such as $x \to \infty$ are, in truth, very good. For instance, in Figure 4, it looks as though the asymptotic approximation (26) for $J_0(x)$ is naturally precise for all $x \geq 1$. However, to approximate $J_0(x)$ over the entire positive domain, in the range from zero to infinity, we can utilize two or three terms of the series (26) for $x \leq 1$ and the asymptotic approximation (28) for $x \geq 1$.

2.3. *Fractal Analysis*

In order to carry out fair comparisons and exemplary discussions in which usual xerogels are involved, some chitosan xerogel explanation samples were processed in the analogous conditions as the hydrogels mentioned in Section 4; all were denoted with the letter F and a certain number of indices [18–20]. In other words, we have obtained the SEM experimentally realized images of several different xerogels, each having diverse degrees of porosity [21]. In the following section, however, we will present only the picture that proved to be the most representative [22–24].

Fractal Analysis of SEM Picture

The SEM picture scale bar/line, on the lower right side of Figure 5a, has a length of 100 microns and 1001× magnification. The value of the high-voltage power supply (HV) is 5 kV [25,26]. The second image, Figure 5b, benefits from a magnification that is five times higher than the first image. We can practically say that we captured a portion of a pore together with its wall.

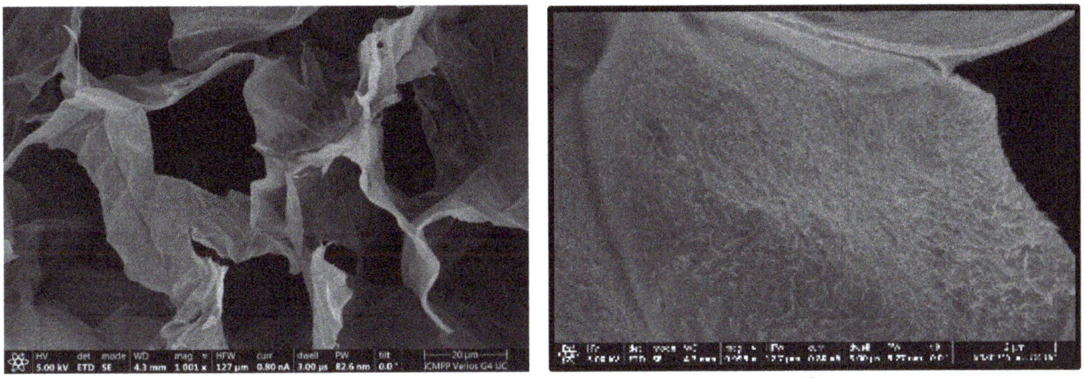

(a) (b)

Figure 5. (**a**) Original SEM image with the material pores evidenced; (**b**) Indoor shot enlarged 5 times compared to the original.

In Figure 6a the FTIR (Fourier transform infrared spectroscopy) spectra of simple xerogel, coded F, is shown, and in Figure 6b, the statistical pore histogram of the L SEM picture is represented.

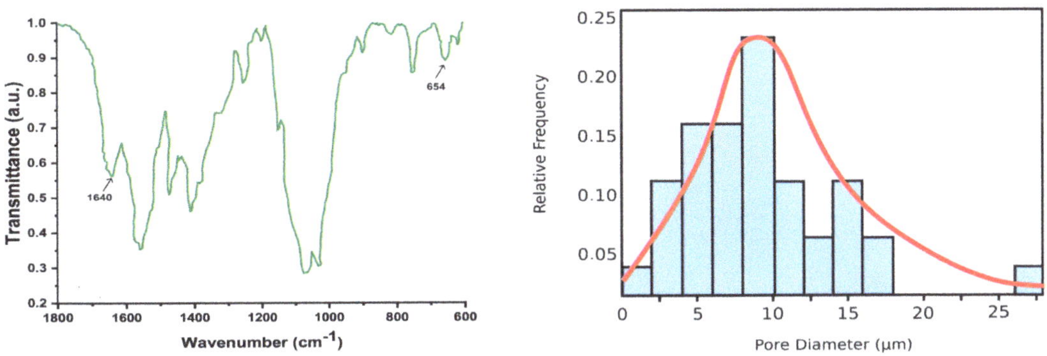

(a) (b)

Figure 6. (**a**) Simple xerogel FTIR spectra; (**b**) Statistical histogram to pore size distribution of L SEM picture.

Fourier transform infrared spectroscopy (FTIR) is a technique used to obtain an infrared spectrum of the absorption or emission of a solid, liquid, or gas. FTIR spectra were produced using a Bruker device, vector 22 (Ettlingen, Germany) at room temperature with a wavenumber resolution of 1 cm^{-1} using KBr pellets in the frequency range of

4000–400 cm^{-1}. In addition, to obtain a good signal-to-noise ratio, 32 scans were run and averaged. Figure 6a shows the obtained FTIR spectra.

The energy-dispersive X-ray analyzer (EDX) has been utilized for the identification of the individual chemical elements in the examined composition for the identification of the investigated xerogel, such as that in Figure 6a. More precisely, in this figure, an electromagnetic emission spectrum is presented. On the Ox-axis (abscissa), the emission energy, is found, representing the energy at which the atoms emit X-rays whenever they enter into an interaction with an electron flow. The respective energy is well established, being the basic characteristic of every atom, and the recorded spectrum marks the presence of these specific values in the evaluated material sample. The counts are compiled on the Oy-axis (ordinate). In principle, EDX analysis is an elemental surface investigation, its purpose being to establish with certainty the categories of atoms that exist in a tested specimen. The histograms of Figure 6b present the pore dimension distribution data obtained from the F-index SEM image. Concerning the xerogel histogram from the L SEM image, we can say that its upper limit is a normal (Gaussian) distribution, that is, the continuous red line in the graph.

In Figure 7a, the initial image filtered with the Wiener technique is found, and in Figure 7b, the image in grayscale version is represented.

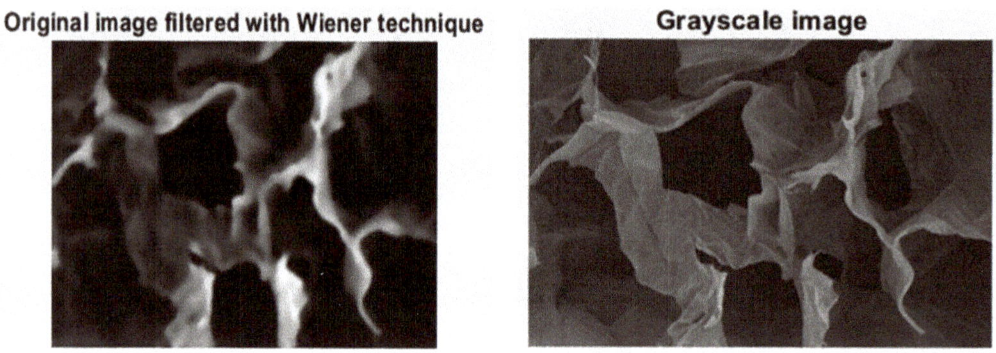

(a) (b)

Figure 7. (**a**) The filtered version of the SEM image; (**b**) The processed grayscale image.

Figure 8a,b present the pictures prepared in the binary version and the fixing of the applied mask, respectively, utilized to compute the lacunarity.

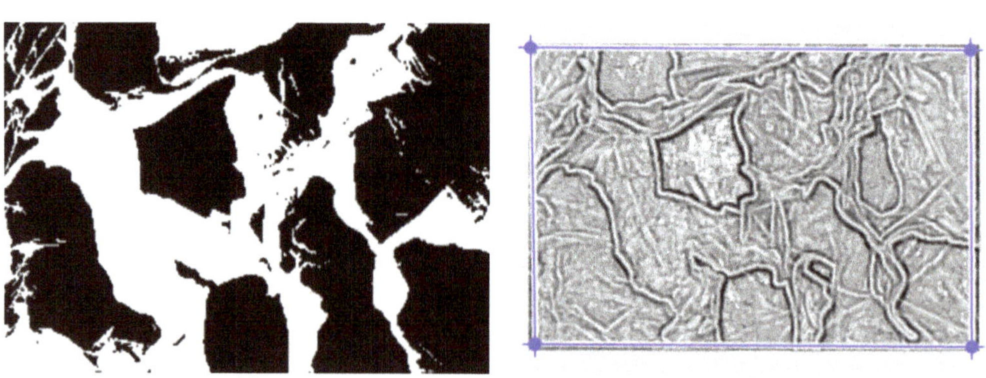

(a) (b)

Figure 8. (**a**) The binary version of the picture; (**b**) The mask made for the lacunarity calculation.

Figure 9 presents the assessed voxels of the L picture, which is a 3D eloquent voxels portrait of the gray-level amount on the Oz-axis. This conforms with the position and number of pixels, labeled on the last two plane axes: the Ox-axis and Oy-axis, respectively.

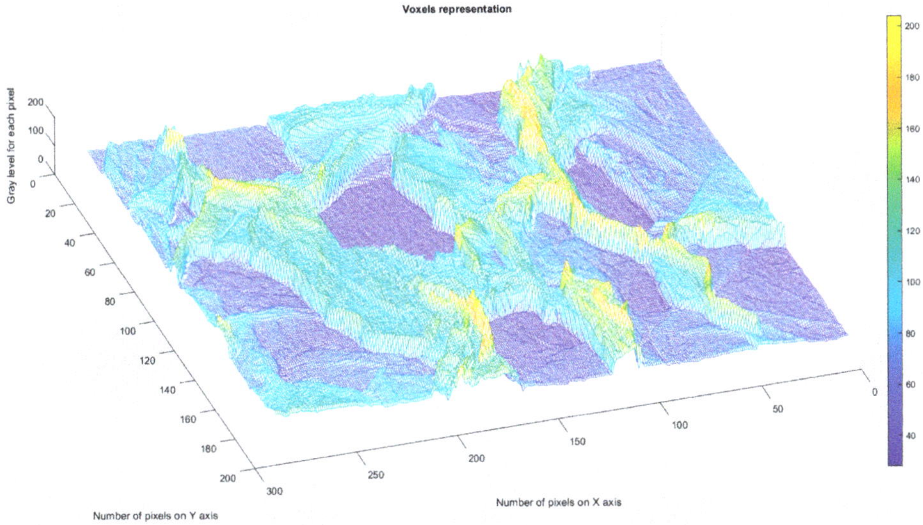

Figure 9. Voxels representation within the evaluated image.

Figure 10 shows an examination of the elected image area; more precisely, this figure shows the fractal dimension calculation as the box size *r* function with dedicated software for the box-counting procedure. The blue vertical bars attached to each point (intersecting the horizontal line) represent the standard deviation (square root of the variance).

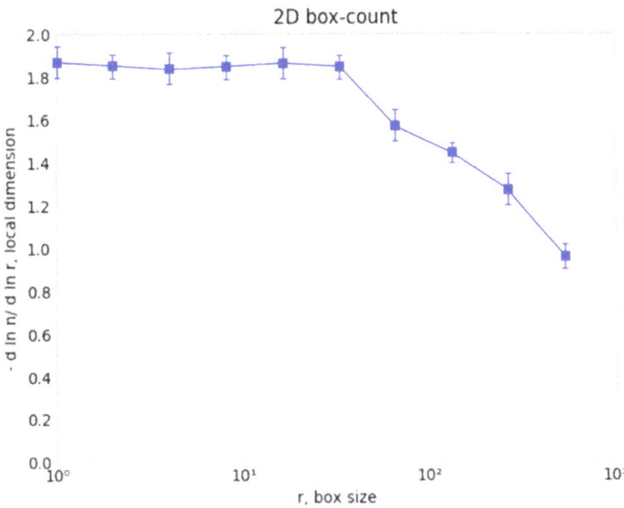

Figure 10. Local fractal dimension evaluated by the box-counting method.

Basically, we have the local fractal dimension computed with the 2D box-computing method.

Figure 11 presents the straight lines achieved by implementing multiple linear regression to certain accurate data collections, considered as checking the function ln($N(n(r))$ depending on the argument ln(r).

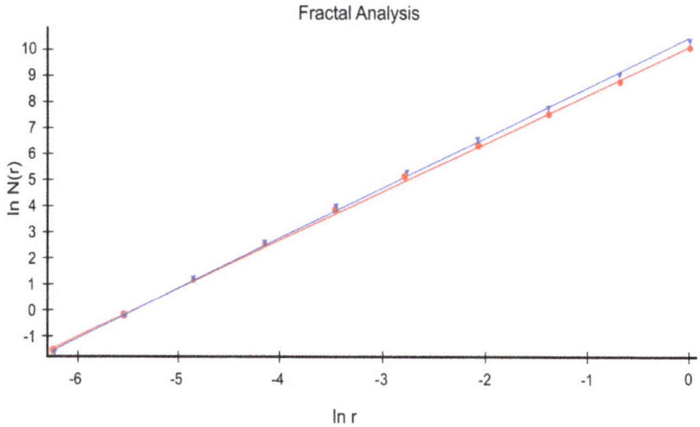

Figure 11. Verification of the results using the HarFA program.

Statistics

Blue regression line: y = 1.9063x + 10.4307, R = 9.99951

Red regression line: y = 1.8503x + 10.1113, R = 0.99939

In Figure 11, the slopes of the straight blue and red lines achieved through multiple linear regression are equal to the suitable fractal dimension of the considered SEM picture [18]. Thus, the fractal dimension values of $d_1 = 1.850$ for the red regression line (square mask) and $d_2 = 1.906$ for the blue regression line (rectangular mask) were obtained. The Hurst index (exponent) $H = 2 - D$ calculated with a square mask is $H_1 = 0.150$ and calculated for the type of rectangular mask is $H_2 = 0.094$.

For the same xerogel SEM image, the fractal parameter values are found in Table 1.

Table 1. Fractal parameters values of L SEM image.

Index	FD_1	SD_1	H_1	FD_2	SD_2	H_2	Lacunarity
L	1.668	±0.3127	0.332	1.615	±0.1445	0.242	0.0526

Table 1 is founded on the fractal parameter values achieved by the complex processing of the original SEM image. In this table, FD_1 is the fractal dimension calculated with a square mask, FD_2 is the fractal dimension calculated with a rectangular mask, SD_1 is the standard deviation calculated with a square mask, while SD_2 is the standard deviation of the fractal dimension calculated for the type of rectangular mask.

The specific fractal parameters calculated in this study are the fractal dimension, the Hurst index, and lacunarity [27,28]. The high accuracy with which these fractal indicators were calculated is owing to the methods and the advanced computational software used. Thus, the box-counting method was used to calculate the fractal dimension (see Figure 10), with different algorithms for a square mask and rectangular mask. As a measure of the particular accuracy obtained, they stand as testimony for the values of standard deviation SD_1 calculated with a square mask [29] and the standard deviation of the fractal dimension calculated for the type of rectangular mask SD_2 [30].

To calculate the lacunarity, the modified SEM figures were used, with the binary version of the picture and the mask with well-defined pore contours (see Figure 8a,b).

Finally, Figure 9 presents the assessed voxels of the SEM picture with a 3D eloquent voxels portrait of the gray-level amount on the Oz-axis, conforming with the position and number of pixels, labeled on the last two plane axes: the Ox axis and Oy axis, respectively. In fact, it shows the 3D vision of the lacunarity due to the clear distinction of the pores in a processed SEM image.

The numerical assessment upshot of the elected F SEM image, effectuated by the fractal characterization software cultivated by the authors, is the fractal dimensional worth $FD_1 = 1.668$ and $FD_2 = 1.615$, standard deviation worth $SD_1 = \pm\sqrt{(\sigma\hat{}2)} = \pm 0.3127$ and $SD_2 = \pm\sqrt{(\sigma\hat{}2)} = \pm 0.1445$, and lacunarity worth $\Lambda = 0.0526$. These are presented in Table 1. The Hurst index (exponent) $H = 2 - D$ calculated with a square mask is $H_1 = 0.332$ and calculated for the type of rectangular mask is $H_2 = 0.242$, [10,11].

Finally, we will carry out a comparison of the obtained results in this paper with those published in the only two papers we know of, belonging to us, where a complete analysis of the morphology details of xerogels using multifractal analysis and scanning electron microscopy images was carried out [31] (see the bibliography). In the first paper, the fractal parameters of the SEM images of 5-fluorouracil released from a chitosan-based matrix were evaluated. The average values were for a fractal dimension of $D = 1.8621 \pm 0.0733$ and a lacunarity value of $\Lambda_1 = 0.0385$ [31]. In the second article, the average value for the fractal dimension of a rectangular mask is $D_1 = 1.604 \pm 0.2798$, the fractal dimension of a square mask is $D_2 = 1.596 \pm 0.0460$, and the lacunarity is $\Lambda_2 = 0.0402$ [32]. The fractal dimension is much smaller, but the lacunarity is greater because the xerogels used in the current/implicated article have more voids/interstices, such that they are able to fix mercury in larger quantities. The material has a high porosity, but the pores are occupied with fixed Hg. In the current article, the fractal dimension of an SEM figure for a rectangular mask is $D_3 = 1.668 \pm 0.3127$ and for a square mask is $D_4 = 1.615 \pm 0.1445$, and the lacunarity is $\Lambda_3 = 0.0526$. The free, unoccupied pores increase the lacunarity value of the SEM figures of the tested materials.

3. Conclusions

This paper depicts some xerogel solid materials, equipped with a multitude of pores, that are statistically represented in their compositional structure, most frequently between the values of zero and 25 microns.

To correctly investigate the porous xerogel dynamic comportment, in vitro kinetic information was considered to be respected by an acoustic equation of propagation. Thus, in a fractal porous media, this fractional differential equation was successfully verified and solved with the help of Bessel functions.

Moreover, the fractal character was demonstrated by utilizing fractal analysis and it has been proven on the evaluated SEM pictures of porous xerogel compounds. The fractal parameters, namely the fractal dimension, the lacunarity, and the Hurst index, were calculated with considerable precision.

The results of the numerical evaluation on the selected L SEM picture, effectuated with the aid of the fractal characterization software developed by the authors, demonstrated that the values of the fractal dimensions and the Hurst exponents were $FD_1 = 1.668 \pm 0.312$ and $H_1 = 0.332$, calculated with a square mask. In addition, $FD_2 = 1.615 \pm 0.1445$ and $H_2 = 0.242$, calculated with a rectangular mask. The lacunarity value of $\Lambda = 0.0526$ was the last fractal parameter computed.

4. Materials and Methods

4.1. The Xerogels Materials and Synthesis

The most important materials utilized for obtaining xerogels were the chitosan with an attenuated molecular mass, tri-ethylene glycol monomethyl ether in a 97% concentration, sodium hydride in a 95% concentration, phenothiazine in a 98% concentration, phosphorus V-oxychloride in a 99% concentration, and magnesium sulfate in a 99.5% concentration. All of these were acquired through the SA Company (Sigma-Aldrich Company, St. Louis,

Missouri, United States), and the fractal dimension was calculated for the type of rectangular mask SD_2 [33]. TEGylated phenothiazine means that the phenothiazine heterocycle is replaced by TEG (tri-ethylene glycol); more precisely, the phenothiazine kernel is equipped with a TEG link that strengthens it. Ultimately, we obtained viscous chitosan, and the results were verified with the help of a calibrated Ubbelohde viscometer [34,35]. A suite of three chitosan-founded hydrogels was obtained by the imination chemical reaction with a formyl derivative of tri-ethylene glycol–phenothiazine. Thus, the chitosan hydrogelation in the company of tri-ethylene glycol–phenothiazine aldehyde may occur due to the imine constituent parts along with the self-assemblage in cluster formations from reticulated nodes.

Chitosan is composed of a long biopolymer chain of N-acetylglucosamine. Nanoscale chitosan has several advantages such as high antibacterial activity, broad spectrum of activity, as well as low toxicity to mammalian cells. Thus, chitosan was chosen for various reasons, many of them unrelated to the scope of its research in the present article. However, upon trying to manufacture it for other purposes, we realized its advantages in the currently investigated field, namely, the fact that it is an unsuspected resource in terms of it having a wide range of porosities.

Aerogel conditioning of the chitosan makes it possible to prepare porous solids of a significant specific surface. The increase in the chitosan concentration or the degree of acetylation decreases the specific surface of the synthesized chitosan gel.

The fractal nature of the porous xerogel compounds and the structural attributes of these materials have been demonstrated by particularly highlighting their porous nature. These attributes were presented through the known analyses performed, such as the SEM surface analysis (see Figure 5) and FTIR spectra of xerogel samples (see Figure 6a), revealing that it was the best candidate. It is also among the few safe materials whose behavior can be associated with an equation for advanced continuum-type mechanics of porous media. It has a generally anisotropic, mathematical product-like fractal geometry. In addition, it could be easily verified as the same material; chitosan xerogel provides an exemplary answer to the solution of the fractional propagation equation type, in fractal porous media. This fact thus becomes the biggest advantage of its use.

4.2. Xerogel Characterization

Low-density xerogels with functional amines and exhibiting very attractive textural properties can be obtained by one-step procedures using low-cost commercial ethyl silicate 40 (ES) instead of TEOS. These materials have notably larger pores, and a larger overall pore volume and specific surface area than the corresponding gels prepared using TEOS or TMOS alone [36,37]. The structural properties of xerogels depend on the type of amino functional silane additive. The use of a compound containing a more reactive methoxy group (ATM), acting as a nucleation agent, resulted in xerogels with larger mesopores and a more robust structure, similar to the results observed before.

In order to conduct fair comparisons and exemplary discussions in which usual xerogels are involved, some chitosan xerogel explanation samples were processed in the analogous conditions as the hydrogels mentioned in Section 4. All of these were denoted with the letter F and a certain number of indices. In other words, the experimentally realized SEM images of the samples are now available, and represent images of several different xerogels, each having diverse degrees of porosity. In the entire discussion held in this article, we presented only the picture that proved to be the most representative, as shown in Figure 5 [38].

4.3. Morphology

The xerogels discussed have a sponge-like morphology and are a declared an absorbent substance, with few isolated pores but rather interconnected pores. Statistically speaking, they seem to have a heterogeneous distribution, but with massive tendencies to approach the limit of a uniform distribution, with the pores formed having a diameter from 2 to

35 μm (Figure 5). Whilst the data in the literature describes the pore diameter increase as a decrease in the crosslinking degree, the same tendency was not observed in the case of our specimens [39]. Certainly, the imination rank in the hydrogel stage was not sufficiently large to manage the final morphology; consequently, the water freezing before the lyophilization enacts the crucial role. Thereby, the displacement of the imination balance (stability) to the resulting item in the lyophilization period totally consolidated the morphology type modeled in the freezing phase. However, the morphology was affected by the different sublimation cost/fee of water/acetone crystals from gelid hydrogels. The reduced density and acetone freezing point caused its rapid sublimation in contrast to pure water, resulting in a hydrophobic phenothiazine component congestion and fashioning a visible thin film at the superficial level in the xerogel [40,41].

4.4. The Morphology Characterization Techniques of a Porous Xerogel

The morphology characterization techniques of the porous xerogel structure were carried out via surface analysis by SEM, spectral analysis by Fourier transform infrared spectroscopy (FTIR) and the energy-dispersive X-ray (EDX) analyzer. The EDX has been utilized for the identification of the individual chemical elements in the examined composition [42]. Later, the SEM images will be evaluated by fractal parameter calculation, including the fractal dimension, the Hurst index, and lacunarity. The high accuracy with which these fractal indicators were calculated is owing to the methods and the advanced computational software used. Thus, the box-counting method was used to calculate the fractal dimension (see Figure 10), with different algorithms for a square mask and rectangular mask. As a measure of the particular accuracy obtained, the standard deviation $SD_1 = \pm 0.3127$, calculated with a square mask. Moreover, the standard deviation of the fractal dimension calculated for the type of rectangular mask was $SD_2 = \pm 0.1445$.

To calculate the lacunarity, the modified SEM figures were used, with the binary version of the picture and the mask with well-defined pore contours (see Figure 8a,b) [43]. In addition, Figure 9 presents the assessed voxels of the SEM picture, with a 3D eloquent voxels portrait of the gray-level amount on the Oz-axis, conforming with the position and number of pixels, labeled on the last two plane axes: the Ox-axis and Oy-axis, respectively. In fact, this represents the 3D vision of the lacunarity due to the clear distinction of the pores in a processed SEM image.

We can carry out the SEM analysis of the xerogel-type materials and if we apply the algorithm for calculating the fractal dimension, this will tell us if we have a uniform material without gaps or a lacunar material with pores of various sizes. This technique not only predicts but precisely tells us the surface morphology of the evaluated xerogel. In our opinion, this supposed generalization presented in this paper is the only one capable of carrying out appropriate comparisons and discussions on the evaluated materials. In conclusion, we can apply the fractal dimension analysis to any type of xerogel system. In the assumption of the correct use of the proposed methodology, by maintaining the proper measurement conditions in the laboratory, the experimental results can be easily reproducible and therefore generalized.

Author Contributions: Conceptualization, V.-P.P. and M.-A.P.; methodology, V.-P.P.; software, V.-A.P.; validation, V.-P.P., M.-A.P. and V.-A.P.; formal analysis, V.-P.P., M.-A.P. and V.-A.P.; investigation, V.-A.P. and M.-A.P.; resources, V.-A.P. and M.-A.P.; data curation, V.-A.P.; writing—original draft preparation, V.-P.P.; writing—review and editing, M.-A.P. and V.-P.P.; visualization, V.-A.P.; supervision, V.-P.P.; project administration, V.-P.P. All authors have read and agreed to the published version of the manuscript.

Funding: This research received no external funding.

Institutional Review Board Statement: Not applicable.

Informed Consent Statement: Not applicable.

Data Availability Statement: The data used to support the findings of this study cannot be accessed due to commercial confidentiality.

Acknowledgments: The co-authors M.A. Paun, V.A. Paun and V.P. Paun would like to thank Jenica Paun for her continuous kind support.

Conflicts of Interest: The authors declare no conflicts of interest.

References

1. Czarnobaj, K. Sol–gel-processed silica/polydimethylsiloxane/calcium xerogels as polymeric matrices for Metronidazole delivery system. *Polym. Bull.* **2011**, *66*, 223–237. [CrossRef]
2. Wang, X.; Ben Ahmed, N.; SAlvarez, G.; VTuttolomondo, M.; Hélary, C.; FDesimone, M.; Coradin, T. Sol-gel encapsulation of biomolecules and cells for medicinal applications. *Curr. Top. Med. Chem.* **2015**, *15*, 223–244. [CrossRef] [PubMed]
3. Khattab, T.A.; Dacrory, S.; Abou-Yousef, H.; Kamel, S. Development of microporous cellulose-based smart xerogel reversible sensor via freeze drying for naked-eye detection of ammonia gas. *Carbohydr. Polym.* **2019**, *210*, 196–203. [CrossRef] [PubMed]
4. Alsheekhhussain, Z.; Moaddy, K.; Shah, R.; Alshammari, S.; Alshammari, M.; Al-sawalha, M.M.; Alderremy, A.A. Extension of the Optimal Auxiliary Function Method to Solve the System of a Fractional-Order Whitham–Broer–Kaup Equation. *Fractal Fract* **2024**, *8*, 1. [CrossRef]
5. Al-Sawalha, M.M.; Mukhtar, S.; Shah, R.; Ganie, A.H.; Moaddy, K. Solitary Waves Propagation Analysis in Nonlinear Dynamical System of Fractional Coupled Boussinesq-Whitham-Broer-Kaup Equation. *Fractal Fract.* **2023**, *7*, 889. [CrossRef]
6. Al-Sawalha, M.M.; Yasmin, H.; Shah, R.; Ganie, A.H.; Moaddy, K. Unraveling the Dynamics of Singular Stochastic Solitons in Stochastic Fractional Kuramoto–Sivashinsky Equation. *Fractal Fract.* **2023**, *7*, 753. [CrossRef]
7. Karperien, A.L.; Jelinek, H.F. Box-Counting Fractal Analysis: A Primer for the Clinician. In *The Fractal Geometry of the Brain*; Di Ieva, A., Ed.; Springer Series in Computational Neuroscience; Springer Science+Business Media: New York, NY, USA, 2016 Second chapter.
8. Karperien, A.; Jelinek, H.F.; Miloševic, N.T. Reviewing Lacunarity Analysis and Classification of Microglia in Neuroscience. In Proceedings of the 8th European Conference on Mathematical and Theoretical Biology, Kraków, Poland, 28 June–2 July 2011; 6p
9. Falconer, K. *Fractal Geometry: Mathematical Foundations and Applications*; John Wiley & Sons: Hoboken, NJ, USA, 2004.
10. Hurst, H.E. Long-term storage of reservoirs: An experimental study. *Trans. Am. Soc. Civ. Eng.* **1951**, *116*, 770–799. [CrossRef]
11. Qian, B.; Rasheed, K. Hurst Exponent and financial market predictability. In Proceedings of the IASTED Conference on "Financial Engineering and Applications" (FEA 2004), Cambridge, MA, USA, 8–10 November 2004; pp. 203–209.
12. Fellah, Z.E.A.; Fellah, M.; Ongwen, N.O.; Ogam, E.; Depollier, C. Acoustics of Fractal Porous Material and Fractional Calculus *Mathematics* **2021**, *9*, 1774. [CrossRef]
13. Fellah, Z.E.A.; Fellah, M.; Lauriks, W. Direct and inverse scattering of transient acoustic waves by a slab of rigid porous material *J. Acoust. Soc. Am.* **2003**, *113*, 61–72. [CrossRef]
14. Carpinteri, A.; Chiaia, B.; Cornetti, O. On the mechanics of quasi-brittle materials with a fractal microstructure. *Eng. Fract. Mech* **2003**, *15*, 2321–2349. [CrossRef]
15. Tarasov, V.E. *Fractional Dynamics: Applications of Fractional Calculus to Dynamics of Particules, Fields and Media*; Springer Berlin/Heidelberg, Germany, 2010.
16. Bowman, F. *Introduction to Bessel Functions*; Dover Books on Mathematics Inc.: New York, NY, USA, 2010.
17. Artin, E. *The Gamma Function*; Dover Books on Mathematics; Dover Publications: Mineola, NY, USA, 2015.
18. Prausnitz, M.R.; Langer, R. Transdermal drug delivery. *Nat. Biotechnol.* **2008**, *26*, 1261–1268. [CrossRef] [PubMed]
19. Patel, A.; Cholkar, K.; Agrahari, V.; Mitra, A.K. Ocular drug delivery systems: An overview. *World J. Pharmacol.* **2013**, *2*, 47–64 [CrossRef] [PubMed]
20. Zhang, L.; He, G.; Yu, Y.; Zhang, Y.; Li, X.; Wang, S. Design of Biocompatible Chitosan/Polyaniline/Laponite Hydrogel with Photothermal Conversion Capability. *Biomolecules* **2022**, *12*, 1089. [CrossRef] [PubMed]
21. Flores-López, S.L.; Karakashov, B.; Ramírez-Montoya, L.A.; Menéndez, J.A.; Fierro, V.; Arenillas, A.; Montes-Morán, M.A.; Celzard, A. Effect of the porosity and microstructure on the mechanical properties of organic xerogels. *J. Mater. Sci.* **2021**, *56*, 10312–10325. [CrossRef]
22. Dórea, J.G. Integrating Experimental (In Vitro and In Vivo) Neurotoxicity Studies of Low-Dose Thimerosal Relevant to Vaccines *Neurochem. Res.* **2011**, *36*, 927–938. [CrossRef] [PubMed]
23. Aragay, G.; Pons, J.; Merkoçi, A. Recent Trends in Macro-, Micro-, and Nanomaterial-Based Tools and Strategies for Heavy-Metal Detection. *Chem. Rev.* **2011**, *111*, 3433–3458. [CrossRef]
24. Mahmoud, M.E.; Abdelwahab, M.S.; Ibrahim, G.A.A. The Design of SnO_2-Crosslinked-Chitosan Nanocomposite for Microwave-Assisted Adsorption of Aqueous Cadmium and Mercury Ions. *Sustain. Chem. Pharm.* **2022**, *28*, 100731. [CrossRef]
25. Zhang, L.; Jiao, X.; Zhang, H.; He, S.; Cheng, X. Novel Chitosan–Naphthalimide–Amino Acid Fluorescent Powder for Selective Detection and Removal of Hg^{2+}/Hg^{+} and Fe^{2+} in Aqueous Solution. *Chem. Pap.* **2022**, *76*, 7037–7049. [CrossRef]
26. Lin, H.; Duan, Y.; Zhao, B.; Feng, Q.; Li, M.; Wei, J.; Zhu, Y.; Li, M. Efficient Hg(II) Removal to Ppb Level from Water in Wider PH Based on Poly-Cyanoguanidine/Graphene Oxide: Preparation, Behaviors, and Mechanisms. *Colloids Surf. A Physicochem. Eng Asp.* **2022**, *641*, 128467. [CrossRef]

27. Saenchoopa, A.; Klangphukhiew, S.; Somsub, R.; Talodthaisong, C.; Patramanon, R.; Daduang, J.; Daduang, S.; Kulchat, S. A Disposable Electrochemical Biosensor Based on Screen-Printed Carbon Electrodes Modified with Silver Nanowires/HPMC/Chitosan/Urease for the Detection of Mercury (II) in Water. *Biosensors* **2021**, *11*, 351. [CrossRef]
28. Michailidou, G.; Koumentakou, I.; Liakos, E.v.; Lazaridou, M.; Lambropoulou, D.A.; Bikiaris, D.N.; Kyzas, G.Z. Adsorption of Uranium, Mercury, and Rare Earth Elements from Aqueous Solutions onto Magnetic Chitosan Adsorbents: A Review. *Polymers* **2021**, *13*, 3137. [CrossRef]
29. Seidi, F.; Reza Saeb, M.; Huang, Y.; Akbari, A.; Xiao, H. Thiomers of Chitosan and Cellulose: Effective Biosorbents for Detection, Removal and Recovery of Metal Ions from Aqueous Medium. *Chem. Rec.* **2021**, *21*, 1876–1896. [CrossRef] [PubMed]
30. Sarkar, N.; Chaudhuri, B.B. An efficient approach to estimate Fractal Dimension of textural images. *Pattern Recognit.* **1992**, *25*, 1035–1041. [CrossRef]
31. Paun, M.-A.; Paun, V.-A.; Paun, V.-P. A Multifractal Vision of 5-Fluorouracil Release from Chitosan-Based Matrix. *Gels* **2022**, *8*, 661. [CrossRef] [PubMed]
32. Paun, M.-A.; Nichita, M.-V.; Paun, V.-A.; Paun, V.-P. Xerogels Morphology Details by Multifractal Analysis and Scanning Electron Microscopy Images Evaluations of 5-Fluorouracil Release from Chitosan-Based Matrix. *Gels* **2022**, *8*, 820. [CrossRef] [PubMed]
33. Li, Y.; Qi, X.; Chen, Y.; Wang, L.; Li, Z.; Sun, J.; Jia, J. Voxel field fusion for 3D object detection. In Proceedings of the IEEE Conference on Computer Vision and Pattern Recognition, New Orleans, LA, USA, 18–24 June 2022; Main Conference; 13p.
34. Nichita, M.V.; Paun, M.A.; Paun, V.A.; Paun, V.P. Image Clustering Algorithms to Identify Complicated Cerebral Diseases. Description and Comparison. *IEEE Access* **2020**, *8*, 88434–88442. [CrossRef]
35. Nichita, M.V.; Paun, M.A.; Paun, V.A.; Paun, V.P. Fractal analysis of brain glial cells. Fractals dimension and lacunarity. *Univ. Politeh. Buchar. Sci. Bull. Ser. A Appl. Math. Phys.* **2019**, *81*, 273–284.
36. Bordescu, D.; Paun, M.A.; Paun, V.A.; Paun, V.P. Fractal analysis of Neuroimagistic. Lacunarity degree, a precious indicator in the detection of Alzheimer's disease. *Univ. Politeh. Buchar. Sci. Bull. Ser. A Appl. Math. Phys.* **2018**, *80*, 309–320.
37. Popovic, N.; Radunovic, M.; Badnjar, J.; Popovic, T. Fractal dimension and lacunarity analysis of retinal microvascular morphology in hypertension and diabetes. *Microvasc. Res.* **2018**, *118*, 36–43. [CrossRef]
38. Pereira, L.M. Fractal Pharmacokinetics. *Comput. Math. Methods Med.* **2010**, *11*, 161–184. [CrossRef]
39. Lungu, R.; Paun, M.-A.; Peptanariu, D.; Ailincai, D.; Marin, L.; Nichita, M.-V.; Paun, V.-A.; Paun, V.-P. Biocompatible Chitosan-Based Hydrogels for Bioabsorbable Wound Dressings. *Gels* **2022**, *8*, 107. [CrossRef] [PubMed]
40. de Melo, R.H.C.; Conci, A. How Succolarity could be used as another fractal measure in image analysis. *Telecommun. Syst.* **2013**, *52*, 1643–1655. [CrossRef]
41. Xia, Y.; Cai, J.; Perfect, E.; Wei, W.; Zhang, Q.; Meng, Q. Fractal dimension, lacunarity and succolarity analyses on CT images of reservoir rocks for permeability prediction. *J. Hydrol.* **2019**, *579*, 124198. [CrossRef]
42. Paun, V.-P. Polymer dynamics simulation at nanometer scale in a 2D diffusion model. *Mater. Plast.* **2007**, *44*, 393–395.
43. Lazar, B.; Sterian, A.; Pusca, S.; Paun, V.; Toma, C. Simulating delayed pulses in organic materials. In Proceedings of the ICCSA 2006, Glasgow, UK, 8–11 May 2006; Lecture Notes in Computer Science; Volume 3980, pp. 779–784.

gels MDPI

Article

Fractal Analysis of Four Xerogels Based on TEGylated Phenothiazine and Chitosan

Maria-Alexandra Paun [1,2], Mihai-Virgil Nichita [3], Vladimir-Alexandru Paun [4] and Viorel-Puiu Paun [5,6,*]

1 School of Engineering, Swiss Federal Institute of Technology (EPFL), 1015 Lausanne, Switzerland; maria_paun2003@yahoo.com
2 Division Radio Monitoring and Equipment, Section Market Access and Conformity, Federal Office of Communications (OFCOM), 2501 Bienne, Switzerland
3 Doctoral School, Faculty of Applied Sciences, University Politehnica of Bucharest, 060042 Bucharest, Romania; mihai_nichita9@yahoo.com
4 Five Rescue Research Laboratory, 75004 Paris, France; vladimir.alexandru.paun@ieee.org
5 Physics Department, Faculty of Applied Sciences, University Politehnica of Bucharest, 060042 Bucharest, Romania
6 Academy of Romanian Scientists, 50085 Bucharest, Romania
* Correspondence: viorel.paun@physics.pub.ro

Abstract: The present article describes novel massive materials (in the solid phase) based on TEGylated phenothiazine and chitosan that possess great capability to recover mercury ions from constituent aqueous solutions. These were produced by chitosan hydrogelation accompanied by formyl subsidiary item of TEGylated phenothiazine, attended by lyophilization. The delineation and structure description of the obtained material or supramolecular assembly were realized by FTIR (Fourier transform infrared) spectroscopy, X-ray diffraction, and POM (Polarized Light Optical Microscopy). The morphology of their texture was kept under observation by SEM (Scanning Electron Microscopy). The obtained SEM images were evaluated by fractal analysis. The fractal parameters of interest were calculated, including the fractal dimension and lacunarity.

Keywords: TEGylated phenothiazine; chitosan; SEM images; fractal analysis; fractal parameters

1. Introduction

The pollution of the ambient environment by heavy metals is a crucial concern worldwide, which seriously impacts animals' and human's general state of health [1]. Heavy metals are considered to be perilous pollutants because they are not biodegradable and pollute the air, water, and soil by contamination. These ones have a high penetration rate in the trophic stereotype (food chain) and, over time, in the human body. While several metals, such as chromium, copper, manganese, molybdenum, selenium, and zinc are simply necessary in daily diet, however in reduced quantity, an overexposure determines major intoxication followed by organ deterioration in the long-term period, particularly for children and adolescents [2]. Average concentrations (μg/gHb) in the erythrocytes (or red blood cells) are equal to 0.32 ± 0.16 (for Cr), 38.8 ± 6.68 (for Cu), 1.45 ± 0.36 (for Mn), 0.3 ± 0.04 (for Mo), 8.63 ± 2.30 (for Se), and 0.65 ± 0.10 (for Zn), in tested human subjects with low physical training levels. Other metals, for example cadmium, lead, mercury, and arsenic, were identified as eminently perilous for human body and health, even in low quantities. The geometric averages of cadmium, lead, mercury and inorganic arsenic in the blood are equal to 0.09 μg/L (max 0.26 μg/L) for Cd, 9.9 μg/L (max 42 μg/L) for Pb, 0.70 (max 2.4 μg/L) for Hg and 6.1 μg/L (max 10 μg/L) for iAs. The Centre for Disease Control (CDC), U.S. Food and Drug Administration, Joint Food, World Health Organization (WHO), Agricultural Organization (FAO) and the US Environmental Protection Agency (EPA) have decided their inclusion on the carcinogenic active agents list by all health agencies [3,4].

As regards, heavy metal contamination is closely related to global industrial development, which cannot be suspended. As a consequence, the world's interstate agencies included these metallic compounds on the problematic chemical substances list, which require propriety monitoring and confirming the recommended maximal levels, both in water and soils [5,6]. Among these toxic/poisonous metals, mercury is particularly most perilous because it sublimates very easily, contaminates the air, is effortlessly stored in potable waters and soil, and is a tenacious contamination source [7]. In living organisms, mercury denatures the proteins and kills living cells, especially nervous system cells, the neurons [8]. For this reason, advisable mercury concentration values are very small; more precisely, they are limited below values of 2 ppb [9]. In the circumstances described above, the attention of specialists was concentrated on the discovery and improvement of materials that have the ability to identify and recuperate mercury from the ambient environment and corporeality (human body) [10–16].

Chemical substances generically named as xerogels are a gel type naturally found in the solid phase, which currently have properties such as a superior porosity and significant surface in coincidence with remissivity of the pore dimensions [17,18]. The present study is based on the new solid materials found on chitosan and TEGylated phenothiazine, which present a large capacity to recuperate mercury ions located in aqueous (water) solutions. Among these chemical substances presented above, we mention with primacy chitosan. Chitosan is a special biopolymer considering that it is abundant in nature, is positively electrically charged (cationic), has very low degree of toxicity, is immunodeficient (non-immunocompetent), and is essentially sustainable in an unlimited manner [19,20].

Fractal analysis is a quantitative method of image evaluation that is based on three established parameters, which are considered to be fractal dimension, lacunarity and succolarity [21]. The effective structure determining the properties of gel morphology is characterized by fractal dimensions deduced from the used theoretical model, which also suggests that the size of primary flocs building fractal structures is one of the important factors that determine the linear viscoelastic properties of the gels. The fractal dimension is the measure that discriminates how much a geometric object fills the space that includes it. Fractal dimension is an established quantity that does not amend with the scale, neither with applied translation or rotation procedure [22]. Lacunarity is the one which determines the measurements of the holes' dimension and frequency on the picture. Succolarity measures in what quantity a well-determined fluid can flow over a picture, considering the set of pixels as a barrier with a definitive exact color (black or white, for example) on 2D picture evaluation.

2. Theoretical Part

In fractal theory, the fractal dimension, lacunarity, and succolarity notions are usually utilized to characterize and evaluate the structural information of the pore/hole system in the material. Still, the practical application and calculation of the three fractal parameters is difficult because of the complex definitions and laborious methods of computation. In this subchapter, we mainly introduce the classic definitions and physical meanings of these fractal structural parameters and the calculation method based on the box-counting procedure from pictures. In addition, several examples of the application of fractal parameters in physical property modeling, natural fracture characterization, and permeability prediction of the analyzed compounds are presented. These results can illustrate well the functions of the fractal dimension, lacunarity, and succolarity for the description of complexity and heterogeneity degree, as well as the anisotropy of the material's body structure and porous media.

2.1. Fractal Dimension

The fractal dimension is the essential fractal parameter to depict a complex system, from the fractal point of view. It is a measure to reflect the space availableness (3D), the coverage with differentiable curves of the contour of some surfaces (2D) or the complex bodies' nonuniformity. Referring to porous media, the fractal dimension is utilized in a quantitative manner, to characterize the statistical repartition of the orifices/holes dimension, the porous surfaces (with pores on them) rugosity, and the streamline outline curvature [22]. We still have to say here that, while distinct definitions generate dissimilar fractal dimension values, the fractal dimension is the ordinary procedure to exactly report the distribution of the pore/particle dimension of the porous media pictures. We will now discuss the basic relationship of the fractal scaling convention among the pore/particle collected number, noted $M(\varepsilon)$, and the pore/particle caliber noted ε.

This can be written as the following relation:

$$M(\varepsilon) \propto \varepsilon^{D_f} \tag{1}$$

wherein D_f is a natural fractal dimension of the considered porous space. The fractal dimension can be considered, from a mathematical point of view, also as a measure of how all details in the fractal change with scale. There is also the problem of knowing, when the fractal object is projected against a grille, how many elements the fractal includes/covers as its elements' numbers grow. We mention that the fractal dimension cannot have an integer value, as it is a fractional number, to be precise.

It is important to remember three important statistics ascertained to be the correct measure of the fractal dimension [23]. These are the information dimension, the correlation dimension, and the box-counting dimension, the last being the most frequently used calculation technique. To calculate the fractal dimension value by the box-counting method, we split the fractal (3D) space into hypercubes with the side lengthiness equal to r. Considering that $N(r)$ is the hypercubes number engaged by the fractal geometric points, the box counting fractal dimension is computed to be [24]:

$$D_{bc} = \lim_{r \to 0} \frac{log N(r)}{log r} \tag{2}$$

2.2. *Lacunarity*

The word "lacunarity" refers in a literary way to a lacuna or hollow, as acquired from the word "lake". However, in lexical consideration, it has been differently defined as being denoted by words such as inhomogeneity, gappiness or translational (2D) and rotational (3D) invariance. Currently denoted as Λ in FracLac software, the lacunarity relates to both holes' evidence and heterogeneity measure, equally [25,26].

Lacunarity and fractal dimension are in close communion with each other, thus making possible a good understanding of the fractal object surface morphology with its pores (holes, orifices), entirely. In particular, it refers to the balance between the homogeneity and inhomogeneity of the texture in an integrative version, with an emphasis on the holes' (pores) statistics and their caliber as repartition function, of all things. In fractal analysis theory, the lacunarity notion construes/renders mathematically the measure of current holes (named porous texture) or "true texture" radiography [27]. We find that the observed inhomogeneity degree as well as rotational (3D) and translational (2D) invariance of the picture surface (where reduced lacunarity assumes the proof of image homogeneity), also confirm that rotating the image amends the given context in a non-significant way.

The mathematical equations that govern this process are presented below.

$$\Lambda(\varepsilon) = \frac{Z^{(2)}}{\left(Z^{(1)}\right)^2} \tag{3}$$

$$Z^{(1)} = \sum_{\varepsilon} P \bullet Q(P, \varepsilon) \tag{4}$$

$$Z^{(2)} = \sum_{\varepsilon} P^2 \bullet Q(P, \varepsilon) \tag{5}$$

$$Q(P,\varepsilon) = \frac{n(P,\varepsilon)}{(M-\varepsilon+1)^2} \quad (6)$$

In the equations highlighted above, the letters specified below signify the quantities described in continuation. The map dimension is M, the box dimension is ε and the box mass is equal to P. The $n(P, \varepsilon)$ is the box number containing P pixels and probability $Q(P, \varepsilon)$ is calculated via Equation (6). At the same time $P \bullet Q(P, \varepsilon)$ is the first moment and $P^2 \bullet Q(P, \varepsilon)$ is the second moment, while $Z^{(1)}$ and $Z^{(2)}$ are the sum of the first and second moments, computed by Equations (4) and (5), respectively. Equation (3) is the lacunarity value $\Lambda(\varepsilon)$ of the box dimension ε dataset [28,29]. In comparison with the other fractal parameters, lacunarity is a counterpart to the fractal dimension, but in conjunction, they offer a complete description of fractal object texture. Its value is directly proportional to the quantity repartition of the gaps/orifices present in the material. In other words, if a fractal has large lacunas or holes, the lacunarity is particularly great. However, one can say that if a fractal is almost 2D translationally invariant, it has reduced lacunarity [30,31].

In conclusion, we can say that lacunarity measures the size and frequency of gaps/holes from a representative image.

2.3. Succolarity

Succolarity estimates the image percolation degree and how much a certain fluid may circulate/run through this picture, taking into consideration the pixels' suite with a defined color (e.g., white or black pixels) as possible obstacles in the surface analysis of 2D images. The principal idea is that succolarity utilization is a necessary characteristic in the pattern recognition affair, especially in order to perceive genuine textures. To evaluate this, let us consider an image that respects the representativeness criteria. Assume that every pixel in its plan position may be regarded/thought about as empty (lack of mass for black pixels) or having an impenetrable mass for white pixels.

To calculate the succolarity value, we use the formula:

$$\sigma(BS(k), dir) = \frac{\sum_{k=1}^{n} OP(BS(k)) \times PR(BS(k), pc)}{\sum_{k=1}^{n} OP(BS(k))xm \times PR(BS(k), pc)} \quad (7)$$

where ($BS(k)$) is box size, k is the number of possible divisions of an image in boxes and dir is a direction, one of the known ones, right and left. $PR(BS(k), pc)$ signifies the pressure above the box k centroid, on the considered scale. This can be achieved using the centroid coordinates, more precisely on x (in the horizontal case) or else y (in vertical case) [32]. Let us do a simulation now of the evacuating or percolation capability of a fluid through the picture. The initial image was explored inundated in vertical plan (from bottom to top and from top to bottom) and in a horizontal plan (from left to right and from right to left). In addition, other directions may be utilized to generate various succolarity values of images, if are representative, naturally.

The importance of succolarity, different from fractal dimension and lacunarity, necessary to highlight different fractal properties, is thus demonstrated. The succolarity [33] denotes a particular flow ability that allows crossing the set. Technically speaking, a succolarity reported on fractal dynamic sets is defined as the number evaluation of filaments that allows the percolation phenomenon or, in the same measure, to flow through. The latter is not suitable for the evaluation of SEM images and therefore will not be used in continuation as a fractal parameter of interest.

Note, the three independent fractal parameters (fractal dimension, lacunarity and succolarity) are important characteristics that examine different picture aspects in a subtle complementary manner. Thus, there can be two images that can set forth the identical fractal dimension, but distinct lacunarity, or identical lacunarity, but distinct succolarity, and even a combination of the outcomes is possible.

3. Results and Discussion

3.1. Morphology Notions

Known under this name in specialized literature, the xerogels manifested a similar texture of sponge-type morphology [34], with interconnected structural orifices (holes) and a polymorphous pores repartition with the included diameter in the interval 2 μm to 35 μm (Figure 1). Whilst the other authors report on the same subject the increase of the hole's diameter as the reticular degree has diminished, no such a tendency was noticed for these specimens. This is most likely due to the fact that the imination degree in the hydrogel situation/condition was not sufficiently large to command the morphology, and thus the water congelation anterior to the lyophilization procedure gamed a decisive role. More, the displacement of the imination equilibrium to the chemical compounds at the time of lyophilization simply consolidated the morphology modeled in the congelation stage. In addition, the morphology was affected by the particular sublimation quota of water/acetone crystallites in the frosty hydrogels. The reduced density and freezing time of acetone, generated its rapid sublimation in comparison to water, dictating congestion of the hydrophobic phenothiazine items on the superficial appearance of xerogel outer veneers, composing a so-called thin film/layer.

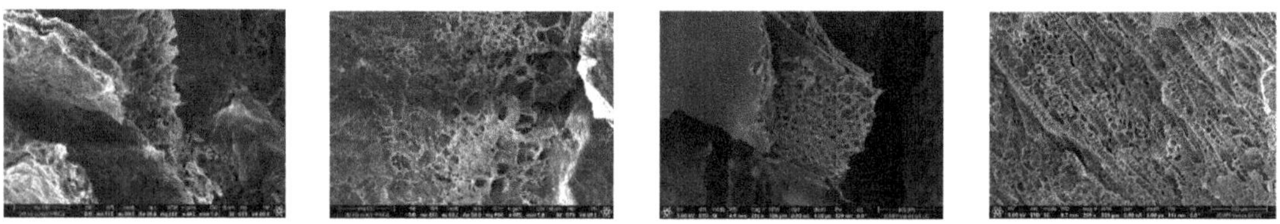

(**a**) 2L (**b**) 4L (**c**) 6L (**d**) CL

Figure 1. Four typical SEM images of the xerogels: (**a**) SEM-2L, (**b**) SEM-4L, (**c**) SEM-6L, (**d**) SEM-CL.

3.2. Fractal Analysis of Scanning Electron Microscope Pictures

We will show the connection between the fractal analysis and the performance of the material in two examples. The first refers to the scaling behavior of gel elasticity. In theory, the gel network is considered a closely packed fractal flocs with the fractal dimension of d. The elastic properties of a floc are dominated by its effective backbone, which can be approximated as a linear chain of springs. The elastic constant (K) of the individual flocs is inversely related to their size (l). Since fractal flocs are considered scale invariant, the size of the flocs l is related to the volume fraction (φ) as $l \propto \varphi^{1/(d-3)}$. The second example refers to cluster–cluster aggregation. The process of colloidal aggregate formation has also been successfully investigated based on fractal considerations. Fractal growth models have been applied to the aggregation process of particles. In the cluster–cluster aggregation process, diffusing particles in a certain medium stick to one another at contact in a random way with probability p.

The achieved hydrogels were proven to be transparent materials, and soft materials, which performed with maximum success in the test of inverted tube, and through the lyophilization process, porous solid materials were produced. The hydrogels acquired were further used, and the corresponding xerogels obtained by lyophilization have been noted with 2L, 4L, and 6L indicators. For a precise comparison with a consecrated xerogel, a chitosan xerogel reference was made ready in the identical conditions as the hydrogels aforesaid and noted with the CL indicator. In Figure 1, representative SEM typical pictures are visible [7]. More precisely, there are four SEM photographic images of four samples from those chemically obtained, representing four distinct xerogels, respectively noted with 2L, 4L, 6L and CL indicators. The SEM pictures scale bars, from the lower right side, in Figure 2 (a) SEM-2L; (b) SEM-4L; (c) SEM-6L and (d) SEM-CL, measure 100 microns for each [7].

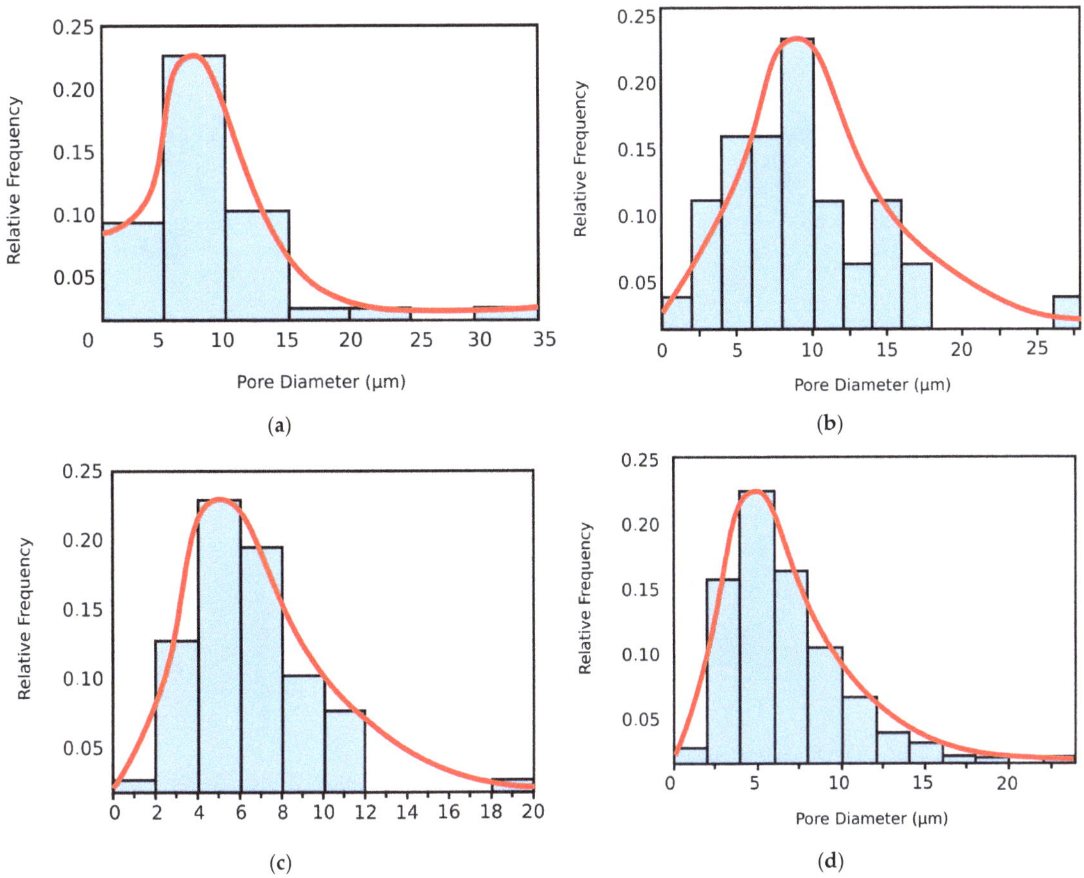

Figure 2. The histograms of pore dimension from SEM images, (**a**) SEM-2L, (**b**) SEM-4L, (**c**) SEM-6L, (**d**) SEM-CL.

Figure 2 shows the histograms of pore dimension (on the abscissa) from SEM images (Figure 1). For the four SEM images, the surface pores diameter is from 2 µm to almost 35 µm. On these PSD histograms are overwritten the PSD curves in red color.

3.2.1. Fractal Parameters of 2L Image

In Figure 3, we have the two phases of 2L original image processing and the fractal analysis techniques, such as the mask image version and binarized version, respectively, for the calculation of the fractal parameters. The threshold, above which the binarization of the 2L image was performed, is 77.

Figure 4 presents the voxels of the evaluated 2L picture, more precisely a 3D graphical portrayal/depiction, with the gray level on the oz axis, while the suitable number of pixels together with their position are marked on the other two plane axes, ox and oy, respectively [35].

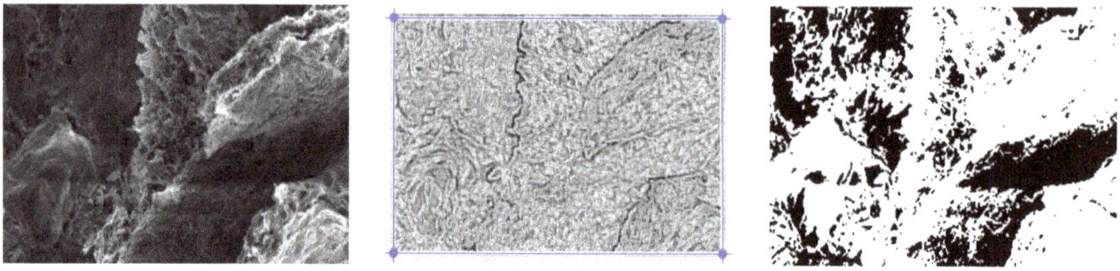

(a) (b) (c)

Figure 3. (**a**) Original 2L image, (**b**) Mask of 2L image, (**c**) The binarized version of the 2L image.

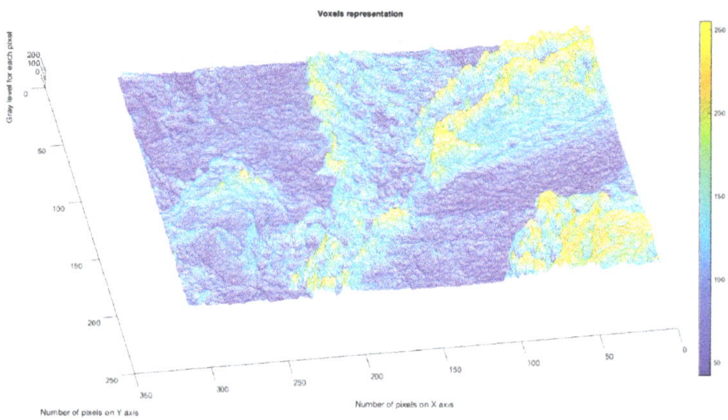

Figure 4. 3D Voxels graphical representation of 2L image.

In Figure 5, we have the two phases of 2L original image processing and the fractal analysis techniques, such as the gray scale with luminance version and the gray scale without luminance version, respectively, utilized for the calculation of the fractal parameters.

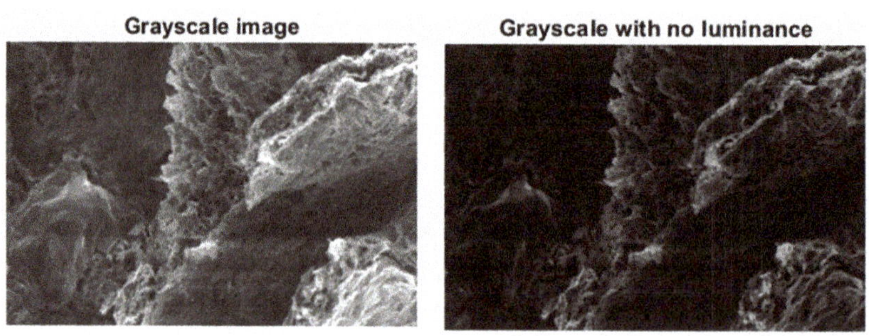

(a) (b)

Figure 5. Grayscale versions of the 2L image: (**a**) with luminance, (**b**) without luminance.

In Figure 6, we have the two phases of 2L original image processing as the filtered image version and the Wiener technique version, respectively, utilized for the calculation of the fractal parameters.

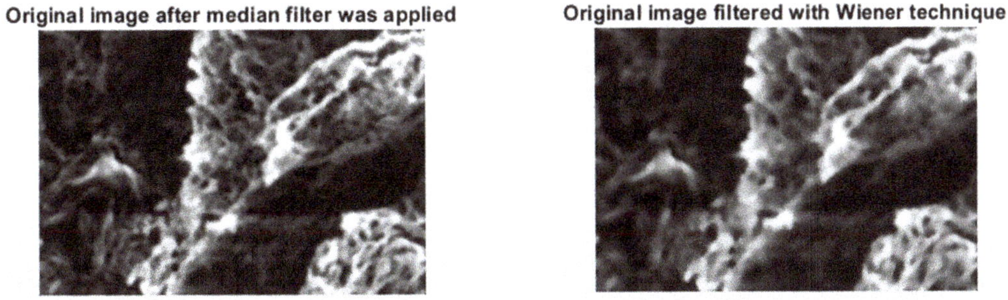

(a) (b)

Figure 6. Filtered version of the 2L image after (**a**) median filter, (**b**) Wiener technique.

In Figure 7, the fractal local dimension by box-counting algorithm in (a) and verification of the results with the HarFA (Harmonic and Fractal Image) program in (b), for the 2L image are presented [36].

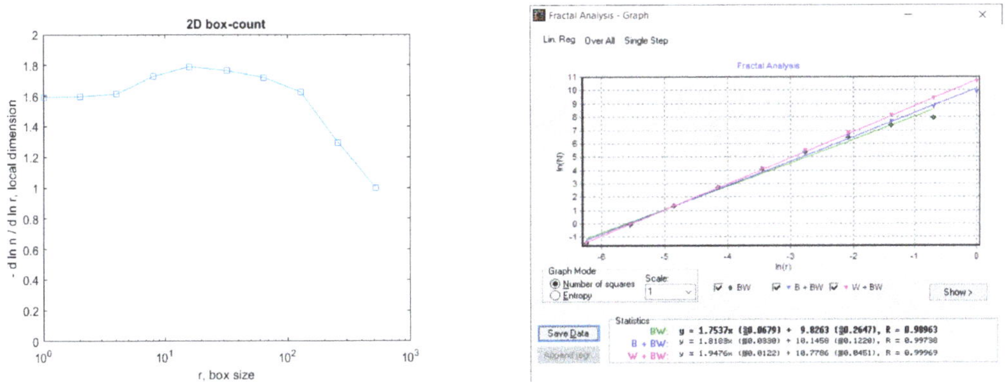

(a) (b)

Figure 7. (**a**) Fractal local dimension for 2L image, (**b**) HarFA program for the 2L image.

In Figure 7a, the (2D) graphic to establish the fractal local dimension for the 2L image, the function of the box size r, by the boxes-counting procedure, is presented.

As a numerical appreciation upshot of the selected 2L picture, performed via the fractal analysis software developed by the authors, the values of Fractal Dimension FD1 = 1.604 and FD2 = 1.596, Standard Deviations $s_1 = \pm\sqrt{(\sigma^{\wedge 2})} = \pm 0.2798$ and $s_2 = \pm\sqrt{(\sigma^{\wedge 2})} = \pm 0.0460$, as well as Lacunarity value $\Lambda = 0.0402$, were estimated, seen in Table 1.

Table 1. Fractal characteristics computation of picture 2L.

FD1	Standard Deviation 1	FD2	Standard Deviation 2	Lacunarity
1.604	±0.27987	1.596	±0.04607	0.0402

In the table above, the following notations were utilized:

- FD1-Fractal dimension with quadratic mask
- Standard deviation 1-Standard deviation with quadratic mask
- FD2-Fractal dimension with a rectangular mask

- Standard deviation 2-Standard deviation with a rectangular mask

Table 1 is a table with all the values of the fractal parameters obtained from the processing of the 2L image.

3.2.2. Fractal Parameters of 4L Image

In Figure 8, we have the two phases of 4L original image processing and the fractal analysis techniques, such as the mask image version and binarized version, respectively, for the calculation of the fractal parameters. The threshold, above which the binarization of the 4L image was performed, is 100.

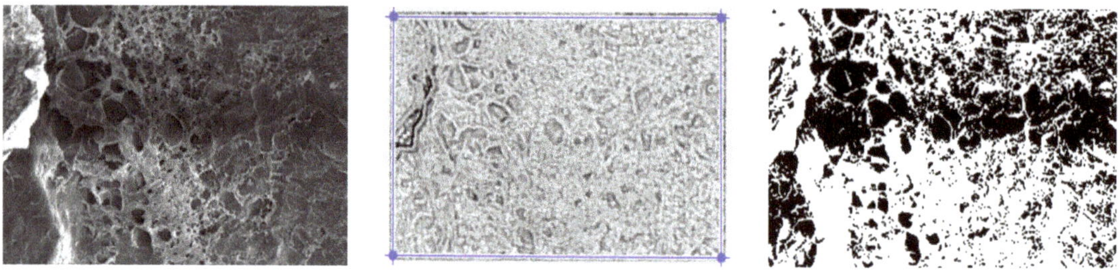

(a) (b) (c)

Figure 8. (**a**) Original 4L image, (**b**) Mask of 4L image, (**c**) The binarized version of the 4L image.

Figure 9 presents the voxels of the evaluated 4L picture, more precisely a 3D graphical portrayal/depiction, with the gray level on the oz axis, while the suitable number of pixels together with their position are marked on the other two plane axes, ox and oy, respectively [35].

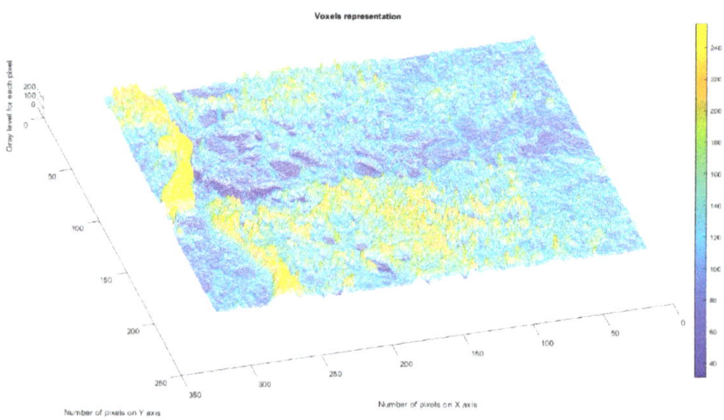

Figure 9. 3D Voxels graphical representation of 4L image.

In Figure 10, we have the two phases of 4L original image processing and the fractal analysis techniques, such as the gray scale with luminance version and the gray scale without luminance version, respectively, utilized for the calculation of the fractal parameters.

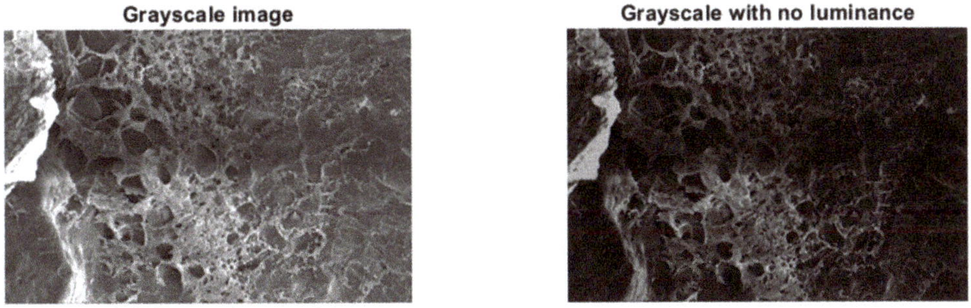

(a) (b)

Figure 10. Grayscale versions of the 4L image: (**a**) with luminance, (**b**) without luminance.

In Figure 11, we have the two phases of 4L original image processing as the filtered image version and the Wiener technique version, respectively, utilized for the calculation of the fractal parameters.

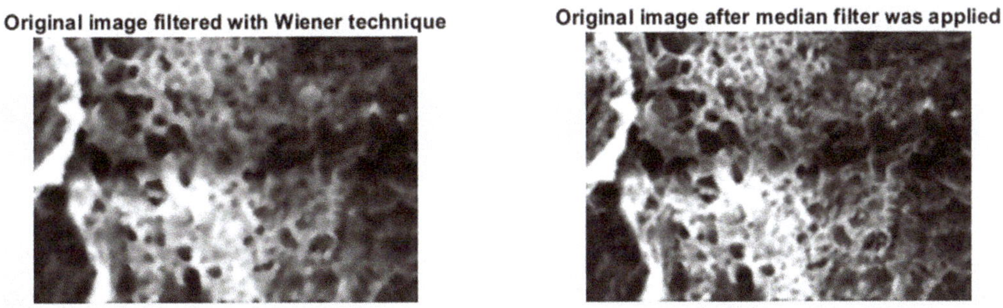

(a) (b)

Figure 11. Filtered version of the 4L image after (**a**) median filter, (**b**) Wiener technique.

In Figure 12, the fractal local dimension by box-counting algorithm in (a) and verification of the results with the HarFA program in (b), for the 4L image are presented [36].

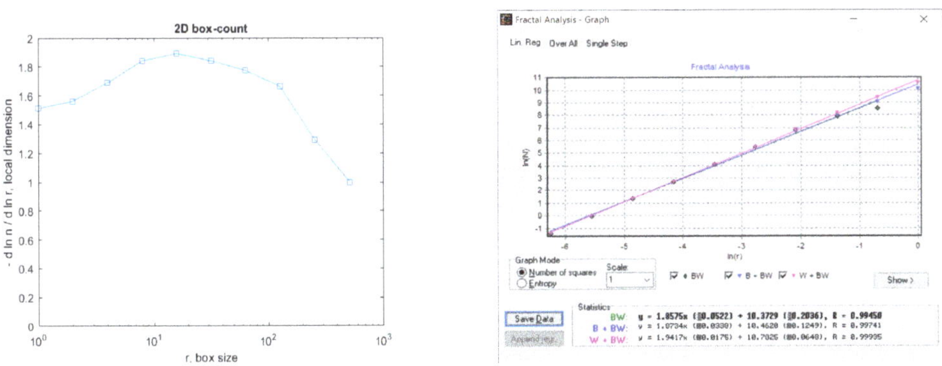

(a) (b)

Figure 12. (**a**) Fractal local dimension for the 4L image, (**b**) HarFA program for the 4L image.

In Figure 12a, the (2D) graphic to establish the fractal local dimension for the 4L image, the function of the box size *r*, by the boxes-counting procedure, is presented.

As a numerical appreciation upshot of the selected 4L picture, performed via the fractal analysis software developed by the authors, the values of Fractal Dimension FD1 = 1.668 and FD2 = 1.615, Standard Deviations $s_1 = \pm\sqrt{(\sigma^{\wedge 2})} = \pm 0.3127$ and $s_2 = \pm\sqrt{(\sigma^{\wedge 2})} = \pm 0.1445$, as well as Lacunarity value $\Lambda = 0.0526$, were estimated, seen in Table 2.

Table 2. Fractal characteristics computation of picture 4L.

FD1	Standard Deviation 1	FD2	Standard Deviation 2	Lacunarity
1.668	±0.3127	1.758	±0.1445	0.0526

Table 2 is a table with all the values of the fractal parameters obtained from the processing of the 4L image.

3.2.3. Fractal Parameters of 6L Image

In Figure 13, we have the two phases of 6L original image processing and the fractal analysis techniques, such as the mask image version and binarized version, respectively, for the calculation of the fractal parameters. The threshold, above which the binarization of the 6L image was performed, is 75.

(**a**) (**b**) (**c**)

Figure 13. (**a**) Original 6L image, (**b**) Mask of 6L image, (**c**) The binarized version of the 6L image.

Figure 14 presents the voxels of the evaluated 6L picture, more precisely a 3D graphical portrayal/depiction, with the gray level on the oz axis, while the suitable number of pixels together with their position are marked on the other two plane axes, ox and oy, respectively [35].

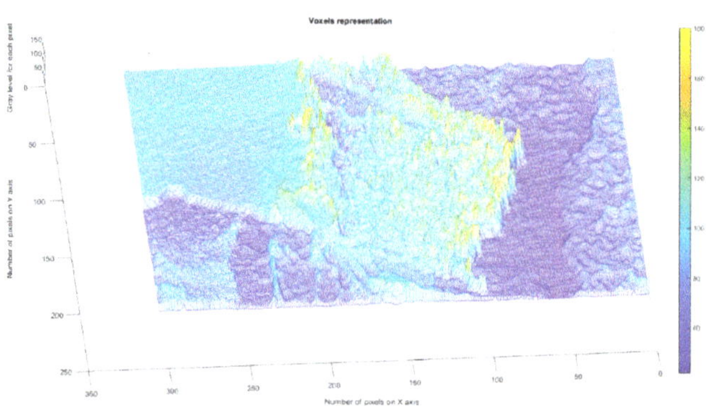

Figure 14. 3D Voxels graphical representation of 6L image.

In Figure 15, we have the two phases of 6L original image processing and the fractal analysis techniques, such as the gray scale with luminance version and the gray scale without luminance version, respectively, utilized for the calculation of the fractal parameters.

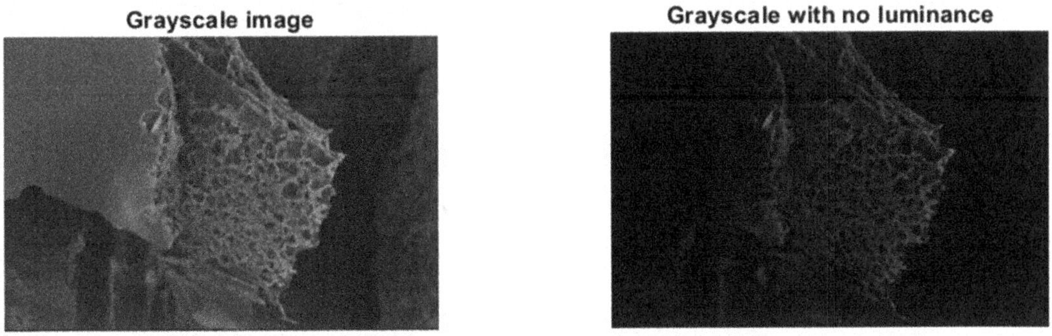

(a) (b)

Figure 15. Grayscale versions of the 6L image: (**a**) with luminance, (**b**) without luminance.

In Figure 16, we have the two phases of 6L original image processing as the filtered image version and the Wiener technique version, respectively, utilized for the calculation of the fractal parameters.

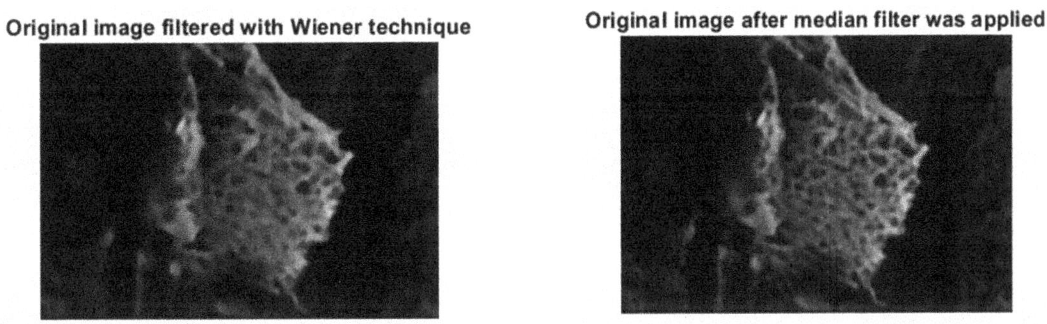

(a) (b)

Figure 16. Filtered version of the 6L image after (**a**) median filter, (**b**) Wiener technique.

In Figure 17, the fractal local dimension by box-counting algorithm in (a) and verification of the results with the HarFA program in (b), for the 6L image are presented [36].

In Figure 17a, the (2D) graphic to establish the fractal local dimension for the 6L image, the function of the box size r, by the boxes-counting procedure, is presented.

As a numerical appreciation upshot of the selected 6L picture, performed via the fractal analysis software developed by the authors, the values of Fractal Dimension FD1 = 1.624 and FD2 = 1.615, Standard Deviations $s_1 = \pm\sqrt{(\sigma^{\wedge 2})} = \pm 0.2947$ and $s_2 = \pm\sqrt{(\sigma^{\wedge 2})} = \pm 0.0298$, as well as Lacunarity value $\Lambda = 0.0381$, were estimated as in Table 3.

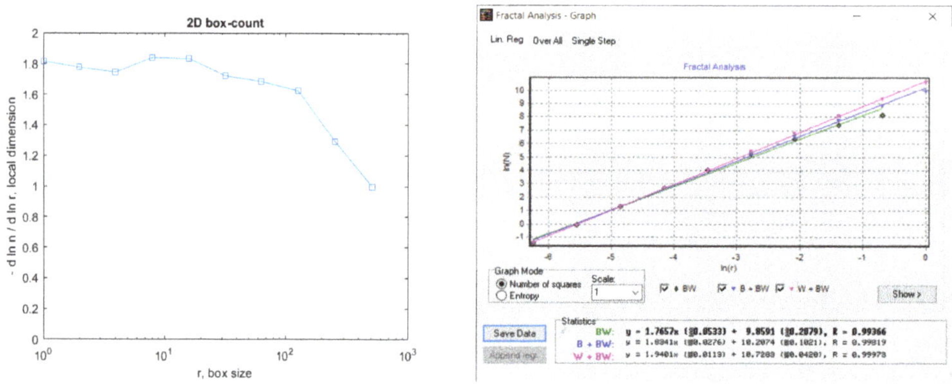

(a) (b)

Figure 17. (**a**) Fractal local dimension for the 6L image, (**b**) HarFA program for the 4L image.

Table 3. Fractal characteristics computation of picture 6L.

FD1	Standard Deviation 1	FD2	Standard Deviation 2	Lacunarity
1.624	±0.2947	1.758	±0.0298	0.0381

Table 3 is a table with all the values of the fractal parameters obtained from the processing of the 6L image.

3.2.4. Fractal Parameters of CL Image

In Figure 18, we have the two phases of CL original image processing and the fractal analysis techniques, such as the mask image version and binarized version, respectively, for the calculation of the fractal parameters. The threshold, above which the binarization of the CL image was performed, is 100.

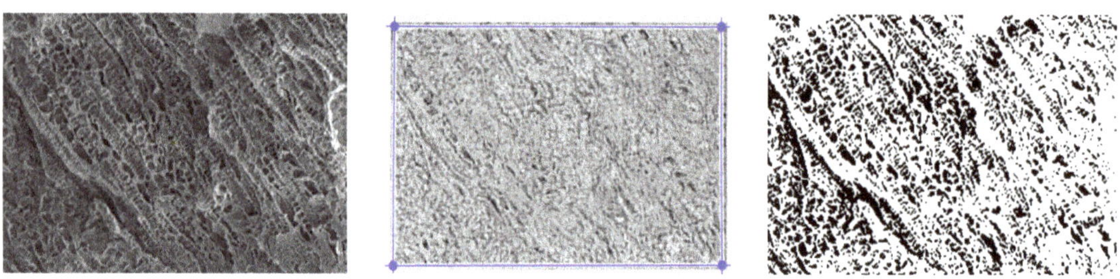

(a) (b) (c)

Figure 18. (**a**) Original CL image, (**b**) Mask of CL image, (**c**) The binarized version of the CL image.

Figure 19 presents the voxels of the evaluated CL picture, more precisely a 3D graphical portrayal/depiction, with the gray level on the oz axis, while the suitable number of pixels together with their position are marked on the other two plane axes, ox and oy, respectively [35].

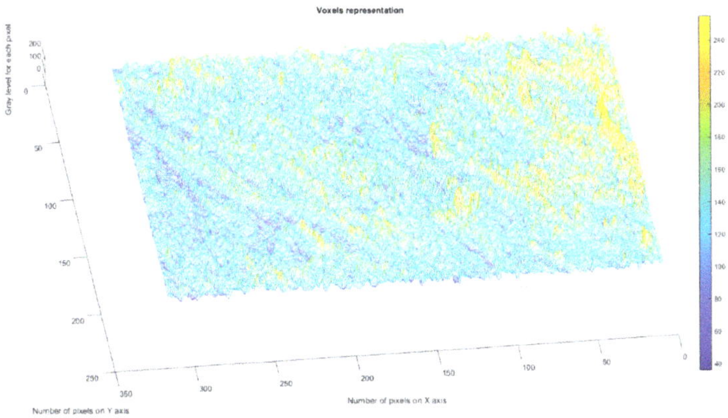

Figure 19. 3D Voxels graphical representation of CL image.

In Figure 20, we have the two phases of CL original image processing and the fractal analysis techniques, such as the gray scale with luminance version and the gray scale without luminance version, respectively, utilized for the calculation of the fractal parameters.

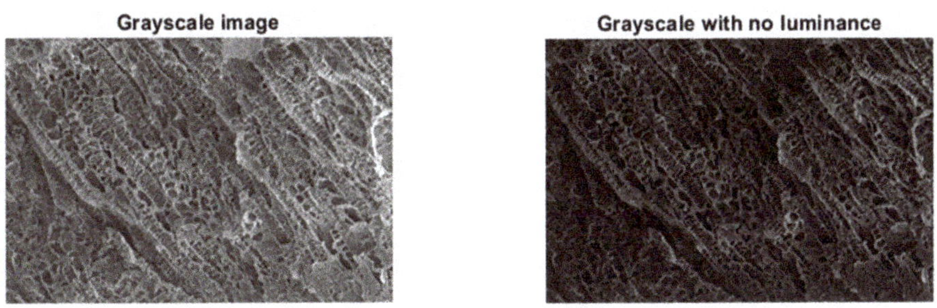

(**a**) (**b**)

Figure 20. Grayscale versions of the CL image: (**a**) with luminance, (**b**) without luminance.

In Figure 21, we have the two phases of CL original image processing as the filtered image version and the Wiener technique version, respectively, utilized for the calculation of the fractal parameters. In Figure 22. the fractal local dimension by box-counting algorithm in (a) and verification of the results with the HarFA program in (b), for the CL image are presented [36].

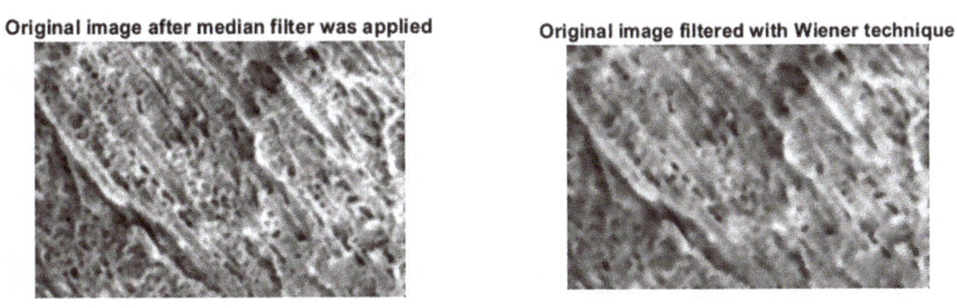

(**a**) (**b**)

Figure 21. Filtered version of the CL image after (**a**) median filter, (**b**) Wiener technique.

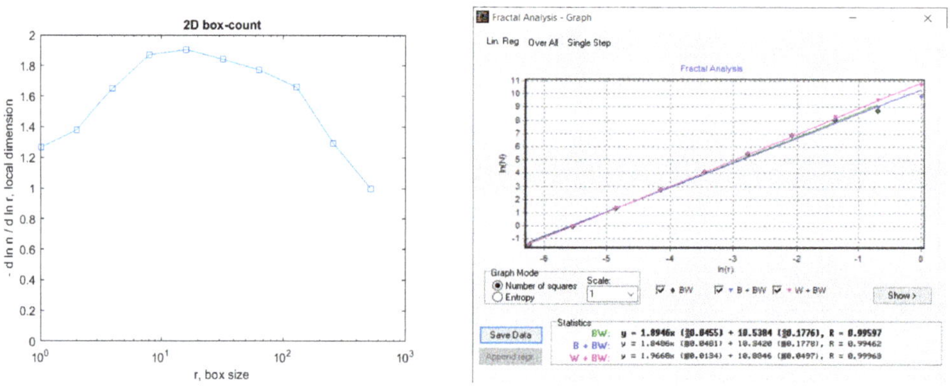

(**a**) (**b**)

Figure 22. (**a**) Fractal local dimension for 6L image, (**b**) HarFA program for the CL image.

In Figure 22a, the (2D) graphic to establish the fractal local dimension for the 2L image, function of the box size *r*, by the boxes-counting procedure, is presented.

As a numerical appreciation upshot of the selected CL picture, performed via the fractal analysis software developed by the authors, the values of Fractal Dimension FD1 = 1.678 and FD2 = 1.518, Standard Deviations $s_1 = \pm\sqrt{(\sigma^{\wedge 2})} = \pm 0.3192$ and $s_2 = \pm\sqrt{(\sigma^{\wedge 2})} = \pm 0.3339$, as well as Lacunarity value $\Lambda = 0.0274$, were estimated, seen in Table 4.

Table 4. Fractal characteristics computation of picture CL.

FD1	Standard Deviation 1	FD2	Standard Deviation 2	Lacunarity
1.678	±0.3192	1.518	±0.3339	0.0274

Table 4 is a table with all the values of the fractal parameters obtained from the processing of the CL image.

3.2.5. Processing of Experimental Results. Discussions

The effective structure determining the properties of gels morphology is characterized by fractal dimensions deduced from the used theoretical model, which also suggests that the size of primary flocs building fractal structures is one of the important factors that determine the linear viscoelastic properties of the gels. Thus, for example, the values of fractal dimension d~1.6–1.75 represent a material's superior porosity, and the fractal dimension d~1.8 is agreed in the case of diffusion-limited cluster-cluster aggregation.

The experimental data obtained, which are the subject of Table 5, were processed with appropriate calculation programs and then were represented graphically [31,37].

Table 5. Fractal characteristics computation of all images.

Index	FD1	Standard Deviation 1	FD2	Standard Deviation 2	Lacunarity
2L	1.604	±0.27987	1.596	±0.04607	0.0402
4L	1.668	±0.3127	1.758	±0.1445	0.0526
6L	1.624	±0.2947	1.758	±0.0298	0.0381
CL	1.678	±0.3192	1.518	±0.3339	0.0274

Table 5 is a table with all the values of the fractal parameters obtained from the processing of every analyzed image.

The histograms of the fractal dimension of four distinct xerogels are presented in Figure 23. In this graph, the error bars of standard deviations for each individual SEM evaluated sample can also be seen.

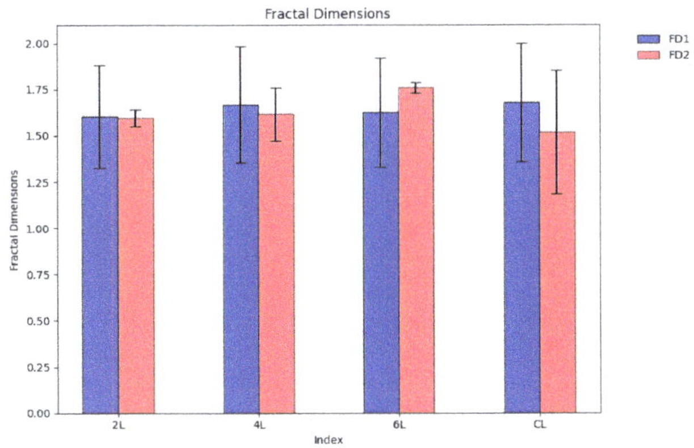

Figure 23. Fractal dimension histograms of the SEM images for four distinct chemical compounds.

The histograms colored in blue marked with FD1 represent Fractal dimension calculation with a quadratic mask, while the histograms colored in dark orange marked with FD2 represent fractal dimension calculation with a rectangular mask. It is observed that for the samples from the SEM images marked with 2L, 4L and CL, the fractal dimension for the calculation with a rectangular mask is smaller than the one calculated with the quadratic mask, while for the sample from the SEM image marked with 6L, the fractal dimension has inverted values for those two types of masks. The calculated values of the fractal dimension are in the range of 1.518 to 1.758, both calculated for a rectangular mask. The different values of the fractal dimension mean a lack of homogeneity of the pores in the four SEM-evaluated samples.

The lacunarity histograms of four distinct xerogels are presented in Figure 24.

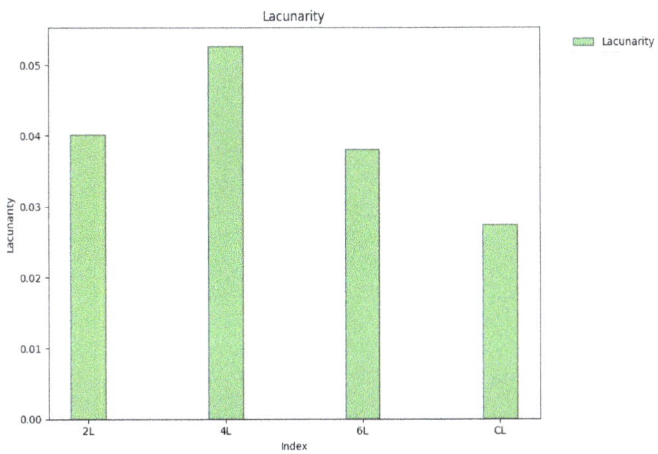

Figure 24. Histograms with the lacunarity value of the SEM images for four distinct chemical compounds.

Note. Therefore, the SEM image of the sample marked CL, which has a lacunarity equal to 0.0274, has on average the smallest pores, while the SEM image of the sample marked 4L has on average the largest pores, at a lacunarity equal to 0.0526.

4. Conclusions

In the current paper, novel massive materials (in the solid phase under normal conditions of temperature and pressure) based on TEGylated phenothiazine and chitosan, some important chemical compounds that show a great capacity to recover the mercury ions from the constitutive aqueous solutions, are presented. The xerogels exhibited a sponge-like morphology type, which works with interconnected pores and a highly heterogeneous pore distribution with diameters ranging from 2 μm to 35 μm. Several important aspects of texture morphology are distinguished inside fractal analysis. Their texture morphology assessments, based on the fractal analysis of the SEM images, were performed accurately. The four SEM pictures indexed 2L, 4L, 6L, and CL of the different chemical formulations have been examined. The obtained results, respectively the values of the calculated fractal parameters, are the subject of Table 5, the two parameters of fractal geometry discussed here being fractal dimension and lacunarity. The fractal dimensions are in the range of 1.518 to 1.758, both values calculated for a rectangular mask. The fractal dimension values d∼1.6–1.75, such as those obtained by us, represent a material with a superior porosity, as expected. The lacunarity values are contained between 0.0274 and 0.0526. Intrinsically determined by the physical presence of pores in the tested samples, they are well surprised by the analyzed SEM images. According to the recognized theoretical assertions, patterns with bigger gaps (or pores) generally prove a higher lacunarity. Based on the values of the fractal parameters presented above, we can thus say that the xerogels obtained experimentally rise to the height of the expected qualities.

The work will be continued with the complex fractal analysis of the SEM images for the same xerogels. Furthermore, a multifractal model will be developed regarding the presence of mercury alongside the basic xerogel as its host, namely a theory of two bodies acting in solidarity (together, as one). All these things will be the subject of a future scientific paper.

5. Materials and Methods

5.1. Materials

The following materials such as reduced molecular weight chitosan, triethylene glycol monomethyl ether 97%, phenothiazine 98%, sodium hydride 95%, phosphorus (V) oxychloride 99%, and magnesium sulfate (MgSO4) 99.5%, have been acquired/bought from the Sigma-Aldrich Company(St. Louis, MO, USA). TEGylated phenothiazine refers to the fact that the phenothiazine heterocycle has been substituted with TEG (Triethylene Glycol), in other words, the phenothiazine core has a TEG chain attached. The chitosan molecular mass (198 kDa) was obtained by viscosity measurement founded on Mark–Houwink formula, with an Ubbelohde type viscometer. The acetylation degree (DA = 18%) was established from 1H-NMR. Acetone, dichloromethane (DCM) 99.5%, and dichloroethane (DCE) 99% were acquired from ROTH Company. Acetic acid and mercury (II) acetate were bought from VWR Company [7]. All resolvents and reagents were utilized as they were received.

5.2. Equipment and Methods

The spectra in the infrared domain were realized with the help of a Spectrometer of type FTIR Bruker Vertex 70 (Bruker Optics Company, 40 Manning Road, Manning Park, Billerica, MA, USA), operating in transmission regime, utilizing KBr granules, at normal temperature and pressure, by 2 cm^{-1} resolution. Origin8 software was utilized to process the recorded spectra. The NMR investigations were executed on the spectrometer of type Bruker Avance Neo (400 MHz) (International Equipment Trading Ltd., 955 Campus Drive, Mundelein, IL, USA) provided with a space probe-type instrument based on four 5 mm diameter cores and unbiased z-axis-gradient detection. The both spectra, photoluminescence, and UV-Vis absorption, were realized on a spectrophotometer of type PerkinElmer LS 55 (International Equipment Trading Ltd., 955 Campus Drive, Mundelein, IL, USA) and a spectrophotometer of type Agilent Cary 60 UV-Vis (Oxford Instruments Company, Abingdon, Oxfordshire, England) respectively, on solid specimens. The SEM pictures were produced with a Scanning Electron Microscope of type SEM EDAX—Quanta 200 (PHILIPS Company, Eindhoven, The Netherlands), at a smaller energy of 20 Kev for the electrons [7].

Author Contributions: Conceptualization, V.-P.P. and M.-A.P.; methodology, V.-P.P.; software, M.-V.N. and V.-A.P.; validation, V.-P.P., M.-A.P. and V.-A.P.; formal analysis, V.-P.P., M.-A.P. and V.-A.P.; investigation, V.-A.P. and M.-A.P.; resources, V.-A.P. and M.-A.P.; data curation, V.-A.P. and M.-V.N.; writing—original draft preparation, V.-P.P.; writing—review and editing, M.-A.P. and V.-P.P.; visualization, M.-V.N. and V.-A.P.; supervision, V.-P.P.; project administration, V.-P.P. All authors have read and agreed to the published version of the manuscript.

Funding: This research received no external funding.

Institutional Review Board Statement: Not applicable.

Informed Consent Statement: Not applicable.

Data Availability Statement: The data used to support the findings of this study cannot be accessed due to commercial confidentiality.

Acknowledgments: The co-authors M.A. Paun, V.A. Paun and V.P. Paun would like to thank Jenica Paun, for her continuous kind support.

Conflicts of Interest: The authors declare no conflict of interest.

References

1. Wu, D.; Sedgwick, A.C.; Gunnlaugsson, T.; Akkaya, E.U.; Yoon, J.; James, T.D. Fluorescent Chemosensors: The Past, Present and Future. *Chem. Soc. Rev.* **2017**, *46*, 7105–7123. [CrossRef]
2. Balali-Mood, M.; Naseri, K.; Tahergorabi, Z.; Khazdair, M.R.; Sadeghi, M. Toxic Mechanisms of Five Heavy Metals: Mercury, Lead, Chromium, Cadmium, and Arsenic. *Front. Pharmacol.* **2021**, *12*, 227. [CrossRef] [PubMed]
3. Tchounwou, P.B.; Yedjou, C.G.; Patlolla, A.K.; Sutton, D.J. Heavy Metal Toxicity and the Environment. *Exp. Suppl.* **2012**, *101*, 133–164. [PubMed]
4. Bansod, B.; Kumar, T.; Thakur, R.; Rana, S.; Singh, I. A Review on Various Electrochemical Techniques for Heavy Metal Ions Detection with Different Sensing Platforms. *Biosens. Bioelectron.* **2017**, *94*, 443–455. [CrossRef]
5. Kinuthia, G.K.; Ngure, V.; Beti, D.; Lugalia, R.; Wangila, A.; Kamau, L. Levels of Heavy Metals in Wastewater and Soil Samples from Open Drainage Channels in Nairobi, Kenya: Community Health Implication. *Sci. Rep.* **2020**, *10*, 8434. [CrossRef]
6. New Limits for Heavy Metals in Food Supplements. Available online: https://www.gmp-compliance.org/gmp-news/new-limits-for-heavy-metals-in-food-supplements (accessed on 11 March 2023).
7. Cibotaru, S.; Ailincai, D.; Andreica, B.I.; Cheng, X.; Marin, L. TEGylated Phenothiazine-Imine-Chitosan Materials as a Promising Framework for Mercury Recovery. *Gels* **2022**, *8*, 692. [CrossRef]
8. Dórea, J.G. Integrating Experimental (In Vitro and In Vivo) Neurotoxicity Studies of Low-Dose Thimerosal Relevant to Vaccines. *Neurochem. Res.* **2011**, *36*, 927–938. [CrossRef]
9. Aragay, G.; Pons, J.; Merkoçi, A. Recent Trends in Macro-, Micro-, and Nanomaterial-Based Tools and Strategies for Heavy-Metal Detection. *Chem. Rev.* **2011**, *111*, 3433–3458. [CrossRef]
10. Fouda, S.R.; El-Sayed, I.E.; Attia, N.F.; Abdeen, M.M.; Abdel Aleem, A.A.H.; Nassar, I.F.; Mira, H.I.; Gawad, E.A.; Kalam, A.; Al-Ghamdi, A.A.; et al. Mechanistic Study of Hg(II) Interaction with Three Different α-Aminophosphonate Adsorbents: Insights from Batch Experiments and Theoretical Calculations. *Chemosphere* **2022**, *304*, 135253. [CrossRef] [PubMed]
11. Mahmoud, M.E.; Abdelwahab, M.S.; Ibrahim, G.A.A. The Design of SnO_2-Crosslinked-Chitosan Nanocomposite for Microwave-Assisted Adsorption of Aqueous Cadmium and Mercury Ions. *Sustain. Chem. Pharm.* **2022**, *28*, 100731. [CrossRef]
12. Zhang, L.; Jiao, X.; Zhang, H.; He, S.; Cheng, X. Novel Chitosan–Naphthalimide–Amino Acid Fluorescent Powder for Selective Detection and Removal of Hg^{2+}/Hg^{+} and Fe^{2+} in Aqueous Solution. *Chem. Pap.* **2022**, *76*, 7037–7049. [CrossRef]
13. Lin, H.; Duan, Y.; Zhao, B.; Feng, Q.; Li, M.; Wei, J.; Zhu, Y.; Li, M. Efficient Hg(II) Removal to Ppb Level from Water in Wider PH Based on Poly-Cyanoguanidine/Graphene Oxide: Preparation, Behaviors, and Mechanisms. *Colloids Surf. A Physicochem. Eng. Asp.* **2022**, *641*, 128467. [CrossRef]
14. Saenchoopa, A.; Klangphukhiew, S.; Somsub, R.; Talodthaisong, C.; Patramanon, R.; Daduang, J.; Daduang, S.; Kulchat, S. A Disposable Electrochemical Biosensor Based on Screen-Printed Carbon Electrodes Modified with Silver Nanowires/HPMC/Chitosan/Urease for the Detection of Mercury (II) in Water. *Biosensors* **2021**, *11*, 351. [CrossRef] [PubMed]

15. Michailidou, G.; Koumentakou, I.; Liakos, E.V.; Lazaridou, M.; Lambropoulou, D.A.; Bikiaris, D.N.; Kyzas, G.Z. Adsorption of Uranium, Mercury, and Rare Earth Elements from Aqueous Solutions onto Magnetic Chitosan Adsorbents: A Review. *Polymers* **2021**, *13*, 3137. [CrossRef]
16. Seidi, F.; Reza Saeb, M.; Huang, Y.; Akbari, A.; Xiao, H. Thiomers of Chitosan and Cellulose: Effective Biosorbents for Detection Removal and Recovery of Metal Ions from Aqueous Medium. *Chem. Rec.* **2021**, *21*, 1876–1896. [CrossRef]
17. Czarnobaj, K. Sol–gel-processed silica/polydimethylsiloxane/calcium xerogels as polymeric matrices for Metronidazole delivery system. *Polym. Bull.* **2011**, *66*, 223–237. [CrossRef]
18. Wang, X.; Ben Ahmed, N.; Alvarez, G.S.; Tuttolomondo, M.V.; Hélary, C.; Desimone, M.F.; Coradin, T. Sol-gel encapsulation of biomolecules and cells for medicinal applications. *Curr. Top. Med. Chem.* **2015**, *15*, 223–244. [CrossRef]
19. Paun, M.-A.; Nichita, M.-V.; Paun, V.-A.; Paun, V.-P. Xerogels Morphology Details by Multifractal Analysis and Scanning Electron Microscopy Images Evaluations of 5-Fluorouracil Release from Chitosan-Based Matrix. *Gels* **2022**, *8*, 820. [CrossRef]
20. Lungu, R.; Paun, M.-A.; Peptanariu, D.; Ailincai, D.; Marin, L.; Nichita, M.-V.; Paun, V.-A.; Paun, V.-P. Biocompatible Chitosan-Based Hydrogels for Bioabsorbable Wound Dressings. *Gels* **2022**, *8*, 107. [CrossRef]
21. Postolache, P.; Borsos, Z.; Paun, V.A.; Paun, V.P. New Way in Fractal Analysis of Pulmonary Medical Images. *Univ. Politeh. Buchar. Sci. Bull. Ser. A Appl. Math. Phys.* **2018**, *80*, 313–322.
22. Husain, A.; Nanda, M.N.; Chowdary, M.S.; Sajid, M. Fractals: An Eclectic Survey, Part-I. *Fractal Fract.* **2022**, *6*, 89. [CrossRef]
23. Nichita, M.V.; Paun, M.A.; Paun, V.A.; Paun, V.P. Fractal analysis of brain glial cells. Fractals dimension and lacunarity. *Univ. Politeh. Buchar. Sci. Bull. Ser. A Appl. Math. Phys.* **2019**, *81*, 273–284.
24. Sarkar, N.; Chaudhuri, B.B. An efficient approach to estimate Fractal Dimension of textural images. *Pattern Recognit.* **1992**, *25*, 1035–1041. [CrossRef]
25. Karperien, A.L.; Jelinek, H.F. Box-Counting Fractal Analysis: A Primer for the Clinician. In *The Fractal Geometry of the Brain*, Second, chapter; Di Ieva, A., Ed.; Springer Series in Computational Neuroscience; Springer Science+Business Media: New York, NY, USA, 2016.
26. Karperien, A.; Jelinek, H.F.; Miloševic, N.T. Reviewing Lacunarity Analysis and Classification of Microglia in Neuroscience. In Proceedings of the 8th European Conference on Mathematical and Theoretical Biology, Kraków, Poland, 28 June–2 July 2011; 6p
27. Paun, M.A.; Nichita, M.V.; Paun, V.A.; Paun, V.P. Fractal analysis in the quantification of medical imaging associated with multiple sclerosis pathology. *Frontiers in Bioscience-Landmark.* **2022**, *27*, 66. [CrossRef]
28. Bordescu, D.; Paun, M.A.; Paun, V.A.; Paun, V.P. Fractal analysis of Neuroimagistic. Lacunarity degree, a precious indicator in the detection of Alzheimer's disease. *Univ. Politeh. Buchar. Sci. Bull. Ser. A Appl. Math. Phys.* **2018**, *80*, 309–320.
29. Popovic, N.; Radunovic, M.; Badnjar, J.; Popovic, T. Fractal dimension and lacunarity analysis of retinal microvascular morphology in hypertension and diabetes. *Microvasc. Res.* **2018**, *118*, 36–43. [CrossRef]
30. Paun, M.A.; Postolache, P.; Nichita, M.V.; Paun, V.A.; Paun, V.P. Fractal Analysis in Pulmonary CT Images of COVID-19-Infected Patients. *Fractal Fract.* **2023**, *7*, 285. [CrossRef]
31. Pereira, L.M. Fractal Pharmacokinetics. *Comput. Math. Methods Med.* **2010**, *11*, 161–184. [CrossRef]
32. De Melo, R.H.C.; Conci, A. How Succolarity could be used as another fractal measure in image analysis. *Telecommun. Syst.* **2013**, *52*, 1643–1655. [CrossRef]
33. Xia, Y.; Cai, J.; Perfect, E.; Wei, W.; Zhang, Q.; Meng, Q. Fractal dimension, lacunarity and succolarity analyses on CT images of reservoir rocks for permeability prediction. *J. Hydrol.* **2019**, *579*, 124198. [CrossRef]
34. Paun, M.-A.; Paun, V.-A.; Paun, V.-P. A Multifractal Vision of 5-Fluorouracil Release from Chitosan-Based Matrix. *Gels* **2022**, *8*, 661. [CrossRef] [PubMed]
35. Li, Y.; Qi, X.; Chen, Y.; Wang, L.; Li, Z.; Sun, J.; Jia, J. Voxel field fusion for 3d object detection. In Proceedings of the IEEE Conference on Computer Vision and Pattern Recognition 2022, New Orleans, LA, USA, 21–24 June 2022; 13p.
36. Available online: http://imagesci.fch.vut.cz/includes/harfa_download.inc.php (accessed on 7 March 2023).
37. Scott, D.W. *Statistics: A Concise Mathematical Introduction for Students, Scientists, and Engineers*; John Wiley & Sons, Inc.: Hoboken, NJ, USA, 2020.

gels

Communication

Study of the Possibility of Using Sol–Gel Technology to Obtain Magnetic Nanoparticles Based on Transition Metal Ferrites

Nina Shabelskaya [1,2], Sergey Sulima [3,*], Elena Sulima [3], Oleg Medennikov [1], Marina Kulikova [1], Tatyana Kolesnikova [1] and Svetlana Sushkova [2]

1 Department of Ecology and Industrial Safety, Faculty of Technology, Platov South-Russian State Polytechnic University (NPI), 346400 Novocherkassk, Russia
2 Laboratory "Agrobiotechnologies for Improving Soil Fertility and Quality of Agricultural Products", D.I. Ivanovsky Academy of Biology and Biotechnology, Southern Federal University, 344090 Rostov-on-Don, Russia
3 Department of Chemical Technologies, Faculty of Technology, Platov South-Russian State Polytechnic University (NPI), 346400 Novocherkassk, Russia
* Correspondence: s_sulima@mail.ru

Abstract: The article presents results for the magnetic nanoparticles sol–gel method synthesis of cobalt (II) ferrite and organic–inorganic composite materials based on it. The obtained materials were characterized using X-ray phase analysis, scanning and transmission electron microscopy, Scherrer, Brunauer–Emmett–Teller (BET) methods. A composite materials formation mechanism is proposed, which includes a gelation stage where transition element cation chelate complexes react with citric acid and subsequently decompose under heating. The fundamental possibility of obtaining an organo–inorganic composite material based on cobalt (II) ferrite and an organic carrier using the presented method has been proved. Composite materials formation is established to lead to a significant (5–9 times) increase in the sample surface area. Materials with a developed surface are formed: the surface area measured by the BET method is 83–143 m^2/g. The resulting composite materials have sufficient magnetic properties to be mobile in a magnetic field. Consequently, wide possibilities for polyfunctional materials synthesis open up for various applications in medicine.

Keywords: sol–gel method; cobalt (II) ferrite; composite materials; magnetic nanoparticles; magnetic composites

1. Introduction

An increase in the technogenic load on an individual in modern society leads to the development of many diseases, including cancer. Research in the field of medicine is becoming more and more interdisciplinary, new methods of treatment are created. Interventional procedures are widely used [1–3], such as transarterial chemoembolization, catheter intra-arterial delivery of nanoparticles, etc. The possibilities of non-surgical and drug-free direct selective influence on individual cells in an external magnetic field by oscillations of magnetic nanoparticles fixed on the cell membranes or introduced into cells have been investigated [4–7]. Study of the processes of synthesis of organo–inorganic composites is an interesting and challenging task. The formation of dual-use materials containing an inorganic magnetic part and an organic one will expand the functional capabilities of substances. Ferrites transition elements are promising candidates for medical applications due to a successful combination of their technical parameters: ease of synthesis and presence of magnetic properties. Iron oxide compounds exhibit antibacterial properties and are non-toxic [8], which allows them to be considered as promising materials for theranostics [9,10]. The nanoparticles' developed surface displays an increased adsorption activity [11], which can be used for organic–inorganic composite materials synthesis [12,13]. One of the promising methods for obtaining magnetic nanoparticles is the sol–gel method [14–16]. Often these processes are lengthy and time-consuming. For example, in [16], transition metal salts were dissolved in deionized water, citric acid was added, and stirred for 1 h. The mixed solution was heated in a water bath at 95 °C for 3 h. Then, the resulting transparent sol was dried at a temperature of 95 °C for 12 h. Finally, the mixture was calcined at 400 °C for 2 h in a muffle furnace. Therefore, the main goal of the study is to investigate the possibility of using simply the sol–gel technology to obtain cobalt (II) ferrite magnetic nanoparticles and composite organo–inorganic materials based on them.

2. Results and Discussion

During the synthesis of cobalt (II) ferrite, when solutions of transition metal nitrates and ammonia were mixed, a brown amorphous precipitate formation was observed, which may be associated with the preparation of a mixture of cobalt (II) and iron (II) hydroxides according to Reactions (1) and (2):

$$Co(NO_3)_2 + 2NH_4OH = Co(OH)_2 + 2NH_4NO_3 \quad (1)$$

$$Fe(NO_3)_3 + 3NH_4OH = Fe(OH)_3 + 3NH_4NO_3 \quad (2)$$

with the introduction of a citric acid solution, the precipitate dissolved. This may be due to the transition metal citrates formation according to Reactions (3) and (4):

$$Co(OH)_2 + C_6H_8O_7 = Co(C_6H_6O_7) + 2H_2O \quad (3)$$

$$Fe(OH)_3 + C_6H_8O_7 = Fe(C_6H_5O_7) + 3H_2O, \quad (4)$$

During further heat treatment, the solution first turned into a viscous gel and then into a solid porous material. Presumably, bulky citrate complexes of transition metals were formed. In this case, citric acid acted in a dual role: as an organic fuel and a dispersing agent that prevented the formation of cobalt (II) ferrite large particles. When the heat treatment stopped, the formation of a black powder was observed.

Figure 1a shows an FC sample X-ray diffraction pattern. Analysis of the X-ray diffraction pattern reveals that the sample is only cobalt (II) ferrite having a cubic spinel structure (PDF Number 010-74-6403). The array parameter a and the crystallite sizes D calculated with the Scherrer formula are given in Table 1. The lines of the diffractogram for pure cobalt (II) ferrite are the clearest and the most pronounced in comparison with similar lines for other synthesized materials (FC/C and FC/S composites). This may be due to the formation of the most defect-free sample.

Table 1. Synthesized materials characteristics.

Sample	a, nm	D_{311}, nm	D_{400}, nm	D_m, nm	S_{BET}, m^2·g^{-1}
FC	0.8386	11.6	8.8	10.2	16
FC/S	0.8386	12.1	5.3	8.7	83
FC/C	0.8381	9.9	8.8	9.4	143

Figure 1b,c show X-ray diffraction patterns of the synthesized composite materials. They are represented by lines characteristic of cobalt (II) ferrite (PDF Number 010-74-3419 for FC/C (Figure 1b), PDF Number 010-74-6403 for FC/S (Figure 1c)) (the carbon part is X-ray amorphous).

Remarkably, cobalt (II) ferrite formation on the surface of an organic support leads to the formation of smaller magnetic particles, as the diffraction pattern peaks are blurred and less clear.

Figure 2a, shows an FC sample micrograph (SEM image). Evidently, a porous material is formed. The specific surface area values (S_{BET}) measured by the BET method are given in Table 1. Figure 3 shows the adsorption–desorption isotherms of N_2.

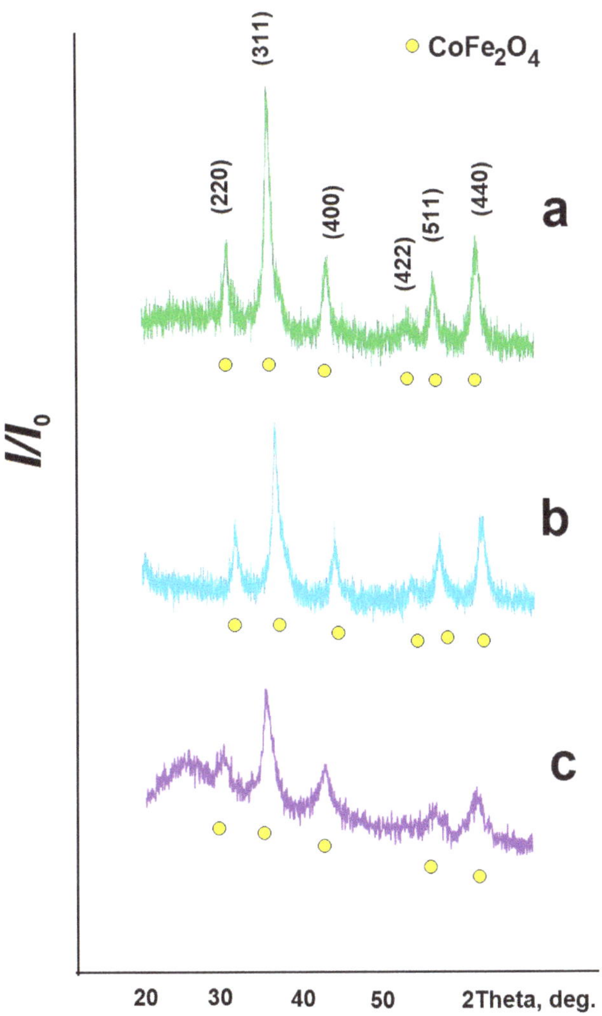

Figure 1. X-ray pattern of synthesized samples: (**a**)—$CoFe_2O_4$, (**b**)—$CoFe_2O_4$/C, (**c**)—$CoFe_2O_4$/S.

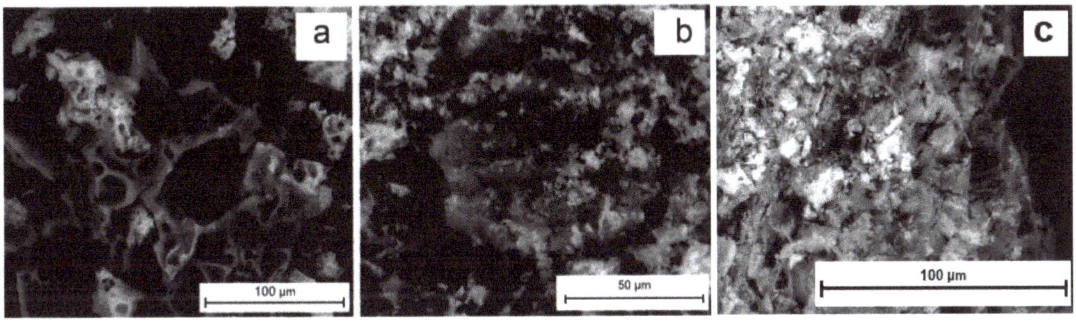

Figure 2. Micrographs of synthesized samples: (**a**)—$CoFe_2O_4$, (**b**)—$CoFe_2O_4$/C, (**c**)—$CoFe_2O_4$/S.

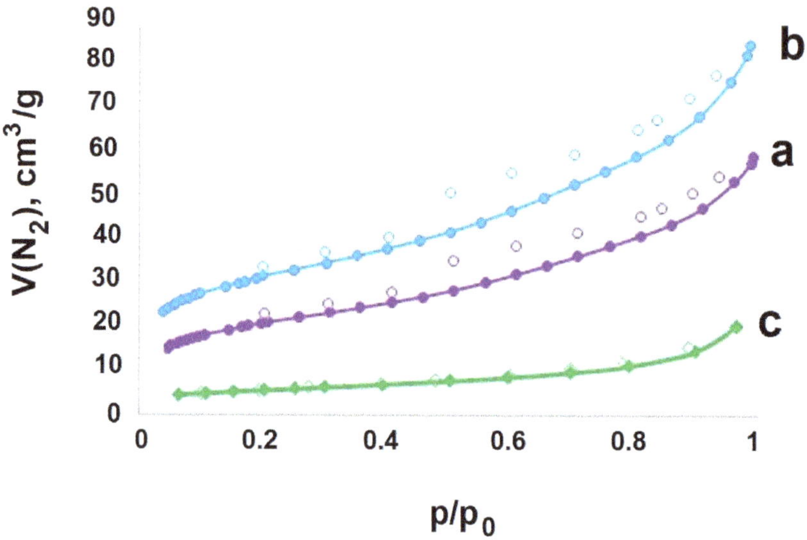

Figure 3. Adsorption–desorption isotherms of N_2 samples: (**a**)—$CoFe_2O_4$/S, (**b**)—$CoFe_2O_4$/C, (**c**)—$CoFe_2O_4$.

The resulting cobalt (II) ferrite exhibits the properties of a magnetic material. Figure 4 shows the magnetic hysteresis loops of synthesized samples. According to the measurements, it demonstrates saturation magnetization of M_S = 44 emu/g (electromagnetic unit/g).

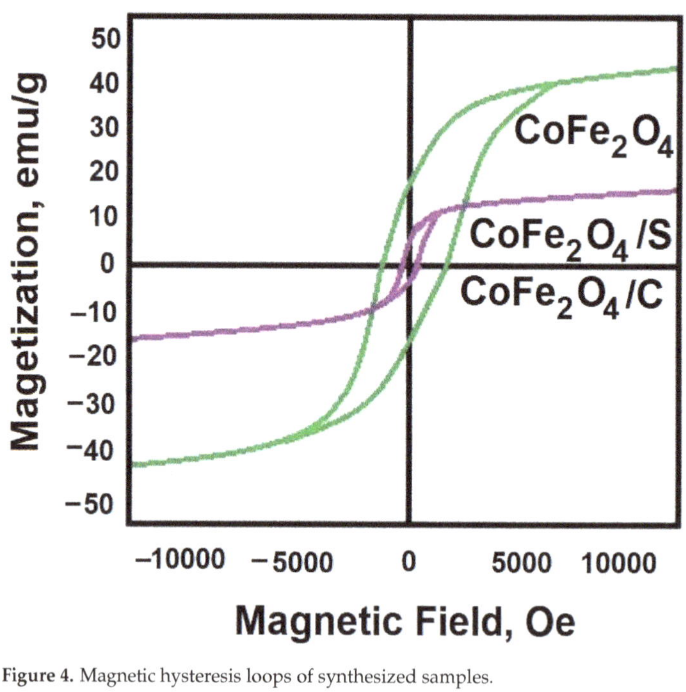

Figure 4. Magnetic hysteresis loops of synthesized samples.

For a possible use as a targeted medication delivery carrier, for example, composite materials with a magnetic and an organic components formation by the sol–gel method were necessary to study. For this purpose, the FC/S and FC/C composite material synthesis was carried out. In the process, all technological operations were carried out similarly to those described above, with one difference: either biochar from sunflower husks or activated carbon from coconut shell, respectively, were initially placed in the reaction vessel.

As a result, an organic–inorganic composite material was successfully obtained, consisting of cobalt (II) ferrite and a carbon-containing component.

Figure 5 shows a schematic representation of citrate complexes formation and cobalt (II) ferrite formation on the surface of a carbonaceous material.

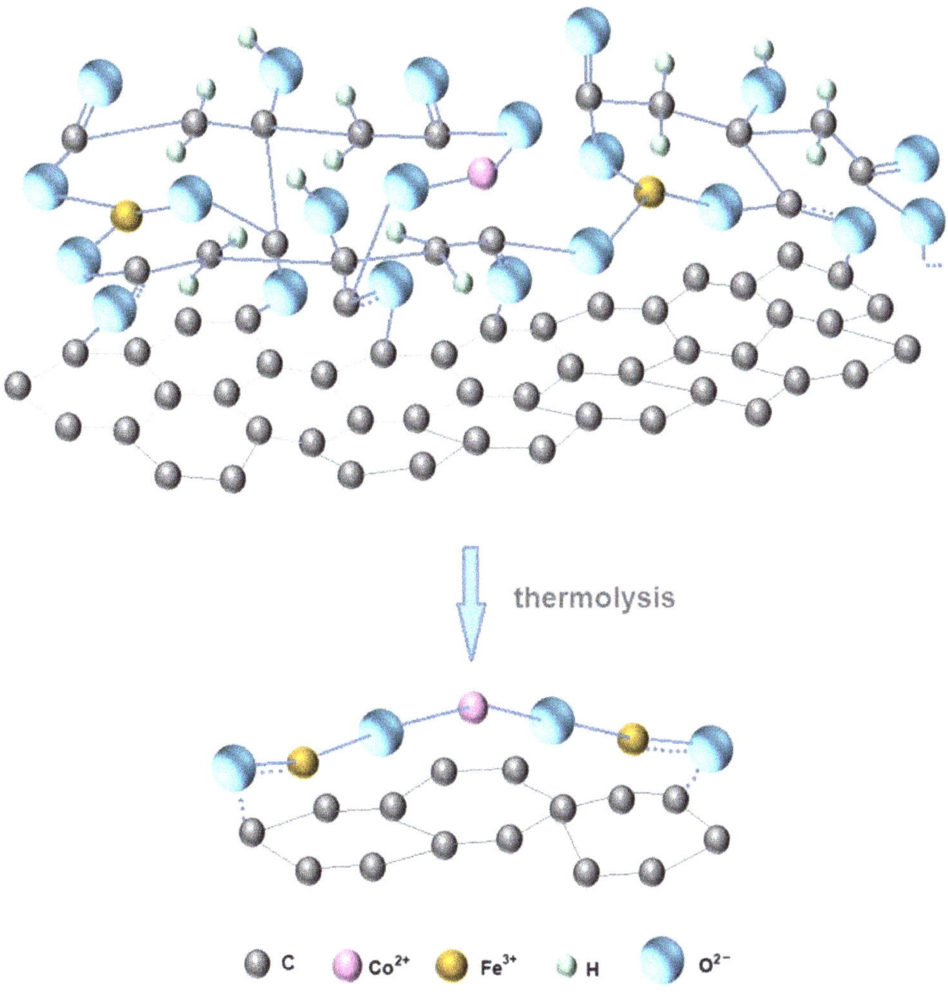

Figure 5. Possible structure of the intermediate citrate complex of transition elements.

Figure 2b shows an FC/C composite material sample micrograph. Cobalt (II) ferrite cluster formation on the surface of a carbon particle is observed. A similar effect is noted for the FC/S composite formation; see Figure 6a for a TEM image of the synthesized material. Regular geometric cobalt (II) ferrite particles in the shape of octahedrons are observed to form on the biochar surface. The particle sizes are shown in Figure 6b. The particle size of cobalt (II) ferrite lies in the range up to −200 nm. The crystallite sizes calculated by the Scherrer method, and the surface area values measured by the BET method, are presented in Table 1.

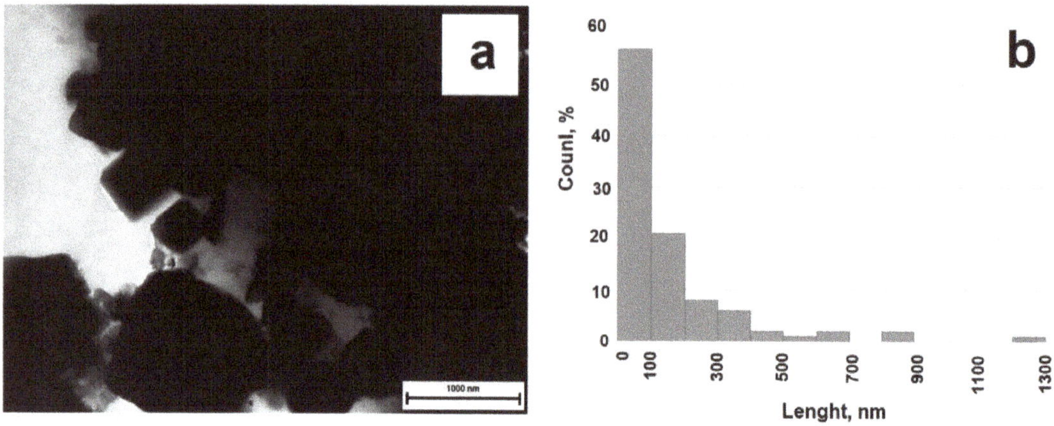

Figure 6. TEM image of the sunflower/$CoFe_2O_4$ biochar sample (**a**), particle size distribution (**b**).

The organic–inorganic composite materials formation is accompanied by a significant (5–9 times) increase in the surface area of the material (see Table 1). This result suggests a higher activity of the synthesized materials in comparison with pure cobalt (II) ferrite in the processes associated with surface activity—adsorption, catalytic. At the same time, the synthesized composites display sufficiently pronounced magnetic characteristics (with a value of M_S = 14 emu./g for $CoFe_2O_4$/S (and $CoFe_2O_4$/C)) to be moved in an aqueous solution by a magnet (see Figure 7).

Figure 7. *Cont.*

Figure 7. Magnetic separation of synthesized material: composites (**a–d**), schematic representation (**e**,**f**).

The synthesized materials were tested in the process of adsorption of chromium (VI) compounds from an aqueous solution. The results are shown in Table 2.

Table 2. Adsorption properties of synthesized materials.

Sample	m_s, g	m_d, mg	N, mg·g^{-1}
FC	0.020	0.0008	0.04
FC/S	0.026	0.1815	6.98
FC/C	0.023	0.0161	0.70

Thus, a simple sol–gel method for the synthesis of low-dimensional cobalt (II) ferrite has been applied. This method makes it possible to synthesize organo–inorganic composites in situ.

The results obtained are promising and could be used as a material synthesis method for a number of applications, such as targeted medication delivery.

3. Conclusions

A simple one-stage method for producing cobalt (II) ferrite has been proposed. The material structure formation mechanism is investigated, including the formation of a gel-like matrix with distributed transition elements cation inclusions in it. During gel thermal decomposition, nanosized ferrite particles are formed.

Using the method described, organo–inorganic composite materials based on cobalt (II) ferrite and an organic carrier are possible to obtain, as the study proved. Unlike existing analogues, the method allows to obtain composites in situ.

Composite materials formation is established to lead to a significant (5–9 times) increase in the surface area of the samples. Materials with a developed surface are formed: the surface area measured by the BET method is 83–143 m^2/g.

The resulting composite materials have sufficient magnetic properties to be mobile in a magnetic field. Hence, wide possibilities open up for the synthesis of polyfunctional materials for various applications.

4. Materials and Methods

4.1. Materials

The following solutions were used as starting materials: cobalt (II) nitrate with the concentration of 183 g/L; iron (III) nitrate with the concentration of 242 g/L; an ammonia solution of 25% (wt.); and a 1200 g/L solution of citric acid. Analytical purity reagents of $Co(NO_3)_2 \cdot 6H_2O$, $Fe(NO_3)_3 \cdot 9H_2O$, $NH_3 \cdot H_2O$, $C_6H_8O_7$ were used.

4.2. Synthesis of Biochar

To test an organic–inorganic composite material synthesis possibility, sunflower husks biochar and activated carbon from coconut shells were used as organic doping agents. The choice is determined by the availability of organic matter, its low cost, as well as the need to dispose of agricultural waste (sunflower husk and coconut peel) to obtain products with high added value. Sunflower husk biomass was produced in a muffle furnace by pyrolysis in a sealed metal vessel at the temperature of 100–700 °C with a temperature change step of 200 °C. The temperature rise rate was 11 °C/min. The holding time at temperatures of 100, 300, 500 °C was 20 min; at a temperature of 700 °C it was 45 min. Ready-made activated charcoal from coconut shells was used.

4.3. Material Synthesis

Cobalt (II) ferrite and composite materials synthesis was carried out in one stage with the formation of CoF_2O_4 and an in situ organic–inorganic composition. The formation of cobalt (II) ferrite on the surface of a carbon-containing substance was carried out according to the method developed by the authors and described in [17,18]. In a typical procedure, 25 mmol of $Co(NO_3)_2$ and 50 mmol of $Fe(NO_3)_3$ were simultaneously placed in a stainless-steel reaction vessel under continuous stirring. Then, 200 mmol of ammonia and 156 mmol of citric acid were introduced. The mixture was then heated until the liquid completely evaporated. Consequently, a viscous gel-like substance was formed, which gradually turned into a solid porous material. With further heating, a process of intense decomposition occurred, accompanied by the release of gaseous substances and the reaction system glowing. Heating continued until gaseous products of metal nitrates thermolysis stopped emitting. To obtain a composite organic–inorganic material, 25 g of biochar were first placed in the reaction vessel. Next, the synthesis procedure described above was carried out. The samples were designated as: FC for cobalt (II) ferrite, FC/S for cobalt (II) ferrite and sunflower husk biochar, FC/C for cobalt (II) ferrite and coconut shell activated carbon.

4.4. Characteristics

The following methods were used to characterize the resulting composite materials: X-ray diffraction (XRD), transmission electron microscopy, Brunauer–Emmett–Teller (BET) analysis, and the Scherrer method.

The phase composition was studied on an ARL X'TRA X-ray diffractometer (Ecublens, Switzerland) (monochromatized Cu-Kα radiation used) by point-by-point scanning (step of 0.01°, accumulation time at a point of 2 s) in the range of 2θ values from 20 °C to 70 °C. The crystallite size was calculated along line 311 using the Scherrer Equation (5):

$$D = 0.94 \cdot \lambda / (B \cdot \cos\theta) \quad (5)$$

where D is the average crystal size, nm, λ is the wavelength of X–ray radiation, nm, B is the value of the peak line width at half its height, rad., cosθ is the cosine value of the angle for the peak.

The calculation was carried by peak 311 (designated D_{311}) and 400 (designated D_{400}), the middle value (D_m).

The surface area was determined on a ChemiSorb 2750 V apparatus. Nitrogen physical adsorption isotherms were obtained at 77 °K. Prior to measurement, the samples were degassed.

Ultrastructural images of the samples were obtained on a Tecnai G12 BioTwins transmission electron microscope (FEI, Philips, Czech Republic, Černovice) in bright field mode at an accelerating voltage of 100 kV.

The magnetic characteristics were measured using a Lake Shore VSM 7404 model magnetometer at the Shared Use Center of the International Research Institute for Intelligent Materials of the Southern Federal University, Russia, at room temperature (300 °K) with a maximum field value of 18 kOe. The test sample was fixed on a quartz rod, placed in an electromagnetic field. The rod with the sample was driven by a mechanical oscillator. As a result, the saturation magnetization M_S of the obtained samples was determined.

The adsorption activity of the synthesized materials was studied on a model solution of potassium dichromate with a concentration of 5 mmol/L. In this case, 5 mL of the initial potassium dichromate solution was passed through a reaction column containing 2 cm^3 of the adsorbent. Next, 5 mL of deionized water was passed through the column, and the content of dichromate ions in the washing solution was determined. The determination was carried out by the intrinsic color of the solution by the photocolorimetric method using a KFK-2-UHL 4.2 device (Yurga, Russia) with a wavelength of 364 nm. The degree of purification (N) was calculated by Equation (6):

$$N = m_d \cdot / m_s \quad (6)$$

where N is the adsorption capacity of the sample, mg/g, m_d is the mass of desorbed $Cr_2O_7^{2-}$ ions, m_s is the mass of the sample, g.

Author Contributions: Conceptualization, N.S. and S.S. (Sergey Sulima); methodology, N.S., S.S. (Sergey Sulima) and O.M.; validation, N.S., S.S. (Sergey Sulima) and S.S. (Svetlana Sushkova); formal analysis, E.S., M.K. and T.K.; data curation, N.S. and S.S. (Sergey Sulima); writing—original draft preparation, N.S. and S.S. (Sergey Sulima); writing—review and editing, S.S. (Sergey Sulima) and E.S.; visualization, O.M.; funding acquisition, S.S. (Sergey Sulima). All authors have read and agreed to the published version of the manuscript.

Funding: The research was financially supported by the Ministry of Science and Higher Education of the Russian Federation project on the development of the Young Scientist Laboratory (no. LabNOTs-21-01AB, FENW-2021-0014).

Institutional Review Board Statement: Not applicable.

Informed Consent Statement: Not applicable.

Data Availability Statement: Not applicable.

Acknowledgments: The authors are grateful to the staff of the Platov South-Russian State Polytechnic University (NPI) Shared Use Center, Yatsenko, A.N. for assistance in capturing and interpreting XRF data.

Conflicts of Interest: The authors declare no conflict of interest. The funders had no role in the design of the study; in the collection, analyses, or interpretation of data; in the writing of the manuscript; or in the decision to publish the results.

References

1. Yang, C.M.; Lee, J.; Lee, S.Y.; Lee, H.; Chathuranga, K.; Lee, J.; Park, W. Silk Fibroin/Tannin/ZnO Nanocomposite Hydrogel with Hemostatic Activities. *Gels* **2022**, *8*, 650. [CrossRef] [PubMed]
2. Friedrich, R.P.; Cicha, I.; Alexiou, C. Iron oxide nanoparticles in regenerative medicine and tissue engineering. *Nanomaterials* **2021**, *11*, 2337. [CrossRef] [PubMed]
3. Fan, X.; Liu, S.; Ruan, K. Application of Magnetic Nanoparticles Fe_3O_4 in the Field of Orthopedics and Medicine. In Proceedings of the 2021 2nd International Academic Conference on Energy Conservation, Environmental Protection and Energy Science, ICEPE 2021, Shanghai, China, 4–5 July 2021. [CrossRef]
4. Koksharov, Y.A.; Khomutov, G.B.; Taranov, I.V.; Gulyaev, Y.V.; Gubin, S.P. Magnetic nanoparticles in medicine: Progress, problems, and advances. *J. Commun. Technol. Electron.* **2022**, *67*, 101–116. [CrossRef]
5. Taheri, M.; Mohebat, R.; Moslemin, M.H. Facile synthesis of ZnO-$H_3PW_{12}O_40$@Fe_3O_4/EN-MIL-101(CR) as magnetic core-shell nanoparticles derived from metal-organic frameworks: Application in medicine and its catalytic activity. *Mol. Cryst. Liq. Cryst.* **2021**, *730*, 1954276. [CrossRef]
6. Ullah Khan, A.; Chen, L.; Ge, G. Recent development for biomedical applications of magnetic nanoparticles. *Inorg. Chem. Commun.* **2021**, *134*, 108995. [CrossRef] [PubMed]
7. Senthilkumar, S.; Dhivya, V.; Sathya, M.; Rajendran, A. Synthesis and characterization of magnetite/hydroxyapatite nanoparticles for biomedical applications. *J. Exp. Nanosci.* **2021**, *16*, 160–180. [CrossRef]
8. Ezealigo, U.S.; Ezealigo, B.N.; Aisida, S.O.; Ezema, F.I. Iron oxide nanoparticles in biological systems: Antibacterial and toxicology perspective. *JCIS Open* **2021**, *4*, 100027. [CrossRef]
9. Sankaranarayanan, S.A.; Thomas, A.; Rengan, A.K.; Revi, N.; Ramakrishna, B. Iron oxide nanoparticles for theranostic applications—Recent advances. *J. Drug Deliv. Sci. Technol.* **2022**, *70*, 103196. [CrossRef]
10. Pawlik, P.; Pruba, M.; Blasiak, B.; Depciuch, J.; Parlinska-Wojtan, M.; Kitala, D.; Vorobyova, S.; Konakov, A.; Ivashkevich, O.; Stec, M.; et al. Application of iron-based magnetic nanoparticles stabilized with triethanolammonium oleate for theranostics. *J. Mater. Sci.* **2022**, *57*, 4716–4737. [CrossRef]
11. Hamza, M.F.; Wei, Y.; Benettayeb, A.; Wang, X.; Guibal, E. Efficient removal of uranium, cadmium and mercury from aqueous solutions using grafted hydrazide-micro-magnetite chitosan derivative. *J. Mater. Sci.* **2020**, *55*, 4193–4212. [CrossRef]
12. Muratov, D.G.; Kozhitov, L.V.; Yakushko, E.V.; Vasilev, A.A.; Popkova, A.V.; Tarala, V.A.; Korovin, E.Y. Synthesis, structure and electromagnetic properties of FeCoAl/C nanocomposites. *Mod. Electron. Mater.* **2021**, *7*, 99–108. [CrossRef]
13. Thamer, B.M.; Abdulhameed, M.M.; El-Newehy, M.H. Tragacanth Gum Hydrogel-Derived Trimetallic Nanoparticles Supported on Porous Carbon Catalyst for Urea Electrooxidation. *Gels* **2022**, *8*, 292. [CrossRef] [PubMed]
14. Yuheng, G.; Haiyan, L.; Shouqiang, L.; Leilei, C.; Zhenhai, L. Study on the Structure, Magnetic Properties and Mechanism of Zn-Doped Yttrium Iron Garnet Nanomaterial Prepared by the Sol-gel Method. *Gels* **2022**, *8*, 325. [CrossRef]
15. Fernández-Hernán, J.P.; Torres, B.; López, A.J.; Rams, J. The Role of the Sol-Gel Synthesis Process in the Biomedical Field and Its Use to Enhance the Performance of Bioabsorbable Magnesium Implants. *Gels* **2022**, *8*, 426. [CrossRef] [PubMed]
16. You, Y.; Shi, Z.; Li, Y.; Zhao, Z.; He, B.; Cheng, X. Magnetic cobalt ferrite biochar composite as peroxymonosulfate activator for removal of lomefloxacin hydrochloride. *Sep. Purif. Technol.* **2021**, *2721*, 118889. [CrossRef]
17. Shabelskaya, N.P.; Egorova, M.A.; Vasileva, E.V.; Polozhentsev, O.E. Photocatalytic properties of nanosized zinc ferrite and zinc chromite. *Adv. Nat. Sci. Nanosci. Nanotechnol.* **2021**, *12*, 015004. [CrossRef]
18. Shabelskaya, N.; Egorova, M.; Radjabov, A.; Burachevskaya, M.; Lobzenko, I.; Minkina, T.; Sushkova, S. Formation of biochar nanocomposite materials based on $CoFe_2O_4$ for purification of aqueous solutions from chromium compounds (VI). *Water* **2023**, *15*, 93. [CrossRef]

gels MDPI

Article

Evaluation of Low-Toxic Hybrid Sol-Gel Coatings with Organic pH-Sensitive Inhibitors for Corrosion Protection of AA2024 Aluminium Alloy

Eva Jaldo Serrano [1], Jesús López-Sánchez [2], Federico García-Galván [3], Aida Serrano [2], Óscar Rodríguez de la Fuente [1,4], Violeta Barranco [5], Juan Carlos Galván [5] and Noemí Carmona [1,4,*]

1 Departamento de Física de Materiales, Facultad de Ciencias Físicas, Universidad Complutense de Madrid, Plaza Ciencias SN, 28040 Madrid, Spain
2 Instituto de Cerámica y Vidrio (ICV-CSIC), C/Kelsen 5, Campus de Cantoblanco, 28049 Madrid, Spain
3 Department of Engineering, School of Architecture, Engineering and Design, Universidad Europea de Madrid, 28670 Villaviciosa de Odón, Spain
4 Instituto de Magnetismo Aplicado (IMA) "Salvador Velayos", A6 Km 22,5, 28230 Las Rozas Madrid, Spain
5 Centro Nacional de Investigaciones Metalúrgicas (CENIM-CSIC), Avda. Gregorio del Amo 8, 28040 Madrid, Spain
* Correspondence: ncarmona@ucm.es

Abstract: Today's environmental needs require the reduction of the weight of vehicles, thus reducing fuel consumption and associated emissions. For this reason, the use of light alloys is being studied, which, due to their reactivity, must be protected before use. In this work, the effectiveness of a hybrid sol-gel coating doped with various organic environmentally friendly corrosion inhibitors applied to an AA2024 lightweight aluminium alloy is evaluated. Some of the inhibitors tested are pH indicators, acting as both corrosion inhibitors and optical sensors for the surface of the alloy. Samples are subjected to a corrosion test in a simulated saline environment and characterised before and after the test. The experimental results regarding their best inhibitor performance for their potential application in the transport industry are evaluated.

Keywords: sol-gel; corrosion inhibitors; protective coating; sensor; chlorophenol red; AA2024 aluminium alloy

1. Introduction

The aerospace, automotive, and marine industries have probably made the greatest anti-corrosion efforts in the last two decades, producing advances that have focused on the substitution of various materials mainly for lightweighting, resulting in fuel and CO_2 emission reductions. In addition, an effort is being made to use more corrosion-resistant materials, better protective finishes and improvements in protective coatings [1].

Light alloys based on aluminium and magnesium are strong candidates for reducing the density of materials used in the transport industry. Their advantages include versatility in production and forming [2–6], a good mechanical strength-to-weight ratio [7], good formability, high fatigue strength [8], and a wide range of possible applications [9]. However, these alloys are very prone to corrosion phenomena, which occur naturally by reacting with oxygen and other gases or ions present in the environment to achieve a state of thermodynamic equilibrium of the material that reduces the internal energetics of the system. In the case of aluminium alloys, the corrosion process is frequent due to the fact that it is a highly reactive element and its alloys contain intermetallic compounds, which can aggravate corrosion phenomena, as they have a different electrochemical potential than the metallic substrate [10,11].

The most common solution used in the past for the protection of metal surfaces was the application of chromium plating treatments, which achieve a good surface finish and improve corrosion behaviour. However, due to the toxicity of hexavalent chromium, international regulations have prohibited its use for some years [12]. Recently, various methods such as painting, lacquering, galvanising and anodising are being developed and are widely used industrially [13]. They are economical but less environmentally friendly. On the other hand, plasma treatment is common, which also improves the hardness and wear resistance of the surface but generates pores that can create areas more prone to corrosion [14–17].

Among the options that have undergone significant development in recent years, the sol-gel process stands out as a very versatile, environmentally friendly and economically viable synthesis method [18,19]. It basically consists of the chemical reaction of precursors in three successive stages: a first hydrolysis stage, followed by polycondensation and finally, thermal densification.

Starting from a colloidal suspension of very small particles in a solvent, the sol is converted into a gel with a two-phase solid-liquid structure. The precursors used in the sol-gel process can be organic or inorganic. If denser coatings are desired, organic-inorganic hybrids can be prepared to form a more barrier-like layer against solvents [20].

The advantages of using the sol-gel method include the following: the high purity of the products obtained due to the high control in the composition of the precursor materials; no pre-treatment is necessary to cover the metal surface; the synthesis can be carried out at room temperature and at atmospheric pressure; the process has low toxicity and is economical; and the coatings show good chemical resistance, constituting a powerful corrosion inhibitor for metals and alloys. In addition, the intrinsic porosity of sol-gel coatings allows them to host other molecules while preserving their properties, acting as active corrosion protection methods [21].

Various corrosion inhibitors can be added to increase the coating's protection. Historically, inorganic salts were used (e.g., Ce^{3+} ions [22], La^{3+} ions [23] or Zr^{3+} ions [24]), and later organic molecules and their derivatives that are less toxic and biodegradable. Most of the early organic inhibitors were weak acids, and some of the more recently used ones are molecules containing several heteroatoms, such as amino acids [25,26]. The use of polymers such as polyaniline or polypyrrole as conductive coatings has also been tested. In this case, the protection mechanism is not due to a barrier effect, but it is believed that a passive oxide film is formed on the metal surface by an anodisation process [27,28], which confers the resulting metal/coating system with self-repairing properties and active corrosion protection [29]. Various types of nanoparticles (NPs) have also been tested [30,31]. TiO_2 NPs have been shown to improve anti-corrosion and antibacterial properties [32]. Graphene and its oxide nanosheets have also been used in sol-gel coatings due to their specific surface areas and impermeability to ions, water, and oxygen. However, they do not provide active protection to the substrate [33,34]. On the other hand, when seeking maximum environmental friendliness, such as toluylalanine they are an effective temporary inhibitor of steel in humid atmospheres [35], and a number of morpholine-Mannich derivatives act as volatile corrosion inhibitors. In the case of seawater, volatile corrosion inhibitors that trap molecules or ions in the medium to prevent their contact with the metal substrate are of interest. Examples of these organic inhibitors that can also be found in the literature are turmeric, L-cysteine or methylene blue, among others [36–38]. Some of the organic corrosion inhibitors can, in some cases, offer a double functionality: an efficient response to corrosion phenomena, thanks to the free radicals in their structure, such as turmeric, dimethyl yellow, chlorophenol red or crystal violet with antioxidant power, and the visual perception of pH changes that can identify corrosion processes as they occur, acting as optical sensors. To the best of the authors' knowledge, this approach has not been developed before.

In this scenario, the aim of this research work is focused on the design, preparation, characterisation, and evaluation of multifunctional sol-gel coatings with different environmentally friendly inhibitors, some of them never used before for this purpose, to be applied on AA2024 aluminium alloy for its corrosion protection.

2. Results and Discussion

2.1. Coatings Characterisation

Table 1 shows prepared coatings with their corresponding nomenclature, preparation and test conditions.

Table 1. Description of the prepared samples.

Sample Name	Sol-Gel Coating	Corrosion Inhibitor Added	Corrosion Test Performed
AL	NO	NO	NO
REF	NO	NO	YES
B	YES	NO	YES
CIS	YES	L-cysteine	YES
CUR	YES	Curcumin	YES
DY	YES	Dimethyl Yellow	YES
CR	YES	Chlorophenol Red	YES
MB	YES	Methylene Blue	YES
CV	YES	Crystal Violet	YES

Coatings thickness was obtained from the interference of reflection spectra in the ultraviolet-visible and near-infrared regions, according to [39]. Table 2 shows specific values for each coating. All coatings had approximate thicknesses between 500 nm and 1 μm.

Table 2. Coatings thickness.

Sample	B	CIS	CUR	DY	CR	MB	CV
Thickness (nm)	$(104 \pm 15)\cdot 10^1$	$(102 \pm 15)\cdot 10^1$	514 ± 95	718 ± 52	$(91 \pm 10)\cdot 10^1$	874 ± 56	719 ± 54

The thickness of the coatings is different due to the different corrosion inhibitors added. The molecules chosen as inhibitors are very different in size and the functional groups they possess. These molecules are responsible for the final pH of the sols and the charge density on the metal, in this case, on the Si ions of the alkoxides. Hydrolysis is facilitated when Si ions are bound to Figure 1 OH groups and inhibited when Si ions are bound to -OR groups. The higher the number of H bound to O, the greater the hydrolysis. In the same way, the literature explains the effect of catalysts. Acidic or basic catalysis influences the hydrolysis and polycondensation rates and thus affects the final gel structure. Acids increase the kinetics of the reaction, and hydrolysis is complete if sufficient water is added. However, high acid concentrations slow down the kinetics of polycondensation and less branched gels are obtained.

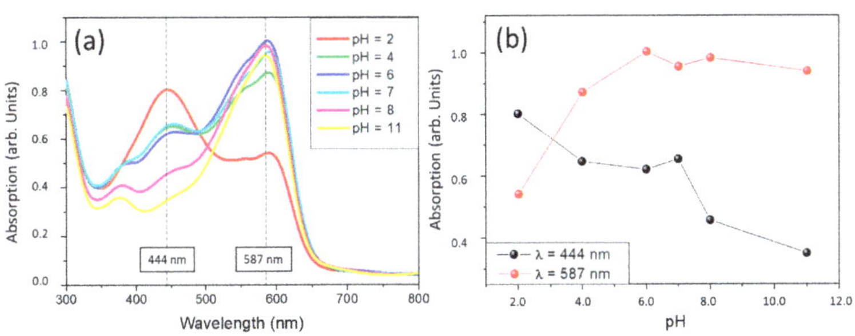

Figure 1. VIS absorption spectra of sample CR on glass substrate at different pHs (**a**). Evolution of maximum absorption spectra at 444 and 587 nm (**b**).

Conversely, alkaline or less acidic conditions result in deprotonation of Si-OH ligands forming Si-O- which slows down the rate of hydrolysis, perhaps because the charge density on the Si ion decreases, and the kinetics of the polycondensation reaction increases [40].

In the sols prepared in this work, differences in pHs from 1 to 4 have been observed, although the only difference in the composition of the sols is the added inhibitor (pH of sols DY, CR, and MB could not be measured due to the intense colour of the solution). It seems that the more acidic the pH of the sol initially is, the lower the thickness of the final coating obtained, which is in agreement with what the literature explains about higher hydrolysis kinetics and lower polycondensation kinetics; that is, the more acidic the sol is. As all the coatings were prepared at the same time and left to stir for the same amount of time, the difference in the degree of polymerisation, viscosity and, therefore, the thickness of the coating can only be due to the catalytic effect of the corrosion inhibitor added, maintaining the pH acidic or lowering it, depending on the molecule.

Some of the inhibitors used have a dual function. On the one hand, they act as corrosion inhibitors; on the other hand, they are pH indicators, changing their colour when the environment becomes more alkaline or acidic. These inhibitors are CV, CR, and DY.

The optical behaviour of the coatings as a function of the environmental pH has been studied by immersion in buffers of different pHs for 1 min. CV and DY samples showed overlapped spectra independently of the pH of the solution. Therefore, it seems that these pH indicator molecules that have been introduced into the sol-gel coating lose their functionality when they are encapsulated. The spectra collected for the CR samples are shown in Figure 1. Results show that the CR samples are sensitive to pH changes, and absorption differences can be observed in the visible region. The molecules of chlorophenol red would act as pH sensors despite being encapsulated within the sol-gel coating and would not lose their properties as pH indicators.

In the case of sample CR, two absorption maxima are observed at 444nm and 587nm at an initial pH = 2, corresponding to violet and yellow wavelength absorption, respectively, which corroborates the visually observed colours. It turns from a reddish colour to an orange-yellow with decreased pH.

Contact angle measurements show a moderate hydrophilic behaviour for all coatings (Figure 2). All obtained values are less than 90°. Lower values have been obtained for the sample REF, which is the uncoated Al alloy, and for the sample CV. The angle between the water and the CV surface is relatively low, so the area in contact with the water is higher in this case, which could be detrimental because water droplets are more likely to remain on the surface. The best results are obtained with B, CIS, and CR samples.

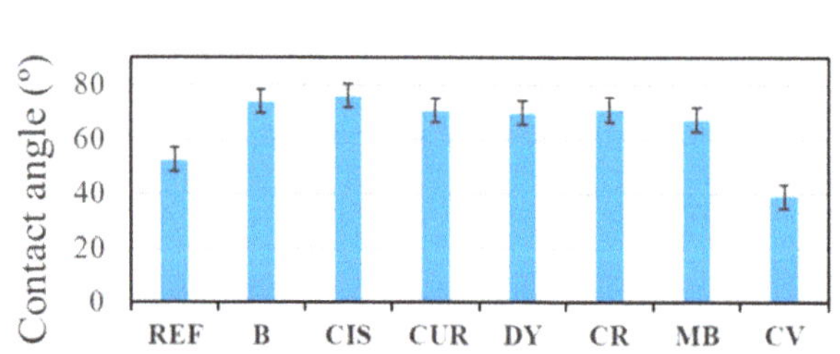

Figure 2. Contact angle values of the samples.

2.2. Weathering Test

At the beginning of the corrosion test, the pH of all 0.6M NaCl solutions was neutral, but after 14 days of sample immersion, the pH of almost all solutions changed to acidic, pH was around 4, or even lower. In samples solution DY and CV, the pH after the test was around 3. This result indicates that the medium has acidified as the corrosion process of the alloys has been carried out, so it seems that protons have been released to the saline medium, and the following reaction may have happened.

$$\downarrow Al + 3H_2O \rightarrow \downarrow Al(OH)_3 + 3\,H^+ \quad (1)$$

All samples show signs of surface deterioration after two weeks of immersion in the corrosive medium. Figure 3 shows the comparison of the samples before and after the corrosion test. Some of them show signs of corrosion, as the REF sample is fully covered with white deposits of oxides well adhered to the surface. Part of the aluminium in the alloy may have been dissolved, as indicated in Equation (1), thus partially compensating the weight loss due to Al dissolution by the formation of $Al(OH)_3$ deposits. Thus, it may explain the lightweight variation of only some milligrams in the samples after the weathering test. On the bottom of some of the sample containers, small black dots are observed together with abundant white crystallisations, which could confirm these statements.

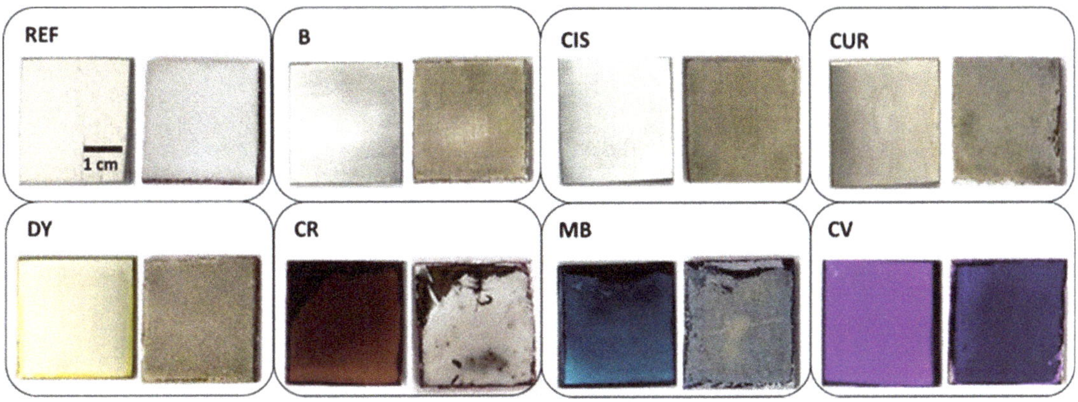

Figure 3. Overview picture of sol-gel coated samples before (**left**) and after (**right**) the corrosion test.

2.3. Optical Microscopy Characterisation

Optical microscopy images of the surface of the samples before and after the corrosion test have been taken to monitor microstructural changes (Figure 4). Before the test, micrographs of the surfaces show homogeneous and transparent coatings colourless for samples B, CIS, CUR, and DY, a red coating for sample CR, blue for sample MB, and violet for sample CV. The polishing lines of the surfaces are also visible. Once the corrosion test was completed, samples were rinsed three times with distilled water to remove surface deposits, and they were again optically characterised.

In the sample without inhibitor (B) and the samples with curcumin (CUR), L-cysteine (CIS), methylene blue (MB), and crystal violet (CV), localised corrosion is observed, which can be in the form of pitting between the grains of alloy AA2024 and oxides resulting from the reaction with the corrosive medium. As for the sample without sol-gel (REF), the sample with dimethyl yellow (DY) and the sample with chlorophenol red (CR), a more uniform distribution of these corrosion products is observed, forming a thick corrosion layer.

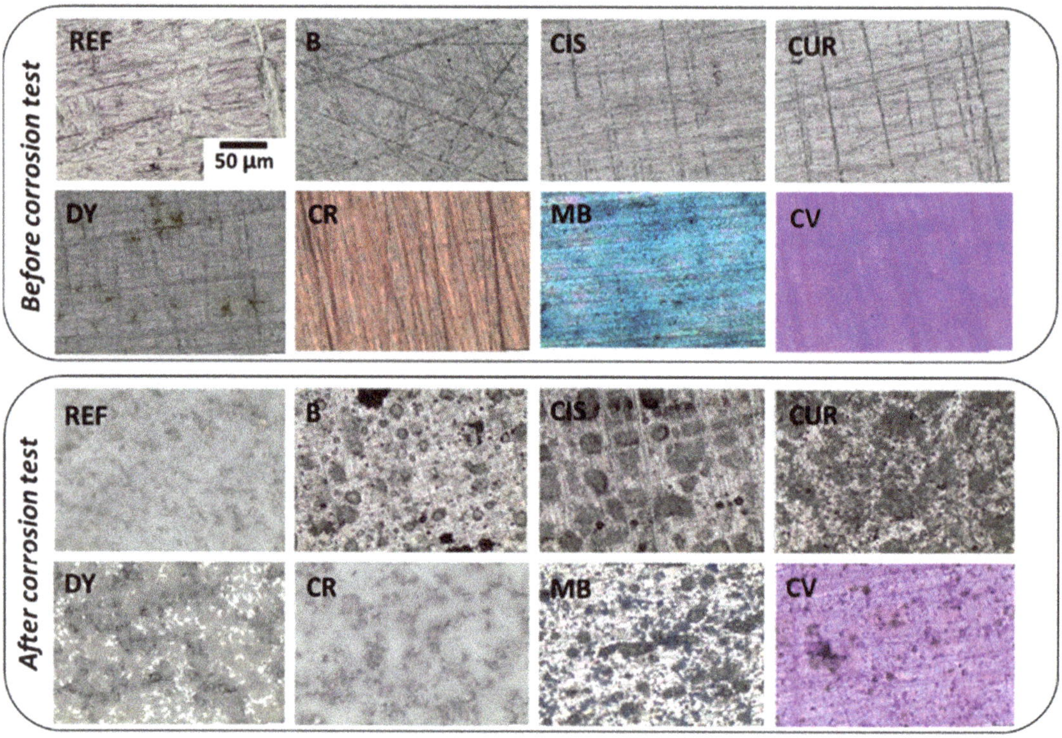

Figure 4. OM images of the sol-gel coated samples: before and after the corrosion test.

2.4. Weight Variation

Each sample was weighed before and after the corrosion test. Once the samples were washed with distilled water three times after the test, they were placed in a desiccator and weighed one week later so that all the samples would lose their residual moisture after the 14 days of testing and washing. Figure 5 shows the weight variation of each sample subjected to the corrosion test (before and after the test). The weight variation is the final weight minus the initial weight.

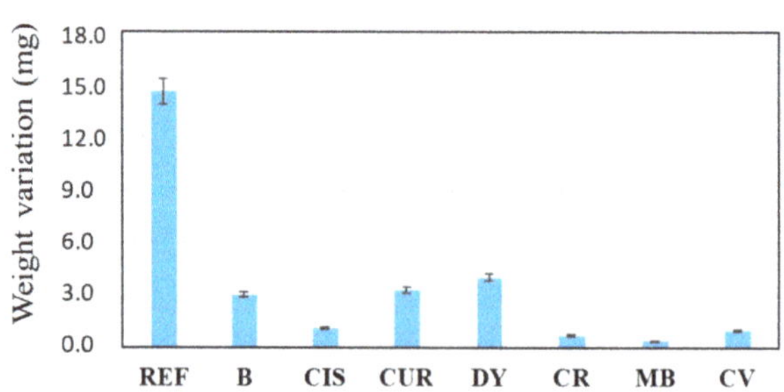

Figure 5. Difference in weight of samples before and after the corrosion test.

An increase in the weight of the samples before and after the corrosion test, in the order of milligrams, is observed. On the one hand, a certain reduction in the weight of the sample was expected due to the loss of metal released by reacting with the ions in the medium. On the other hand, the same metal ions have most likely reacted with OH- ions, forming metal hydroxides deposited on the surface again, increasing the final weight of the samples.

In the weight variation, the uncoated sample (REF) stands out with an increase of almost 15 mg in weight. On the other hand, the rest of the samples can be grouped into two blocks, i.e., those that have barely modified their weight (<2 mg), CIS, CR, MB and CV; on the other hand, those that are in an intermediate situation between the REF sample and the previous ones: B, CUR and DY. A priori, it is logical to think that the thicker the coating, the greater its protective effect on the alloy. However, sol-gel coatings have an inherent nanoporosity in their structure and do not exert a barrier effect, such as other organic polymers (i.e., epoxy resins or polyurethanes). In the coatings presented in this work, the corrosion-inhibiting effect of the chosen organic molecules must also be taken into account. Thus, the CUR and DY coatings, although they have a lower thickness than the sol-gel coating without inhibitor, sample B, the weight variation observed in Figure 5 is similar, so it is thought that their protective effect is also similar. The coatings of similar thickness to the B sample (reference sample coated but without inhibitor), such as the CIS, CR, MB, and CV samples, show significantly better results against corrosion. Thus, it seems that the thickness of the coating affects the barrier effect, and furthermore, the addition of inhibitor molecules improves the anti-corrosion effect of the coating. The presence of heterogroups and double bonds in the inhibitor molecules seems to play a crucial role in the corrosion inhibitory effect of the coatings.

2.5. *Surface Characterisation by SEM*

SEM/EDX characterisation allowed a compositional and morphological study of the corrosion products, the surface of the metal alloy, and the sol-gel coatings. Figure 6 shows the surface of sample REF in secondary electrons (SE), corresponding to the alloy subjected to the ageing test without sol-gel coating. It shows a completely covered surface with white deposits. At higher magnification (Figure 6c), some cracks appear on the surface of the alloy.

Figure 6. SEM images of the surface of sample REF. (**a**) 10 µm; (**b**) 5 µm; (**c**) 1 µm.

An EDX analysis of the surface of Figure 6a shows exclusively O and Al, probably in the form of Al oxides or hydroxides. If these deposits had good adhesion with the substrate, they could act as protectors, but in this case, part of the product is easily lifted, which may indicate that they could be mostly hydroxides. Similarly, the samples that have been coated with the sol-gel are shown in backscattered electrons (BSE) in Figure 7 [41,42]. In comparison, no cracks or corrosion products such as oxides/hydroxides appear in the sample without protective coating from the crosssection view (sample AL of Figure 7). Sample REF, corresponding to the weathered alloy, shows flakes partially detached from the bulk (sample REF in Figure 7). All coated samples show the sol-gel coatings deposited on the alloys. The lighter and brighter grey colour of the SEM images corresponds to the aluminium alloy substrate (left side of the images). The sol-gel coating with the corresponding inhibitor is marked with a red double arrow on all images, and the darkest grey colour corresponds to the organic epoxy resin of the embedding (right side of the images).

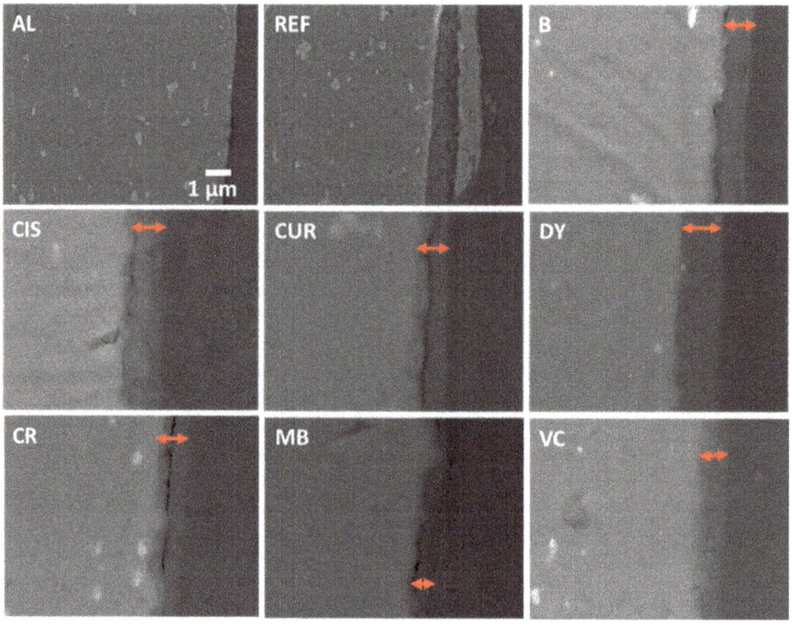

Figure 7. SEM images of a crosssection of samples AL, REF, B, CIS, CUR, DY, CR, MB, and CV.

The coatings have remained unaltered during the accelerated salt weathering and are visible in the SEM images. Their thicknesses have remained constant and in good agreement with the values obtained from the VIS reflectance measurements (Table 2). EDX analyses of the images show a composition similar to that of the nominal AA2024 alloy for the inside of the REF sample, where the O content increases when approaching the outer crust (Table 3).

Table 3. EDX analysis of the crosssection of the weathered samples.

Sample	C (wt. %)	O (wt. %)	Mg (wt. %)	Al (wt. %)	Si (wt. %)	Cl (wt. %)	Cu (wt. %)
REF int		6.1 ± 1.2	1.38 ± 0.28	89.1 ± 1.4			3.41 ± 0.81
B coat	58.5 ± 2.9	23.3 ± 2.7	-	7.84 ± 0.69	6.04 ± 0.60	4.32 ± 0.51	-
CIS coat	37.5 ± 2.6	42.1 ± 2.2	-	16.78 ± 0.91	1.43 ± 0.28	2.15 ± 0.28	-
CUR coat	64.1 ± 2.1	19.6 ± 2.0	-	6.68 ± 0.46	5.93 ± 0.45	3.72 ± 0.35	-
DY coat	61.7 ± 2.5	21.3 ± 2.4	-	2.91 ± 0.32	8.24 ± 0.62	5.86 ± 0.50	-
CR coat	58.7 ± 2.7	26.7 ± 2.5	-	5.64 ± 0.51	5.23 ± 0.49	3.66 ± 0.42	-
MB coat	63.4 ± 1.8	23.2 ± 1.7	-	5.56 ± 0.35	4.85 ± 0.34	2.93 ± 0.26	-
CV coat	57.6 ± 2.5	25.5 ± 2.3	-	6.39 ± 0.51	6.28 ± 0.52	4.25 ± 0.42	-

The "int" means the interior of the substrate; the "coat" means the coating.

The analyses of the sol-gel coatings from the crosssection of samples in Figure 7 shows the presence of Si and O in the coating and C from the organic part of the precursor alkoxides and the inhibitors. Variable amounts of Cl can also be observed, probably from the NaCl solution in which the samples have been immersed and have not been removed by the washes with distilled water. Finally, small amounts of Al appear from the substrate. No major differences are observed between the sol-gel coatings with the different inhibitors.

3. Conclusions

Seven hybrid sol-gel coatings with different corrosion inhibitors have been prepared, applied on AA2024 alloy samples, and characterised before and after a corrosion test in a simulated saline environment. The prepared coatings have proved to be effective against corrosion phenomena in a saline environment, and the observed corrosion was more severe in the uncoated sample. In this sample, numerous whitish deposits were visible on its surface, and some cracks when approaching a neat area. On the coated samples, no cracks or very significant deterioration were observed.

Weight variations of tens of milligrams before and after the corrosion test are measured in the uncoated sample, and slight increases of several milligrams for the coated samples so that the coating plays a fundamental role in inhibiting the formation of superficial aluminium hydroxides.

The thicknesses of the coatings calculated by UV-VIS reflectance spectra vary from 500 nm to 1 µm and are of the same order as the ones observed in the SEM images of the crosssections of the samples subjected to the corrosion test. This demonstrates the effectiveness of the coating and the absence of deterioration. The addition of a different corrosion inhibitor molecule to a sol of the same composition varies the pH of the solution and the charge density of the Si^{4+} ions of the alkoxides. This fact varies the speeds of the hydrolysis and polycondensation reactions, which are accelerated or slowed down. Therefore the viscosity of the sol at the time of its application by dip-coating may be different, thereby obtaining coatings of different thicknesses by varying the inhibitor molecule of the sol-gel coating.

When comparing the results of the coated samples with the selected inhibitors (CIS, CUR, DY, CR, MB, and CV) and the sample coated only with sol-gel (B), with optical microscopy, it has been possible to confirm the reduction of the hydroxides and other corrosion products deposited. This may be due to plugging the pores with the inhibitor when the corrosion process begins on the substrate. The weight increase of the coatings after the test is very small for the CIS, CR, MB, and CV samples, confirming its effectiveness in inhibiting degradation on the surface of the AA2024 alloy. However, there is no direct relation between the effectiveness of protection against corrosion and the coating thickness. This is because corrosion protection is achieved through active protection due to the action of the inhibitors and not only due to a barrier effect. Additionally, the CR coating, which has been optically characterised in pH buffer solutions, has demonstrated a dual functionality as a pH indicator and corrosion inhibitor. The colour change as a function of pH variation is visible by UV-VIS spectrophotometry and to the naked eye, which is a very significant result for future industrial applications.

4. Materials and Methods

4.1. Aluminium Alloys

Commercial AA2024 aluminium alloy (Cu 4.5 and Mg 1.5 in wt%) was selected as a substrate for this work. Plane sheets with a thickness of 0.30 cm were cut into squares of 3×3 cm^2, exposing a total area of 21.6 cm^2. They were all polished with 1200 SiC paper, rinsed with ethanol and dried at 25 °C prior to coating.

4.2. Sol-Gel Coatings

Tetraethyl orthosilicate (TEOS, Sigma Aldrich, 99%), 3-(trimethoxysilyl) propyl methacrylate (Sigma Aldrich, 98%) and 3-(cloropropil)trietoxisilano (Sigma Aldrich, 95%) were used as precursors (molar ratio 2:1:1). Absolute ethanol was employed as solvent. Distilled water was added to promote the complete hydrolysis of the alkoxides, and nitric acid was added as a catalyst. The molar ratio of precursors: EtOH:H_2O:HNO_3 was 1:8:3:0.01.

Prior to the sol hydrolysis process, an environmentally friendly corrosion inhibitor was added to the sol. For this purpose, six organic molecules have been selected, i.e., L-cysteine (CIS), curcumin (CUR), dimethyl yellow (DY), chlorophenol red (CR), crystal violet (CV) and methylene blue (MB) (Figure 8).

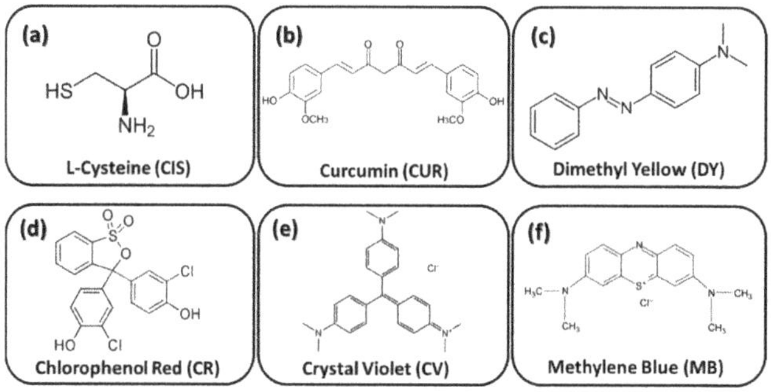

Figure 8. Corrosion inhibitors added into the sol-gel coatings: (**a**) L-cysteine; (**b**) curcumin; (**c**) dimethyl yellow; (**d**) chlorophenol red; (**e**) crystal violet; and (**f**) methylene blue.

The inhibitors were chosen due to their properties as biocompatible molecules, non-toxic and environmentally friendly. Specifically, molecules with conjugated bonding and heteroatoms have been shown to be more effective as corrosion inhibitors [43].

Coatings were performed by dip-coating on AA2024 aluminium alloy and commercial glass slides at a withdrawal rate of 5 $mm \cdot s^{-1}$. Final thermal treatment was performed at 80 °C for 24 h to complete the densification of the resulting coatings.

4.3. Corrosion Test Solution

To evaluate the effectiveness of the coatings doped with several molecules after their application to the Al alloy, a corrosion test in a 0.6 M NaCl solution simulating marine medium was employed [44]. Samples suspended on a fishing line were immersed into new and sterilised single hermetic containers of 150 mL for 14 days and stored in a UV hood [45]. UV irradiation was considered sufficient to prevent bacterial growth, and the salt solution was not changed periodically during the duration of the corrosion test [46]. At the end of the test, samples were washed 3 times by immersion in distilled water. The weight of the coated samples before and after the corrosion test was registered.

4.4. Characterisation Techniques

The macroscopic information of the coatings surfaces on the Al alloy before and after the degradation test was obtained by optical microscopy (OM) in reflection mode with an Olympus CX41 optical microscope.

UV-VIS optical absorption spectra of the coatings applied on glass slides at different pHs were registered with a Shimadzu UV-3100 (UV VIS-NIR) spectrophotometer. It is equipped with an integrating sphere which enables reflection measurements that allow the measurement of the thickness of the coating.

Contact angle measurements were carried out to determine the hydrophilic or hydrophobic behaviour of the different coatings. A Kruss EasyDrop shape analyser was used at room temperature. Doses of 2µL of distilled water were used for each measurement.

Surface characterisation and semi-quantitative chemical composition analyses of the corroded surfaces and small deposits that remained adhered to the surface of the samples were carried out by SEM with an electronic microscope model JEOL JSM 6335F operating at an acceleration voltage of 20 kV. An energy-dispersive X-ray (EDX) detector was employed for the elemental analysis of the surfaces. All samples were previously coated with 5 nm of Au to avoid charging effects.

Author Contributions: Ó.R.d.l.F., V.B., J.C.G. and N.C. conceived the work and coordinated the research. E.J.S., F.G.-G., A.S. and J.L.-S. prepared the samples and carried out the weathering experiment. J.L.-S., A.S., Ó.R.d.l.F. and N.C. carried out the structural characterisation. J.C.G. and V.B. were responsible of the funding. E.J.S. and N.C. wrote the original draft. All authors have read and agreed to the published version of the manuscript.

Funding: This research was funded by the Spanish Ministry of Science and Innovation (MICINN) for financing the projects RTI2018-095856-B-C21; and by the Spanish Ministry of Economic Affairs and Digital Transformation (MINECO) through the projects PID2020-114192RB-C41, PID2019-104717RB-I00 and PID2021-122980OB-C51. A.S. acknowledges the financial support from MCIN for a "Ramón y Cajal" contract RYC2021-031236-I.

Institutional Review Board Statement: Not applicable.

Informed Consent Statement: Not applicable.

Data Availability Statement: Not applicable.

Acknowledgments: The authors acknowledge the CNME Unit of the UCM for SEM/EDX measurements.

Conflicts of Interest: The authors declare no conflict of interest.

References

1. Davis, J.R. (Ed.) *Corrosion: Understanding the Basics;* Chapter 1: "The Effects and Economical Impact of Corrosion"; ASM International: Almere, The Netherlands, 2000.
2. Zhang, C.; Xue, J.-L.; Liu, X.; Li, X.; Zhu, J.; Liu, Q.-C.; Qian, Y. Production of aluminium alloys in electrolysis cells based on Hall-Héroult process: A review. *Chin. J. Eng.* **2019**, *41*, 835–846. [CrossRef]
3. Luo, A.; Sachdev, A.; Apelian, A.K. Alloy development and process innovations for light metals casting. *J. Mater. Process. Technol.* **2022**, *306*, 117606. [CrossRef]
4. Tonelli, L.; Laghi, V.; Palermo, M. AA5083 (Al–Mg) plates produced by wire-and-arc additive manufacturing: Effect of specimen orientation on microstructure and tensile properties. *Prog. Addit. Manuf.* **2021**, *6*, 479–494. [CrossRef]
5. Jurczak, W.; Trzepieciński, T.; Kubit, A.; Bochnowski, W. Static and dynamic properties of al-mg alloys subjected to hydrostatic extrusion. *Materials* **2022**, *15*, 1066. [CrossRef] [PubMed]
6. Liew, Y.; Wijesinghe, S.; Blackwood, D.J. Investigation of the electrochemical Breakdown Response in Sensitised AA5083 Aluminium Alloy. *Sustainability* **2021**, *13*, 7342. [CrossRef]
7. Zhang, H.; Nan, Y.; Guo, C.; Cui, J. Age hardening and intergranular corrosion behavior of new type al-4.5Mg-0.6Zn-0.5Cu-XAg(wt%) alloy. *J. Alloys Compd.* **2022**, *16*, 910. [CrossRef]
8. Zhou, C.; Yang, X.; Luan, G. Effect of root flaws on the fatigue property of friction stir welds in 2024-T3 aluminum alloys. *Mater. Sci. Eng. A* **2006**, *418*, 155–160. [CrossRef]
9. Dobrotă, D.; Racz, S.G.; Oleksik, M.; Rotaru, I.; Tomescu, M.; Simion, C.M. Smart Cutting Tools Used in the Processing of Aluminum Alloys. *Sensors* **2022**, *22*, 28. [CrossRef]
10. Mu, D.K.Q.; Jiang, Z.H.; Zhang, Z. Microstructural evolution, precipitation and mechanical properties of ultrafine-grained and coarse-grained 2024 aluminum alloys fabricated by powder metallurgy. *J. Mater. Res.* **2022**, *37*, 692–704. [CrossRef]
11. Mansurov, Y.N.; Rakhmonov, J.U.; Letyagin, N.V.; Finogeyev, A.S. Influence of impurity elements on the casting properties of al-mg based alloys. *Non-Ferr. Met.* **2018**, *44*, 24–29. [CrossRef]
12. Panda, S.K.; Choudhury, S. Chromium stress in plants. *Braz. J. Plant Physiol.* **2005**, *17*, 95–102. [CrossRef]
13. Twite, R.L.; Bierwagen, G.P. Review of Alternatives to Chromate for Corrosion Protection of Aluminum Aerospace Alloys. *Prog. Org. Coat.* **1998**, *33*, 91–100. [CrossRef]
14. Rincón, R.; Andrés, I.; García Vergara, S.J.; Skeldon, P. Estudio de películas anódicas de aluminio formadas en ácido crómico por espectroscopía de impedancia electroquímica. *Rev. ION* **2012**, *25*, 45–51. Available online: http://www.scielo.org.co/scielo.php?script=sci_arttext&pid=S0120-100X2012000300007&lng=en&tlng=es (accessed on 23 November 2011).
15. Nadimi, M.; Dehghanian, C.; Etemadmoghadam, A. Influence of SiO_2 nanoparticles incorporating into ceramic coatings generated by PEO on aluminium alloy: Morphology, adhesion, corrosion, and wear resistance. *Mater. Today Commun.* **2022**, *31*, 103587. [CrossRef]
16. Guo, Y.; Rogov, A.; Hird, A.; Mingo, B.; Matthews, A.; Yerokhin, A. Plasma electrolytic oxidation of magnesium by sawtooth pulse current. *Surf. Coat. Technol.* **2022**, *429*, 127938. [CrossRef]
17. Javidi, M.; Fadaee, H. Plasma electrolytic oxidation of 2024-T3 aluminum alloy and investigation on microstructure and wear behavior. *Appl. Surf. Sci.* **2013**, *286*, 212–219. [CrossRef]
18. Horváth, M.; Sinkó, K. Hieratical porous SiO_2 cryogel via sol-gel process. *Gels* **2022**, *8*. [CrossRef] [PubMed]
19. Mauri, E.; Giannitelli, S.M.; Trombetta, M.; Rainer, A. Synthesis of nanogels: Current trends and future outlook. *Gels* **2021**, *7*, 36. [CrossRef]
20. Rodríguez-Alonso, L.; López Sánchez, J.; Serrano, A.; Rodríguez de la Fuente, O.; Galván, J.C.; Carmona, N. Hybrid sol-gel coatings doped with non-toxic corrosion protection on AZ61 magnesium alloy. *Gels* **2022**, *8*, 34. [CrossRef]
21. Akbarzadeh, S.; Sopchenski Santos, L.; Vitry, V.; Paint, Y.; Olivier, M. Improvement of the corrosion performance of AA2024 alloy by a duplex PEO/clay modified sol-gel nanocomposite coating. *Surf. Coat. Technol.* **2022**, *434*, 128168. [CrossRef]

22. Lakshmi, R.V.; Aruna, S.T.; Sampath, S. Ceria nanoparticles vis-à-vis cerium nitrate as corrosion inhibitors for silica-alumina hybrid sol-gel coating. *Appl. Surf. Sci.* **2017**, *393*, 397–404. [CrossRef]
23. Carbonell, D.J.; García Casas, A.; Izquierdo, J.; Souto, R.M.; Galván, J.C.; Jiménez-Morales, A. Scanning electrochemical microscopy characterization of sol-gel coatings applied on AA2024-T3 substrate for corrosion protection. *Corros. Sci.* **2016**, *111*, 625–636. [CrossRef]
24. Abuín, M.; Serrano, A.; Llopis, J.; García, M.A.; Carmona, N. Silica doped with lanthanum sol-gel thin films for corrosion protection. *Thin Solid Films* **2012**, *520*, 5267–5271. [CrossRef]
25. Barranco, V.; Carmona, N.; Galván, J.C.; Grobelny, M.; Kwiatkowski, L.; Villegas, M.A. Electrochemical study of tailored sol-gel thin films as pre-treatment prior to organic coating for AZ91 magnesium alloy. *Prog. Org. Coat.* **2010**, *68*, 347–355. [CrossRef]
26. Jafarzadeh, S.; Claesson, P.M.; Sundell, P.E.; Tyrode, E.; Pan, J. Active corrosion protection by conductive composites of polyaniline in a UV-cured polyester acrylate coating. *Prog. Org. Coat.* **2016**, *90*, 154–162. [CrossRef]
27. Bazlia, L.; Yusuf, M.; Farahani, A.; Kiamarzi, M.; Seyedhosseini, Z.; Nezhadmansari, M.; Aliasghari, M.; Iranpoor, M. Application of composite conducting polymers for improving the corrosion behavior of various substrates: Review. *J. Compos. Compd.* **2020**, *2*, 228–240. [CrossRef]
28. Siva, T.; Ramadoss, A.; Sathiyanarayanan, S. Emerging Action of Corrosion Prevention Based on Sustained Self-healing Coatings. *Surf. Interfaces* **2021**, *26*, 101440. [CrossRef]
29. Ashraf, M.A.; Liu, Z.; Peng, W.X.; Yoysefi, N. Amino acid and TiO_2 nanoparticles mixture inserted into sol-gel coatings: An efficient corrosion protection system for AZ91 magnesium alloy. *Prog. Org. Coat.* **2019**, *136*, 105296. [CrossRef]
30. Borisova, D.; Möhwald, H.; Shchukin, D.G. Mesoporous silica nanoparticles for active corrosion protection. *ACS Nano* **2011**, *5*, 1939–1946. [CrossRef]
31. Liu, J.; Lou, Y.; Zhang, C.; Yin, S.; Li, H.; Sun, D.; Sun, X. Improved corrosion resistance and antibacterial properties of composite arch-wires by N-doped TiO_2 coating. *RSC Adv.* **2017**, *7*, 43938–43949. [CrossRef]
32. Xiong, L.; Liu, J.; Li, Y.; Li, S.; Yu, M. Enhancing corrosion protection properties of sol-gel coating by pH-responsive amino-silane functionalized graphene oxide-mesoporous silica nanosheets. *Prog. Org. Coat.* **2019**, *135*, 228–239. [CrossRef]
33. Maeztu, J.D.; Rivero, P.J.; Berlanga, C.; Bastidas, D.M.; Palacio, J.F.; Rodríguez, R. Effect of Graphene oxide and fluorinated polymeric chains incorporated in a multilayered sol-gel nanocoating for the design of corrosion resistant and hydrophobic surfaces. *Appl. Surf. Sci.* **2017**, *419*, 138–149. [CrossRef]
34. Saurbier, K.; Mendorf, V.; Schultze, J.W.; Geke, J.; Penninger, J.; Roβmaier, H. Toluylalanine as an inhibitor of atmospheric corrosion. *Corros. Sci.* **1992**, *33*, 1351–1359. [CrossRef]
35. Vuorinen, E.; Ngobeni, P.; Van der Klashorst, G.H.; Skinner, W.; de Wet, E.; Ernst, W.S. Derivatives of cyclobexylamine and morpholine as volatile corrosion inbibitors. *Br. Corros. J.* **1994**, *29*, 120–121. [CrossRef]
36. Hernández, L.; Veleva, L.; García-Galván, F.; Galván, J.C. Effect of ZrO2 and L-cys nanoparticles as dopants in sol-gel of mesoporous silica coating for corrosion protection of AZ61 magnesium alloy. *Revista De Metalurgia* **2019**, *55*, e155. [CrossRef]
37. Derong, L.; Lijuan, X.; Wen, Q.; Douglas, A.; Zhijun, W.; Hong, C.; Zhang, Q. Preparation, characterization and antioxidant properties of curcumin encapsulated chitosan/lignosulfonate micelles. *Carbohydr. Polym.* **2022**, *281*, 119080. [CrossRef]
38. Oguzie, E.E. Corrosion inhibition of mild steel in hydrochloric acid solution by methylene blue dye. *Mater. Lett.* **2005**, *59*, 1076–1079. [CrossRef]
39. Montero, E.F.; García, M.A.; Villegas Broncano, M.A.; Llopis, J. Estudio de las propiedades ópticas de recubrimientos Sol-Gel dopados con fluoresceína en función de la concentración y del pH. *Boletín Soc. Española Cerámica Vidr.* **2004**, *43*, 8–11. [CrossRef]
40. Brinker, C.J.; Shcherer, G.W. *The Physics and Chemistry of Sol-Gel Processing*; Academic Press Ltd.: Cambridge, MA, USA, 1990.
41. Pogrebnjak, A.D.; Webster, R.F.; Tilley, R.D.; Buranich, V.V.; Ivashchenko, V.I.; Takeda, Y.; Oyoshi, K.; Sakenova, R.; Piotrowska, K.; Zukowski, P.; et al. Formation of Si-Rich Interfaces by Radiation-Induced Diffusion and Microsegregation in CrN/ZrN Nanolayer Coating. *ACS Appl. Mater. Interfaces* **2021**, *13*, 16928–16938. [CrossRef]
42. Xu, X.; Su, F.; Li, Z. Tribological properties of nanostructured TiAlN/W2N multilayer coating produced by PVD. *Wear* **2019**, *430–431*, 67–75. [CrossRef]
43. Palanisamy, G. *Corrosion Inhibitors*; Ambrish, S., Ed.; IntechOpen: London, UK, 2019.
44. *ASTM Standard G31-72*; Standard Practice for Laboratory Immersion Corrosion Testing of Metals (Reapproved 1990); Annual Book of ASTM Standards. ASTM International: West Conshohocken, PA, USA, 1990.
45. Galván, J.C.; Larrea, M.T.; Braceras, I.; Multigner, M.; González-Carrasco, J.L. In vitro corrosion behaviour of surgical 316LVM stainless steel modified by Si+ ion implantation—An electrochemical impedance spectroscopy study. *J. Alloys Compd.* **2016**, *676*, 414–427. [CrossRef]
46. Pacha-Olivenza, M.A.; Gallardo-Moreno, A.M.; Vadillo-Rodríguez, V.; González-Martín, M.L.; Pérez-Giraldo, C.; Galván, J.C. Electrochemical analysis of the UV treated bactericidal Ti6Al4V surfaces. *Mater. Sci. Eng. C Mater. Biol. Appl.* **2013**, *33*, 1789–1794. [CrossRef] [PubMed]

gels

MDPI

Article

Chemiresistors with In_2O_3 Nanostructured Sensitive Films Used for Ozone Detection at Room Temperature

Mariana Chelu [1], Paul Chesler [1,*], Cristian Hornoiu [1,*], Mihai Anastasescu [1], Jose Maria Calderon-Moreno [1], Daiana Mitrea [1], Costin Brasoveanu [2], Carmen Moldovan [2] and Mariuca Gartner [1]

1 "Ilie Murgulescu" Institute of Physical Chemistry—Romanian Academy, Splaiul Independentei 202, 060021 Bucharest, Romania
2 National Institute for Research and Development in Microtechnologies, Strada Erou Iancu Nicolae 126A, 077190 Voluntari, Romania
* Correspondence: pchesler@icf.ro (P.C.); chornoiu@icf.ro (C.H.)

Abstract: Detection of greenhouse gases is essential because harmful gases in the air diffuse rapidly over large areas in a short period of time, causing air pollution that will induce climate change with catastrophic consequences over time. Among the materials with favorable morphologies for gas detection (nanofibers, nanorods, nanosheets), large specific surfaces, high sensitivity and low production costs, we chose nanostructured porous films of In_2O_3 obtained by the sol-gel method, deposited on alumina transducers, with gold (Au) interdigitated electrodes (IDE) and platinum (Pt) heating circuits. Sensitive films contained 10 deposited layers, involving intermediate and final thermal treatments to stabilize the sensitive film. The fabricated sensor was characterized using AFM, SEM, EDX and XRD. The film morphology is complex, containing fibrillar formations and some quasi-spherical conglomerates. The deposited sensitive films are rough, thus favoring gas adsorption. Ozone sensing tests were performed at different temperatures. The highest response of the ozone sensor was recorded at room temperature, considered to be the working temperature for this specific sensor.

Keywords: In_2O_3 nanostructured films; sol-gel; Au IDEs; Pt heater; ozone detection; room temperature

1. Introduction

Stratospheric ozone exists naturally in the upper atmosphere and protects us from damage caused by the sun's ultraviolet rays, having a beneficial effect on the planet [1].

Tropospheric ozone (ground-level ozone or surface ozone), hereafter referred to as ozone (O_3), is a toxic air pollutant that can harm human health and the environment, being the main component of "smog" [2]. O_3 is a secondary pollutant that is produced when primary pollutants, such as volatile organic compounds (VOCs) and nitrogen oxides (NO_x) from stationary air, chemically react in the presence of sunlight [3]. When the level of pollutants emitted by automobiles, refineries, power plants, industrial facilities or factories is very high during sunny periods, there is a very high level of ozone in those areas. It can also have a negative impact on vegetation and ecosystems [4,5].

Breathing atmosphere containing high levels of ozone affects certain individuals, such as asthmatics, the elderly, children and those who perform extended outdoor activities. For this reason, OSHA (Occupational Safety and Health Administration) recommends a human exposure threshold of 0.1 ppm O_3 in air for a maximum of 8 h, as higher concentrations (>50 ppm) can cause serious illness (both healthy adults and asthmatics may experience significant reduction in lung function and inflammation of the upper airways) or even death. At the same time, the European Air Quality Directive 2008/50/EC sets a target value of 120 $\mu g/m^3$ O_3 for the average daily concentration of 8 h of exposure, which can be exceeded up to a maximum of 25 days per year on average over 3 years [6,7]. An extensive study analyzed the high concentrations of surface ozone in the area of Bucharest, the capital of Romania, using both data measured at monitoring stations and data obtained from satellites. These were correlated with high NO_x emissions due to road traffic in the urban area of the capital, compared to other regions of the country [8]. Consequently, to effectively control pollutant emissions, reliable and accurate air quality data are needed. In this sense, there is an increased interest in the development of high-performance air quality monitoring systems, based on low-cost sensors, which are miniaturized and easy to implement in portable devices [9].

Among various sensor systems for the detection of gaseous pollutants, metal oxide semiconductors (SMOX) are the most studied materials as O_3 sensors, having detection limits in line with OSHA recommendation [10]. The operating principle of chemiresistive gas sensors (chemiresistors) is based on the variation of the resistance of the SMOX sensitive layer, depending on the concentration of the target gas in the analyzed atmosphere, which generates a proportional current flow in the detection device circuit [11–13]. Several types of SMOX have been comprehensively investigated for O_3 detection [14–17]. For example, commercial sensors contain sensitive layers in the form of thick or thin films of indium and tin oxide (ITO), tungsten oxide (WO_3), zinc oxide (ZnO) or titanium dioxide (TiO_2) [18–20]. They are generally manufactured by sputter deposition or other high-cost methods, involving high energy consumption or expensive laboratory equipment [21–25]. In addition, the influence of the environment (temperature, humidity or interfering chemical species) contributes to a variation in the performance of these commercial sensors, in terms of the collected data [26].

Recently, different types of sensors have been reported for O_3 detection that operate at a temperature lower than 100 °C with good sensor response [27–30]. Based on the density functional theory, theoretical studies were made on the interaction of the O_3 molecule with group III nitride as well as Al and Ga doped $B_{12}N_{12}$. Exploration of the structural, thermodynamic, electronic, dipole moment and NBO (natural born orbital) charge transfer properties revealed a promising sensing behavior for the O_3 molecule [31]. Using a simple one-step solution combustion method, *p*-type $CuCo_2O_4$ spinel nanomaterials rich in oxygen vacancies were synthesized. The $CuCo_2O_4$-based O_3 sensor was investigated at a temperature of 90 °C, and showed a high response (R_s = 27)—for 1 ppm O_3 under 70% relative humidity (RH), together with sensor chemical stability and very good sensor selectivity [32].

For monitoring tropospheric ozone in ambient air, two different types of SMOX sensors based on nanostructured sensing layers of ZnO:Ga, pure (SnO_2) and graphene-doped tin dioxide (SnO_2:G) nanofibers were manufactured by magnetron sputtering and electrospinning. The materials were tested and compared with commercial SMOX sensors under both controlled laboratory conditions and real-life conditions. Promising results were obtained, which showed that the prepared sensors performed similarly to the commercial sensors and exhibited excellent performance compared to the usual commercial sensors [33].

The highly sensitive photoluminescence emission of ZnO nanoparticles (NPs) at room temperature was explored for O_3 detection. Both commercially available luminescent ZnO NPs and ZnO/polymer nanohybrids were investigated. Polymer matrices of the type poly (poly (ethylene glycol) methyl ether methacrylate) (PPEGMA) and polydimethylsiloxane (PDMS) were used. The materials were investigated by exposure to ozone gas at room temperature over a range of concentrations between 1600–50 ppb. The ZnO/PDMS-based material showed an ozone detection limit of 50 ppb, for a response/recovery time in the range of 50–100 s [34].

In this study, we report the fabrication of a room temperature-operated sensor for ozone detection, fabricated by a simple/low-cost strategy based on the preparation of the sensitive layer of In_2O_3 using the sol-gel method. The versatility and advantages of this ecological method consist in the ease of synthesizing the sensitive material with controlled morphology, favorable for the detection of gaseous ozone under laboratory conditions. The results presented in this paper proved the detection performance of the obtained sensor, which exhibited a limit of detection (LOD) at room temperature of 0.08 ppm O_3. The response/recovery time of the sensor was fast; sensor signal has low-noise; and sensor recovery was complete after each sensing test, allowing for a new sensing cycle without sensor replacement.

2. Results and Discussions

2.1. Sensor Characterization Methods

2.1.1. Atomic Force Microscopy (AFM)

Atomic force microscopy (AFM) measurements were performed in "non-contact" mode with an XE-100 device from Park Systems in order to minimize the probe (tip)–surface interaction. The XE100 microscope is equipped with decoupled XY and Z scanners, thus eliminating unwanted artifacts (bow effect, for example). All AFM measurements were made with NCHR tips (Nanosensors™), having a typical radius of curvature of ~8 nm, length of ~125 μm, width of 30 μm, spring constant of ~42 N/m and resonance frequency of ~330 kHz. After recording, the images were processed with the XEI program (v 1.8.0), produced by the same company (Park Systems).

The AFM figures are presented in the so-called "enhanced contrast" mode, in which the color of a pixel is determined by its variation in comparison to the neighboring pixels. Therefore, unlike the classic color mode used in AFM imaging, where the gradient of a single color is used, in the "enhanced contrast" mode, there is not a linear dependence in color with the height difference. On the other hand, the information of the profile lines (line scans) is not perturbed in the z-axis; instead, the morphology of the sample can be better visualized (enhanced) in the case of flat or rough samples.

Figure 1 presents AFM images of the sample tested from the point of view of sensor activity: a film of In_2O_3 obtained by 10 repetitive depositions on Au-IDE/Alumina. From Figure 1 it can be observed that the In_2O_3 film having 10 layers deposited onto alumina substrates (alumina transducers) is continuous, adherent to the substrate and has a complex morphology, showing the presence of both fibrillar formations and quasi-spherical conglomerates on the surface. The profile line in Figure 1a reflects a large difference in height, of about 1 μm, probably also influenced by the corrugation of the alumina ceramic substrate (which exhibit large particles on the surface—not shown here).

Thus, at the scale of (8×8) μm^2 the In_2O_3/alumina sample exhibits a RMS (Root Mean Square)–roughness (a representation of surface roughness, calculated as deviations from the mean plane within the sampling area, using measured microscopic peaks and valleys) of 181 nm and a peak-to-valley parameter of 1185 nm. At smaller scales, scanned over (2×2) μm^2, there are smoother parcels/areas, as highlighted by the profile line from Figure 1b, which shows a difference in z axis of about 60 nm and an increased repeatability/uniformity. Globally, at the scale of (2×2) μm^2, a RMS roughness of 19.6 nm and a peak-to-valley parameter of 229 nm were estimated.

2.1.2. Scanning Electron Microscopy (SEM)

Scanning electron microscopy (SEM) revealed the microstructure and distribution of the active phases deposited as films on the transducers, both on the alumina wafer and on the gold interdigital electrodes. The microstructure of the samples was evaluated by SEM using a Quanta 3D field emission microscope equipped with an energy dispersive X-ray (EDX) spectrometer, operating at voltages between 10 and 30 kV and at a voltage of 5 kV using secondary electron (SE) or backscattered electron (BSE) images.

Figure 2 shows the transparent In_2O_3 film (gold IDEs are visible through the film) deposited on the alumina transducer. It can be seen that through successive sol-gel depositions followed by subsequent heat treatment, both the alumina substrate grains and the interdigitated Au electrodes were completely covered with a film of In_2O_3. Another observable feature is that the In_2O_3 film is continuous and folds to the characteristic surface morphology of the polycrystalline support with micron-sized alumina grains. After the deposition of the In_2O_3 film, the surface morphology becomes more complex, also evidenced by the AFM images, with fibrillar formations as well as some quasi-spherical conglomerates. These morphological features are probably the result of the release of gas bubbles during the drying of the sensitive layer prepared by sol-gel and heat treatment.

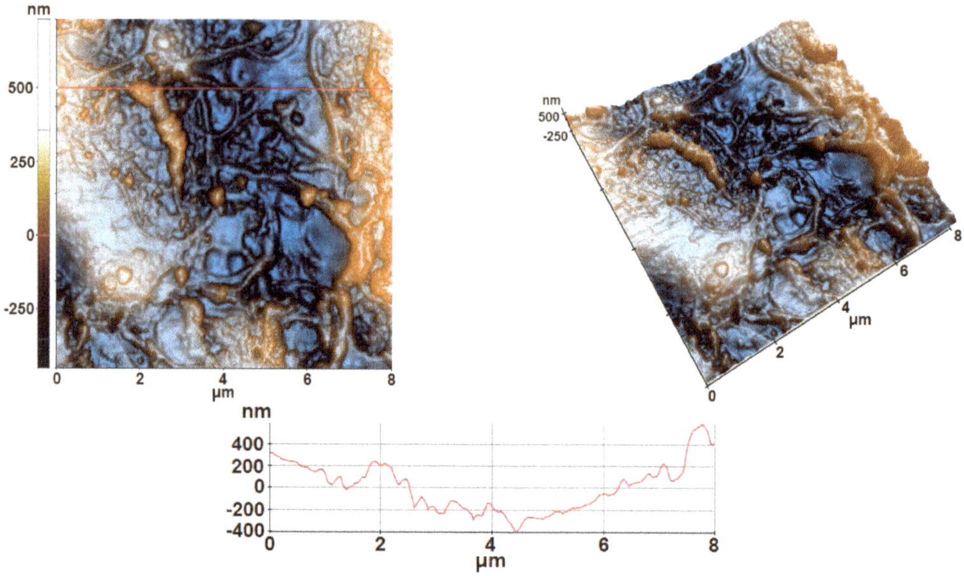

(a)

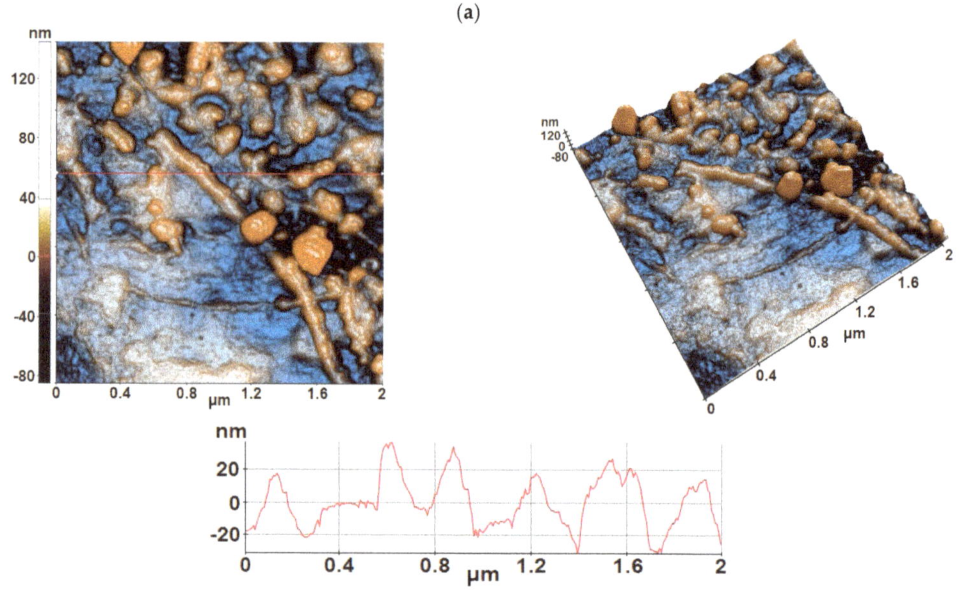

(b)

Figure 1. AFM images of the 10-layered In_2O_3 sensitive film deposited on Au-IDE/Alumina (alumina ceramic support with interdigitated gold electrodes) at the scale of: (**a**) (8 × 8) μm^2; (**b**) (2 × 2) μm^2.

Figure 2. Top view SEM micrographs of the 10-layered In_2O_3 sensitive film of the fabricated sensor (having alumina wafer with Au IDEs), at different magnifications.

Cross-sectional SEM images (Figure 3) of a fractured sample show that the In_2O_3 film is about 2 microns thick, yet the film conforms to the roughness of the alumina layer, which consists of alumina grains of several microns. The film is continuous, filling the intergranular step spaces and the voids between the alumina grains; therefore, the film is not of uniform thickness and can be considerably thicker at intergranular voids. The characteristics of the substrate-film interface can be observed in the cross-sectional images in Figure 3, which show close contact between the alumina substrate and the In_2O_3 film. The presence of porosity is clear in the cross-sectional SEM images—two types of pores exist: a fine internal porosity, which is observed throughout the cross-section and larger rounded pores on the surface with diameters similar to the thickness of the film.

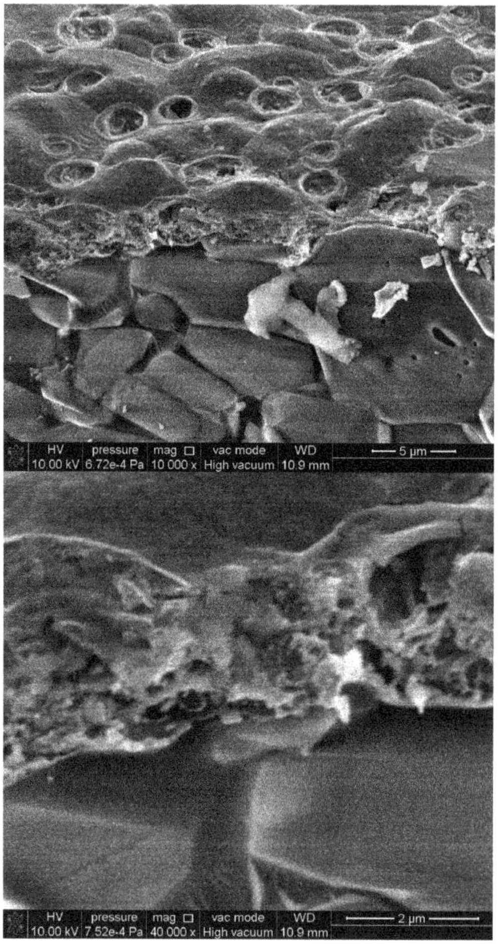

Figure 3. Cross-sectional SEM micrographs of the sensor's In_2O_3 sensitive film at different magnifications.

Higher magnification SEM micrographs taken in tilted view of the In_2O_3 film (Figure 4) show the surface features and the distribution of rounded pores along the film surface. The EDX spectra (Figure 4) show only the presence of indium and oxygen. The absence of carbon indicates that the organic parts were removed during the heat treatment. The rounded-shape pores observed on the surface of the films indicates the formation of gas bubbles during the heat treatment and densification of the In_2O_3 film. SEM and EDX observations confirm the absence of any carbon-containing residues in the film. It should be noted that the formation of the film takes place through 10 successive depositions and, therefore, 10 successive intermediate thermal treatments, followed by a final heat treatment at a slightly higher temperature. We have used these thermal treatment conditions to remove any remaining acetylacetonate from the film, in accordance with the literature [35,36] and also with our previous studies.

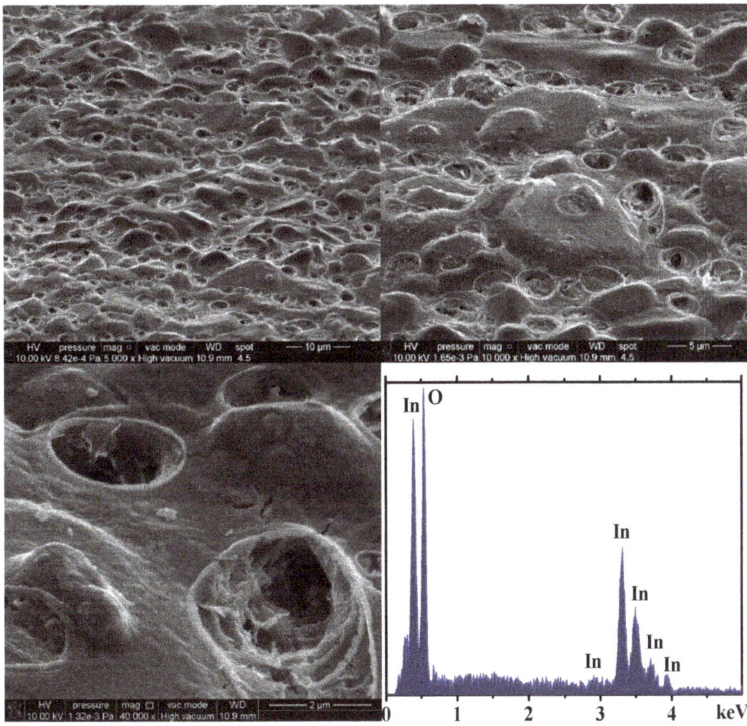

Figure 4. Tilted view SEM micrographs of the sensor's 10-layered In_2O_3 sensitive film at different magnifications, together with EDX spectra.

2.1.3. X-ray Diffraction (XRD)

X-ray Diffraction (XRD) was used to examine the structure of the sensitive film. A Rigaku Ultima IV X-ray diffractometer was used to acquire the diffractograms. The equipment was set in thin film geometry, with parallel beam (PB) optics, operated at 40 kV and 30 mA, using Cu Kα (λ = 1.5406 Å) radiation. Measurements were performed in continuous scan mode, with a scan speed of 2°/min and a step width of 0.02° (2θ), at a fixed incidence angle, α, of 0.5°. Data were collected over the angular range of 15–80° (2θ). To determine the structure of the sensor sample, X-ray diffraction was used (Figure 5).

Three compounds, α-Al_2O_3, Au and In_2O_3, were identified in the X-ray diffractogram. α-Al_2O_3 and Au belong to the sensor substrate (transducer), only In_2O_3 lines being assigned to the sensitive film. In_2O_3 was identified using the ICDD file no. 6-0416. In_2O_3 crystallized in a cubic crystal system, with space group Ia3. The lattice constant of the cubic In_2O_3 crystals, calculated from the whole pattern fitting, was 10.120(4) Å, very close to the indexed value (10.1180 Å in the ICDD file no. 6-0416). The crystallite size was around 17 nm, as determined by applying the Scherrer formula for the crystal plane (222) only.

2.2. Gas Sensing Measurements

The gas sensing experiments were performed using own-design sensing cell and transducers. The transducer model was used previously by our research group to detect small concentrations of dangerous gases [37,38]. The alumina transducer sized at 20 mm × 10 mm × 0.2 mm contains Au interdigital electrodes (IDE) with 53 pairs of interdigits, 6400 μm in length, 25 μm wide, with 25 μm spacing gap and a Pt heating circuit (1200 μm width).

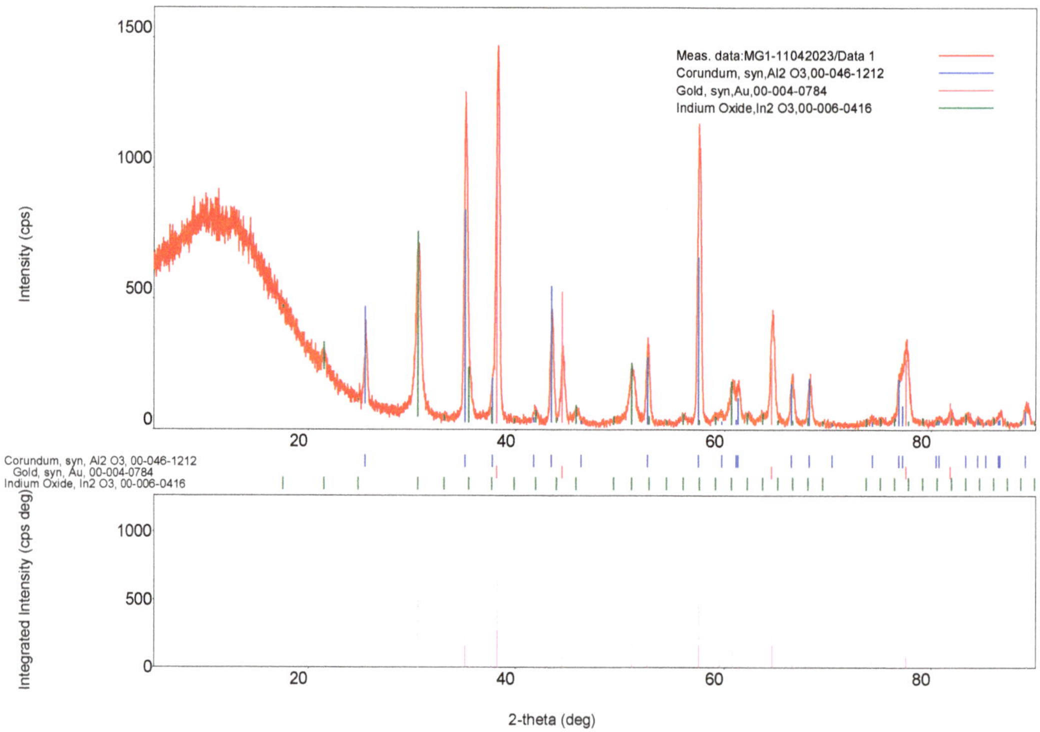

Figure 5. XRD pattern of the sensor sample.

To provide the target gas (different concentrations of ozone), an ozone generator was coupled with the setup gas lines (Figure 6), using a glass mixing vessel, which combines the target and the carrier gas before sensing cell entry. A commercial Dräger X-am 5000 gas detector was used to check the ozone concentration in air at the exit of the mixing vessel. The sensing measurements were performed in triplicates under laboratory conditions (the gas cylinders contain high purity dry gases, as provided by the manufacturers). To ensure the necessary working temperature, a direct current voltage generator was used to provide stable current for the contact pads of the platinum heater circuit on the back side of the sensor. The sensor operating voltage applied to the IDE pads was set at 1.5 V direct current (DC).

The fabricated sensors were tested for ozone detection in the 0.08–1.0 ppm range, considering OSHA recommendations for ozone exposure in the work environment [39], higher ozone concentrations having inflammatory effect over the human airways, as mentioned in the Section 1. The tested working temperatures for the fabricated sensors were situated in the range between room-temperature and 100 °C, because above 105 °C ozone is known to detonate [40]. Sensor response to the target gas (ozone) was defined as the ratio between sensor electrical resistance in the target gas and carrier gas (air), and it was permanently monitored and automatically recorded using a GPIB interface connected to the RLC bridge and the data acquisition computer. No additional signal filtering was involved in the sensing measurements.

From Figure 7, it can be observed that with increasing working temperatures, the response of the sensor for ozone decreases, in accordance with those mentioned above (the ozone molecule destabilizes with the increase in the temperature, until the total decomposition by explosion at 105 °C). The highest sensor response was recorded for the tested concentration of 0.08 ppm, decreasing in intensity with the increase in the concentration of the target gas (towards 1.0 ppm). At the maximum tested concentration of 1.0 ppm, the sensor generates the lowest response for ozone. Considering the room temperature testing conditions, the decrease in the sensor response with increasing tested ozone concentrations can be explained as follows: the adsorption centers on the surface of the sensitive layer are gradually saturated with ozone molecules, as the tested concentrations for the target gas increase. The desorption of gas molecules from these active centers is made difficult by the low temperature conditions; consequently, active centers on the surface of the sensitive film are more and more occupied, leading to a decrease in the yield of the chemical reaction taking place on the surface of the sensor, which occurs with charge transfer, leading to a change in sensor electrical resistance—captured by the RLC bridge connected to the experimental setup. A smaller resistance variation is reflected by a weaker sensor signal.

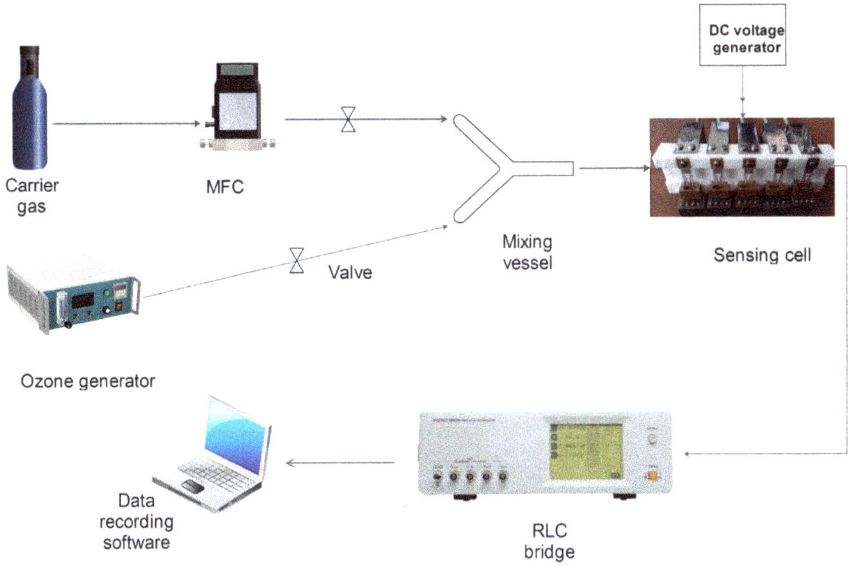

Figure 6. Experimental setup used for ozone sensing measurements.

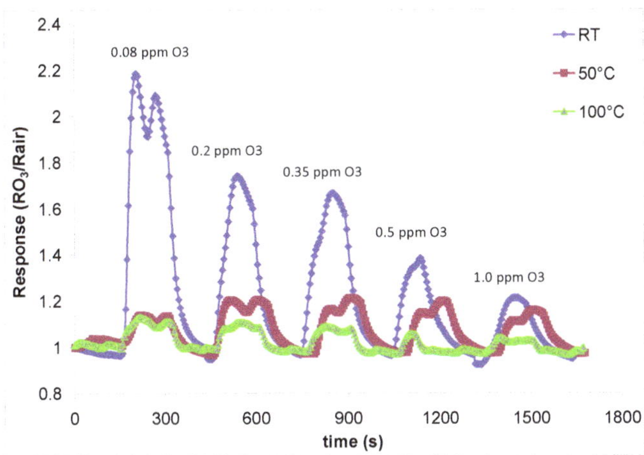

Figure 7. Sensor response/recovery characteristics for ozone detection at different tested concentrations and working temperatures.

Sensor sensitivity S to the target gas was defined as

$$S = \frac{R_{gas} - R_{air}}{R_{air}} \times 100\% \quad (1)$$

with R_{gas} = sensor electrical resistance when exposed to target gas, and R_{air} = sensor electrical resistance when exposed to carrier gas.

In Figure 8, we can observe sensor sensitivity to ozone at working temperatures ranging between room temperature and 100 °C. As in the case of sensor response as emphasized in Figure 7, sensor sensitivity to ozone decreases with increasing target gas concentration. Maximum sensor sensitivity is achieved for the lowest ozone tested concentration (0.08 ppm) at room temperature.

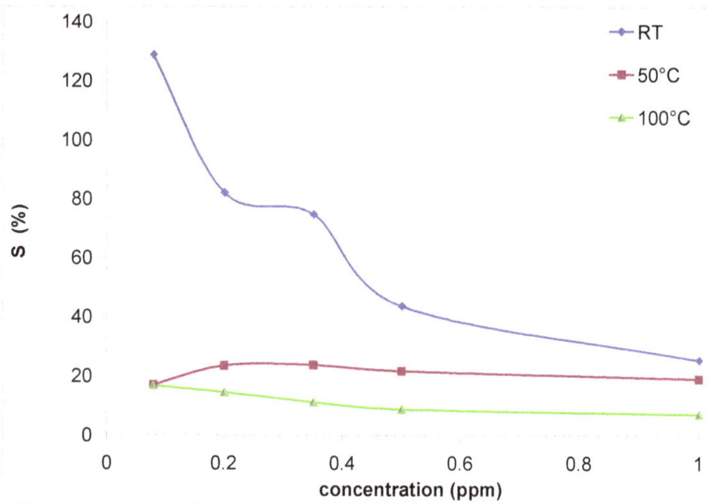

Figure 8. Sensor sensitivity (S) to ozone at different working temperatures.

Regarding the sensing mechanism involved in ozone detection for this specific case, according to dos Santos Silva et al. [41], for SMOX materials, the interaction between the indium oxide and the ozone molecules (oxidizing gas) leads to ionic species adsorbed on the In_2O_3 surface, accompanied by electron consumption, consequently increasing the depletion layer of the oxide material and its electrical resistance (Equation (2)). This was confirmed by the obtained experimental results, in which sensor electrical resistance increases every time ozone is injected in the sensing cell.

$$O_3 + e^-_{(surface)} \rightarrow O_{2(gas)} + e^-_{(surface)} \quad (2)$$

According to the same cited work [41], a higher porosity of the sensitive film promotes sensor sensitivity/performance, due to a higher surface area which leads to more available adsorption sites for target gas molecules. The ozone sensor used in this work uses a porous alumina wafer. Together with the surface roughness/porosity characteristics described by the AFM and SEM investigations we can state that the high sensitivity of the investigated sensor at room temperature is in close connection with the porosity of both, the sensitive layer and alumina wafer.

We can state that the sensor signal is clean (very low levels of noise) and its recovery is total, an important sensor characteristic when performing multiple sensing cycles without sensor replacement. For 0.08 ppm ozone in air, similar response values were obtained for

3 consecutive sensing cycles (Figure 5). This sensor characteristic is also linked with sensor sample porosity, according to ref. [41].

In Figure 9, we can observe that reproducible results were obtained for 3 consecutive ozone sensing tests conducted at room temperature for the lowest tested ozone concentration of 0.08 ppm. Fast sensor response, followed by a complete sensor recovery (the electrical resistance value of the sensor returns to the baseline), was also recorded under these testing conditions. The sensor signal was very clean, with negligible levels of noise. Important sensor characteristics were determined using Figure 9, such as sensor response time (t_{resp} = 180 s) and sensor recovery time (t_{rec} = 425 s), both considered as short, implying that the studied sensor is quite responsive to the tested target gas.

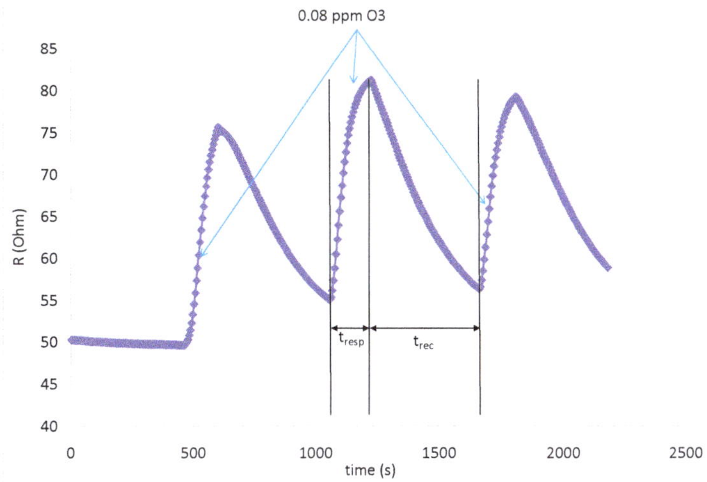

Figure 9. Sensor response/recovery characteristics for 3 consecutive tests with 0.08 ppm ozone in air at room temperature.

Regarding sensor stability, it must be mentioned that sensing measurements were performed several months after sensor fabrication with excellent results (sensor response was ranged between 1.3 and 2.2), so the sensitive film is considered to be chemically stable.

The state of the art regarding SMOX-based ozone sensors, comprised in Table 1, shows that other research groups obtained in some cases successful ozone detection in the ppb range but with higher working temperatures, which involves additional electronics necessary for the detection device, an impediment from the economical perspective [42–46] when mass-production is to be considered.

Table 1. State of the art regarding ozone detection with SMOX-based sensors.

Materials	Concentration	Operating Temperature (°C)	Sensor Response	LOD	Ref.
Al-doped NiO films	4% O_2 in plasma	80	2.54	10 ppb	[16]
WO_3	0.5–2.0 ppm	150	3.9–46.7	-	[47]
ZnO	0.3–5 ppm	200	10.7–26	-	[48]
In_2O_3/ZnO		150	14.4	500 ppb	[49]

Table 1. *Cont.*

Materials	Concentration	Operating Temperature (°C)	Sensor Response	LOD	Ref.
$ZnCo_2O_4$	80–890 ppb	200	71	890 ppb	[50]
$CuWO_4$ nanoparticles	15–50 ppb	250	4.2	15 ppb	[51]
α-Fe_2O_3 nanorods	10–570 ppb	150	-	-	[52]
ZnO sputtered films	0.13 ppm	300	-	-	[53]
In_2S_3 thin films	40 ppm	160	-	-	[54]
In_2O_3 nanostructured films	0.8–1.0 ppm	RT	1.3–2.2	0.08 ppm	This work

3. Conclusions

A nanostructured In_2O_3 film (with 10 layers) obtained by the sol-gel method was used as a sensitive layer in the fabrication of an own-designed chemiresistor, based on an alumina transducer containing Au IDEs and a Pt heating circuit. The deposited sensitive film was characterized using AFM and SEM as continuous, porous and transparent. The sensor responded to low concentrations of ozone (0.08 ppm–1.0 ppm), in compliance with OSHA requirements for the workplace environment (maximum 0.1 ppm for an 8 h exposure time). The sensor response (ranged between 1.3–2.2) was fast (180 s), stable (low-noise and full sensor recovery recorded in 425 s) and reproducible (measurements were performed in triplicates yielding similar ozone detection results) at room temperature, a great advantage from the economical point of view, with no additional heating/electronics being necessary when considering large-scale fabrication. A sensing mechanism was formulated, in agreement with the literature and obtained results, which states that a chemical reaction accompanied by charge transfer occurs on the surface of the In_2O_3 film. The sensitive film porosity and also the porosity of the sensor's alumina wafer both promote gas adsorption and, consequently, sensor performance under low temperature operating conditions. The sensitive layer was considered chemically stable given the fact that the sensing tests were performed with reproducible results several months after sensor fabrication. The low operating temperature of the sensor together with the sol-gel method of obtaining the sensitive layer represents the novelty factor of this research. The good results obtained for the In_2O_3-based sensor for ozone detection at room temperature can be used for the future development of a commercial indoor ozone detector.

4. Materials and Methods

All chemicals were of analytical grade and used as received without further purification. Indium (III) nitrate hydrate [In $(NO_3)_3$ •xH_2O], absolute ethanol (CH_3CH_2OH) and acetylacetone ($CH_3COCH_2COCH_3$) were purchased from Sigma Aldrich; ammonia solution 25% EPR was purchased from Labchem.

A solution of 0.2 M In_2O_3 was prepared by dissolving indium hydrate nitrate in absolute ethanol at room temperature along with stirring. After dissolving the precursor salt, acetylacetonate (AA) is added dropwise to the solution, as a chelating agent, obtaining a transparent solution. The molar ratio between AA:In (III) was 2:1. After 30 min, 25% ammonia solution was added dropwise and kept under continuous magnetic stirring for an hour. The molar ratio between of NH_3:In (III) was 2:1. A homogenous, clear and transparent solution was obtained which was stored (for ageing) at room temperature for 6 days.

In order to obtain a good adhesion of the sensitive film to the support, the transducers were previously cleaned with acetone, deionized water and alcohol and dried at room temperature. The obtained precursor solution was deposited onto transducers by dip coating, with an immersion speed of 50 mm/min for 1 min. Ten layers were deposited consecutively, with an intermediate heat treatment applied after each layer at 300 °C for 10 min and a final sensor heat treatment at 350 °C for one hour, a relatively low temperature [55], with a heating rate of 5 °C/minute.

The sensitive layer was deposited onto the alumina transducer in November 2021. Sensing measurements over the fabricated sensor were performed during June–July 2022.

Author Contributions: Conceptualization, P.C. and C.H.; methodology, P.C., C.H. and C.B.; software, P.C. and C.H.; validation, P.C. and C.H.; formal analysis, P.C. and C.H.; investigation, P.C., C.H., M.A. and J.M.C.-M.; resources, C.B., M.C. and D.M.; data curation, P.C. and C.H.; writing—original draft preparation, P.C. and M.C.; writing—review and editing, P.C.; visualization, P.C.; supervision, M.G. and C.M.; project administration, M.G.; funding acquisition, M.G. All authors have read and agreed to the published version of the manuscript.

Funding: This research was funded by Romanian National Authority for Scientific Research on Innovation, CCCDI-UEFISCDI, grant number "PN-III-P2-2.1-PED-2019-2073, Contract 308PED/2020".

Institutional Review Board Statement: Not applicable.

Informed Consent Statement: Not applicable.

Data Availability Statement: The datasets generated during and/or analyzed during the current study are available from the corresponding author on reasonable request.

Acknowledgments: The paper was carried out within the research program "Science of Surfaces and Thin Layers" of the "Ilie Murgulescu" Institute of Physical Chemistry of the Romanian Academy. The contribution of Ion Stan to the development of the sensing cell is greatly acknowledged.

Conflicts of Interest: The authors declare no conflict of interest.

References

1. Langematz, U. Stratospheric ozone: Down and up through the anthropocene. *ChemTexts* **2019**, *5*, 8. [CrossRef]
2. Jadoon, S.; Nawazish, S.; Majeed, Z.; Baig, A.; Bukhari, S.M.; Faiz, A.u.H.; Ghanim, A.A.J.; Irfan, M.; Rahman, S.; Ullah, F. Health Impacts of Surface Ozone in Outdoor and Indoor Environments of Hattar Industrial Units, KPK, Pakistan. *Atmosphere* **2022**, *13*, 2002. [CrossRef]
3. Muñoz, A.; Borrás, E.; Vera, T.; Colmenar, I.; Ródenas, M.; Gimeno, C.; Fuentes, E.; Coscollá, C.; Calvete-Sogo, H. Atmospheric degradation of two pesticides mixed with volatile organic compounds emitted by citrus trees. Ozone and secondary organic aerosol production. *Atmos. Environ.* **2023**, *295*, 119541. [CrossRef]
4. Al-Qassim, M.; Al-Salem, S.M. Ozone (O_3) ambient levels as a secondary air borne precursor in Fahaheel urban area, the State of Kuwait. *Atmos. Sci. Lett.* **2020**, *21*, 983. [CrossRef]
5. Gupta, P.; Payra, S.; Bhatla, R.; Verma, S. An evaluation of long-term gridded datasets of total columnar ozone retrieved from MERRA-2 and AIRS over the Indian region. *Environ. Sci. Pollut. Res.* **2023**, *30*, 43586–43603. [CrossRef]
6. Available online: https://www.europarl.europa.eu/meetdocs/2014_2019/plmrep/COMMITTEES/ENVI/DV/2021/01-14/Air_quality_in_Europe-2020_report_EN.pdf (accessed on 14 March 2023).
7. Erickson, L.E.; Newmark, G.L.; Higgins, M.J.; Wang, Z. Nitrogen oxides and ozone in urban air: A review of 50 plus years of progress. *Environ. Prog. Sustain. Energy* **2020**, *39*, 13484. [CrossRef]
8. Grigoraş, G.; Ştefan, S.; Rada, C.; Grigoraş, C. Assessing of surface-ozone concentration in Bucharest, Romania, using OML and satellite data. *Atmos. Pollut. Res.* **2016**, *7*, 567–576. [CrossRef]
9. Park, B.; Kim, S.; Park, S.; Kim, M.; Kim, T.Y.; Park, H. Development of Multi-Item Air Quality Monitoring System Based on Real-Time Data. *Appl. Sci.* **2021**, *11*, 9747. [CrossRef]
10. Mirzaei, A.; Ansari, H.R.; Shahbaz, M.; Kim, J.-Y.; Kim, H.W.; Kim, S.S. Metal Oxide Semiconductor Nanostructure Gas Sensors with Different Morphologies. *Chemosensors* **2022**, *10*, 289. [CrossRef]
11. John, A.T.; Murugappan, K.; Nisbet, D.R.; Tricoli, A. An Outlook of Recent Advances in Chemiresistive Sensor-Based Electronic Nose Systems for Food Quality and Environmental Monitoring. *Sensors* **2021**, *24*, 2271. [CrossRef]
12. Goel, N.; Kunal, K.; Kushwaha, A.; Kumar, M. Metal oxide semiconductors for gas sensing. *Eng. Rep.* **2022**, 12604. [CrossRef]
13. Sharma, A.K.; Mahajan, A. Chapter 8—Potential applications of chemiresistive gas sensors. In *Micro and Nano Technologies, Carbon Nanomaterials and Their Nanocomposite-Based Chemiresistive Gas Sensors*; Dhall, S., Ed.; Elsevier: Amsterdam, The Netherlands, 2023; pp. 223–245, ISBN 9780128228371. [CrossRef]
14. Peterson, P.J.D.; Aujla, A.; Grant, K.H.; Brundle, A.G.; Thompson, M.R.; Vande Hey, J.; Leigh, R.J. Practical Use of Metal Oxide Semiconductor Gas Sensors for Measuring Nitrogen Dioxide and Ozone in Urban Environments. *Sensors* **2017**, *17*, 1653. [CrossRef]
15. Avansi, W.J.; Catto, A.C.; da Silva, L.F.; Fiorido, T.; Bernardini, S.; Mastelaro, V.R.; Aguir, K.; Arenal, R. One-Dimensional V_2O_5/TiO_2 Heterostructures for Chemiresistive Ozone Sensors. *ACS Appl. Nano Mater.* **2019**, *2*, 4756–4764. [CrossRef]

16. Paralikis, A.; Gagaoudakis, E.; Kampitakis, V.; Aperathitis, E.; Kiriakidis, G.; Binas, V. Study on the Ozone Gas Sensing Properties of RF-Sputtered Al-Doped NiO Films. *Appl. Sci.* **2021**, *11*, 3104. [CrossRef]
17. Huang, C.-Y.; He, X.-R.; Jhang, J.-J.; Wu, J.-H.; Wu, T.-H.; Lin, T.-Y. A self-powered dual-functional hybrid Cu_2O/SiNWs heterojunction with applications in broadband photodetectors and ozone gas sensors. *Sens. Actuators A Phys.* **2022**, *347*, 113992. [CrossRef]
18. Sironi, S.; Capelli, L.; Spinelle, L.; Gerboles, M.; Aleixandre, M.; Bonavitacola, F. Evaluation of Metal Oxides Sensors for the Monitoring of O_3 in Ambient Air at Ppb Level. *Chem. Eng. Trans.* **2016**, *54*, 319–324. [CrossRef]
19. Wang, C.Y.; Becker, R.W.; Passow, T.; Pletschen, W.; Köhler, K.; Cimalla, V.; Ambacher, O. Photon Stimulated Sensor Based on Indium Oxide Nanoparticles I: Wide-Concentration-Range Ozone Monitoring in Air. *Sens. Actuators B Chem.* **2011**, *152*, 235–240. [CrossRef]
20. Sá, J.P.; Chojer, H.; Branco, P.T.B.S.; Alvim-Ferraz, M.C.M.; Martins, F.G.; Sousa, S.I.V. Two step calibration method for ozone low-cost sensor: Field experiences with the UrbanSense DCUs. *J. Environ. Manag.* **2023**, *328*, 116910. [CrossRef]
21. Baloria, V.; Yadav, A.; Singh, P.; Gupta, G. Chapter 5—Semiconductor oxide based chemiresistive gas sensors. In *Micro and Nano Technologies, Carbon Nanomaterials and Their Nanocomposite-Based Chemiresistive Gas Sensors*; Dhall, S., Ed.; Elsevier: Amsterdam, The Netherlands, 2023; pp. 133–167. [CrossRef]
22. Aguir, K.; Lemire, C.; Lollman, D.B.B. Electrical Properties of Reactively Sputtered WO3 Thin Films as Ozone Gas Sensor. *Sensors Actuators B Chem.* **2002**, *84*, 1–5. [CrossRef]
23. de Lima, B.S.; Komorizono, A.A.; Silva, W.A.S.; Ndiaye, A.L.; Brunet, J.; Bernardi, M.I.B.; Mastelaro, V.R. Ozone detection in the ppt-level with rGO-ZnO based sensor. *Sens. Actuators B Chem.* **2021**, *338*, 129779. [CrossRef]
24. Boulmani, R.; Bendahan, M.; Lambert-Mauriat, C.; Gillet, M.; Aguir, K. Correlation between rf-sputtering parameters and WO_3 sensor response towards ozone. *Sens. Actuators B Chem.* **2007**, *125*, 622–627. [CrossRef]
25. Bendahan, M.; Boulmani, R.; Seguin, J.L.; Aguir, K. Characterization of ozone sensors based on WO_3 reactively sputtered films: Influence of O_2 concentration in the sputtering gas, and working temperature. *Sens. Actuators B Chem.* **2004**, *100*, 320–324. [CrossRef]
26. Hasan, M.H.; Yu, H.; Ivey, C.; Pillarisetti, A.; Yuan, Z.; Do, K.; Li, Y. Unexpected Performance Improvements of Nitrogen Dioxide and Ozone Sensors by Including Carbon Monoxide Sensor Signal. *ACS Omega* **2023**, *8*, 5917–5924. [CrossRef] [PubMed]
27. Klaus, D.; Klawinski, D.; Amrehn, S.; Tiemann, M.; Wagner, T. Light-activated resistive ozone sensing at room temperature utilizing nanoporous In_2O_3 particles: Influence of particle size. *Sens. Actuators B Chem.* **2015**, *217*, 181–185. [CrossRef]
28. Kumar, U.; Li, Y.-N.; Deng, Z.-Y.; Chiang, P.-C.; Yadav, B.C.; Wu, C.-H. Nanoarchitectonics with lead sulfide quantum dots for room-temperature real-time ozone trace detection with different light exposure. *J. Alloy. Compd.* **2022**, *926*. [CrossRef]
29. de Palma, J.V.N.; Catto, A.C.; de Oliveira, M.C.; Ribeiro, R.A.P.; Teodoro, M.D.; da Silva, L.F. Light-assisted ozone gas-sensing performance of SnO_2 nanoparticles: Experimental and theoretical insights. *Sens. Actuators Rep.* **2022**, *4*, 100081. [CrossRef]
30. Bernardini, S.; Benchekroun, M.H.; Fiorido, T.; Aguir, K.; Bendahan, M.; Dkhil, S.B.; Gaceur, M.; Ackermann, J.; Margeat, O.; Videlot-Ackermann, C. Ozone Sensors Working at Room Temperature Using Zinc Oxide Nanocrystals Annealed at Low Temperature. *Proceedings* **2017**, *1*, 423. [CrossRef]
31. Roy, D.; Hossain, M.R.; Hossain, M.K.; Hossain, M.A.; Ahmed, F. Density functional theory study of the sensing of ozone gas molecules by using fullerene-like Group-III nitride nanostructures. *Phys. B Condens. Matter* **2023**, *650*, 414553. [CrossRef]
32. Zhao, Z.; Deng, Z.; Zhang, R.; Klamchuen, A.; He, Y.; Horprathum, M.; Chang, J.; Mi, L.; Li, M.; Wang, S.; et al. Sensitive and selective ozone sensor based on $CuCo_2O_4$ synthesized by a facile solution combustion method. *Sens. Actuators B Chem.* **2023**, *375*, 132912. [CrossRef]
33. Gómez-Suárez, J.; Arroyo, P.; Cerrato-Álvarez, M.; Hontañón, E.; Masa, S.; Menini, P.; Presmanes, L.; Alfonso, R.; Pinilla-Gil, E.; Lozano, J. Development and Field Validation of Low-Cost Metal Oxide Nanosensors for Tropospheric Ozone Monitoring in Rural Areas. *Chemosensors* **2022**, *10*, 478. [CrossRef]
34. Christaki, E.; Vasilaki, E.; Gagaoudakis, E.; Binas, V.; Vamvakaki, M.; Klini, A. Room temperature optical detection of ultra-low ozone concentration using photoluminescent ZnO nanohybrids. *Sens. Actuators B Chem.* **2022**, *359*, 131614. [CrossRef]
35. Al-Resheedi, A.S.; Alhokbany, N.S.; Mahfouz, R.M. Radiation Induced Synthesis of In_2O_3 Nanoparticles-Part II: Synthesis of In_2O_3 Nanoparticles by Thermal Decomposition of Un-Irradiated and γ-Irradiated Indium Acetylacetonate. *Mater. Res.* **2015**, *18*, 931–938. [CrossRef]
36. Gallagher, D.; Scanlan, F.; Houriet, R.; Mathieu, H.J.; Ring, T.A. Indium-Tin Oxide Thin Films by Metal-Organic Decomposition. *J. Mater. Res.* **1993**, *8*, 3135–3144. [CrossRef]
37. Chesler, P.; Hornoiu, C.; Mihaiu, S.; Vladut, C.; Moreno, J.M.C.; Anastasescu, M.; Moldovan, C.; Firtat, B.; Brasoveanu, C.; Muscalu, G.; et al. Nanostructured SnO_2-ZnO Composite Gas Sensors for Selective Detection of Carbon Monoxide. *Beilstein J. Nanotechnol.* **2016**, *7*, 2045–2056. [CrossRef] [PubMed]
38. Firtat, B.; Moldovan, C.; Brasoveanu, C.; Muscalu, G.; Gartner, M.; Zaharescu, M.; Chesler, P.; Hornoiu, C.; Mihaiu, S.; Vladut, C.; et al. Miniaturised MOX Based Sensors for Pollutant and Explosive Gases Detection. *Sens. Actuators B Chem.* **2017**, *249*. [CrossRef]
39. Available online: https://www.osha.gov/laws-regs/standardinterpretations/1988-02-11#:~:text=The%20current%20OSHA%20permissible%20exposure,over%20an%20eight%2Dhour%20workshift (accessed on 14 March 2023).
40. Thorp, C.E. *Bibliography of Ozone Technology, Volume II*; Armour Research Foundation, Illinois Institute of Technology: Chicago, IL, USA, 1955; p. 40.

41. dos Santos Silva, W.A.; de Lima, B.S.; Bernardi, M.I.B.; Mastelaro, V.R. Enhancement of the Ozone-Sensing Properties of ZnO through Chemical-Etched Surface Texturing. *J. Nanoparticle Res.* **2022**, *24*, 96. [CrossRef]
42. Chesler, P.; Hornoiu, C. MOX-Based Resistive Gas Sensors with Different Types of Sensitive Materials (Powders, Pellets, Films), Used in Environmental Chemistry. *Chemosensors* **2023**, *2023 11*, 95. [CrossRef]
43. Chesler, P.; Hornoiu, C.; Bratan, V.; Munteanu, C.; Gartner, M.; Ionescu, N.I. Carbon Monoxide Sensing Properties of TiO_2. *Rev. Roum. Chim.* **2015**, *60*, 227–232.
44. Chesler, P.; Hornoiu, C.; Bratan, V.; Munteanu, C.; Postole, G.; Ionescu, N.I.; Juzsakova, T.; Redey, A.; Gartner, M. CO Sensing Properties of SnO_2–CeO_2 Mixed Oxides. *React. Kinet. Mech. Catal.* **2016**, *117*, 551–563. [CrossRef]
45. Chesler, P.; Hornoiu, C.; Mihaiu, S.; Munteanu, C.; Gartner, M. Tin–Zinc Oxide Composite Ceramics for Selective CO Sensing. *Ceram. Int.* **2016**, *42*, 16677–16684. [CrossRef]
46. Chelu, M.; Chesler, P.; Anastasescu, M.; Hornoiu, C.; Mitrea, D.; Atkinson, I.; Brasoveanu, C.; Moldovan, C.; Craciun, G.; Gheorghe, M.; et al. ZnO/NiO Heterostructure-Based Microsensors Used in Formaldehyde Detection at Room Temperature: Influence of the Sensor Operating Voltage. *J. Mater. Sci. Mater. Electron.* **2022**, *33*, 19998–20011. [CrossRef]
47. Huang, J.; Wang, X.; Gong, Y.; Liu, Y.; Zhou, P.; Suo, X.; Zeng, D.; Li, H. Construction of WO_3 coatings with mi-cro-nano hybrid structures by liquid pre-cursor flame spray for enhanced sensing performances to sub-ppm ozone. *Mater. Lett.* **2017**, *205*, 106–109. [CrossRef]
48. Acuautla, M.; Bernardini, S.; Gallais, L.; Fiorido, T.; Patout, L.; Bendahan, M. Ozone flexible sensors fabricated by photolithography and laser ablation processes based on ZnO nanoparticles. *Sens. Actuators B Chem.* **2014**, *203*, 602–611. [CrossRef]
49. Zhang, D.; Yang, Z.; Li, P.; Zhou, X. Ozone gas sensing properties of metal-organic frameworks-derived In_2O_3 hollow microtubes decorated with ZnO nanoparticles. *Sens. Actuators B Chem.* **2019**, *301*, 127081. [CrossRef]
50. Joshi, N.; da Silva, L.F.; Jadhav, H.S.; Shimizu, F.M.; Suman, P.H.; M'Peko, J.C.; Orlandi, M.O.; Seo, J.G.; Mastelaro, V.R.; Oliveira, O.N., Jr. Yolk-shelled $ZnCo_2O_4$ microspheres: Surface properties and gas sensing application. *Sens. Actuators B Chem.* **2018**, *257*, 906–915. [CrossRef]
51. Catto, A.C.; Fiorido, T.; Souza, É.L.S.; Avansi, W.; Andres, J.; Aguir, K.; Longo, E.; Cavalcante, L.S.; da Silva, L.S. Improving the ozone gas-sensing properties of $CuWO_4$ nanoparticles. *J. Alloy. Compd.* **2018**, *748*, 411–417. [CrossRef]
52. Catto, A.C.; Bernardini, S.; Aguir, K.; Longo, E.; da Silva, L.F. In-situ hydrothermal synthesis of oriented hematite nanorods for sub-ppm level detection of ozone gas. *J. Alloy. Compd.* **2023**, *947*, 169444. [CrossRef]
53. Colmenares, Y.N.; Correr, W.; Lima, B.S.; Mastelaro, V.R. The effect of morphology on the ozone-gas sensing properties of zinc oxide sputtered films. *Thin Solid Film.* **2020**, *703*, 137975. [CrossRef]
54. Souissi, R.; Bouguila, N.; Bendahan, M.; Aguir, K.; Fiorido, T.; Abderrabba, M.; Hali-dou, I.; Labidi, A. Ozone sensing study of sprayed β-In_2S_3 thin films. *J. Alloy. Compd.* **2022**, *900*, 2163513. [CrossRef]
55. Gaponenko, N.V.; Mudryi, A.V.; Sergeev, O.V.; Borisenko, V.E.; Pivin, J.C.; Baran, A.S.; Stepanova, E.A.; Ratko, A.I.; McGilp, J.F. Erbium and Terbium Luminescence from Sol–Gel Derived In_2O_3 Films on Porous Silicon. *Phys. Status Solidi* **1998**, *165*, 131–134. [CrossRef]

gels

MDPI

Review

An Elucidative Review of the Nanomaterial Effect on the Durability and Calcium-Silicate-Hydrate (C-S-H) Gel Development of Concrete

Farqad Yousuf Al-saffar [1,2,*], Leong Sing Wong [1,2,*] and Suvash Chandra Paul [3]

1 Department of Civil Engineering, College of Engineering, Universiti Tenaga Nasional, Jalan IKRAM-UNITEN, Kajang 43000, Selangor, Malaysia
2 Institute of Energy Infrastructure, Universiti Tenaga Nasional, Jalan IKRAM-UNITEN, Kajang 43000, Selangor, Malaysia
3 Department of Civil Engineering, International University of Business Agriculture and Technology, 4 Embankment Drive Road, Sector 10, Dhaka 1230, Bangladesh
* Correspondence: farqad.yousuf@yahoo.com (F.Y.A.-s.); wongls@uniten.edu.my (L.S.W.)

Abstract: Concrete as a building material is susceptible to degradation by environmental threats such as thermal diffusion, acid and sulphate infiltration, and chloride penetration. Hence, the inclusion of nanomaterials in concrete has a positive effect in terms of promoting its mechanical strength and durability performance, as well as resulting in energy savings due to reduced cement consumption in concrete production. This review article discussed the novel advances in research regarding C-S-H gel promotion and concrete durability improvement using nanomaterials. Basically, this review deals with topics relevant to the influence of nanomaterials on concrete's resistance to heat, acid, sulphate, chlorides, and wear deterioration, as well as the impact on concrete microstructure and chemical bonding. The significance of this review is a critical discussion on the cementation mechanism of nanoparticles in enhancing durability properties owing to their nanofiller effect, pozzolanic reactivity, and nucleation effect. The utilization of nanoparticles enhanced the hydrolysis of cement, leading to a rise in the production of C-S-H gel. Consequently, this improvement in concrete microstructure led to a reduction in the number of capillary pores and pore connectivity, thereby improving the concrete's water resistance. Microstructural and chemical evidence obtained using SEM and XRD indicated that nanomaterials facilitated the formation of cement gel either by reacting pozzolanically with portlandite to generate more C-S-H gel or by functioning as nucleation sites. Due to an increased rate of C-S-H gel formation, concrete enhanced with nanoparticles exhibited greater durability against heat damage, external attack by acids and sulphates, chloride diffusion, and surface abrasion. The durability improvement following nanomaterial incorporation into concrete can be summarised as enhanced residual mechanical strength, reduced concrete mass loss, reduced diffusion coefficients for thermal and chloride, improved performance against sulphates and acid attack, and increased surface resistance to abrasion.

Keywords: nanomaterial; concrete; durability; energy savings; C-S-H gel

1. Introduction

Improving concrete durability was and is still a subject of interest for scientists and engineers [1,2]. Many approaches and methods were used to increase concrete's resistance against threats from natural weathering, chemical attacks, surface abrasion, and fire exposure. Concrete durability problems compromise its structural integrity, leading to economic and human losses [3]. Several measures were taken to alleviate the problems [4–6], including but not limited to the following: a decrease in the water-to-cement ratio (w/c), incorporation of supplementary cementitious materials (SCMs) to partially replace cement, application of polymers, and utilization of surface coating substances. By reducing porosity, creating discontinuous pore networks, and strengthening the interfacial transition zone (ITZ) in concrete [7,8], there is a significant enhancement in the durability of concrete. Furthermore, enhancing the mechanical characteristics of concrete and compacting its microstructure can significantly decrease its permeability and ability to absorb water [9]. Pore interconnectivity directly determines the permeability of concrete; the more interconnected the pore system, the greater the permeability and water diffusion. Pore connectivity likely decreases with increases in the degree of cement hydration, owing to a growing solid phase in the cement matrix, which is represented by hydration products (e.g., calcium silicate hydrate (C-S-H), portlandite (CH), ettringite (AFt), and monosulfoalmonate (AFm)). According to Hewlett and Liska [10], the gel that contributes to the strength characteristics of concrete is a porous and amorphous compound called calcium silicate hydrate (C-S-H). The nanostructured, gel-like hydrate bridges the unreacted cement particles as a connected network of solids into the empty spaces (pores) and is thereby considered the principal binding phase among other hydrates. The development of C-S-H gel greatly reduces the permeability of concrete and, as a result, enhances its endurance.

Incorporating ultra-fine materials (e.g., 1–100 nm) such as nanoparticles is among the most influential methods to increase the mechanical strength of concrete and perfect its microstructure [11–14]. Recent studies and published papers demonstrated that nanomaterials play an advanced role in improving concrete's compressive strength, tensile strength, and flexural strength [15,16]. For example, nanosilica [17], which is regarded as the earliest applied and most common nanomaterial in concrete, had a remarkable effect by enhancing the performance of the material with high mechanical and durability properties. This can be attributed to its high surface area, which provides seeds for hydration products, and to its nano-filling effect on reducing the pores in concrete. Thereby, nanosilica can dramatically reduce the permeability of concrete. Nanosilica materials enhance the microstructure not only because of the nanofiller effect but also through pozzolanic reactivity, which in turn promotes cement hydration, formatting a higher amount of C-S-H gel [18]. Other nanomaterials such as nanoclay, titanium oxide, nanometakaolin, nanoC-S-H, nano$CaCO_3$, carbon nanotube, etc., improve concrete microstructure with filling and seeding effects [11–14]. According to the literature, nanoparticles enhance the formation of C-S-H gel with a compact structure by filling its nanopores, as discussed in Section 2. Some nanoparticles can provide nucleation sites owing to their super fineness, thus accelerating the precipitation and formation of C-S-H gel. Owing to the high efficiency of nanomaterials in improving the mechanical properties of cementitious materials, the applications of nanoparticles were broadened to include ultra-high performance concrete [13], self-compacted concrete [19], reactive powder concrete [20], heavyweight concrete [21], marine concrete [22], shotcrete [23], green concrete [24], recycled aggregate concrete [25], lightweight concrete [26], high-volume fly ash concrete [27], carbon fibre-reinforced concrete [28], and foam concrete [29].

The impact of numerous nanomaterials, such as nanosilica, nanoZnO, nanoFe_2O_3, nanoAl_2O_3, and nanoTiO_2, on the properties of concrete was discussed in several published works [11–14]. The results confirmed the ability of these ultra-fine materials to ameliorate the cementitious materials' macroscopic characteristics and functioning. However, due to the agglomeration effect of nanomaterials, these nanoparticles need to be optimised to reduce their negative impact and efficiency in cementitious materials. Hence, several studies were conducted to determine the optimum dosage for better performance in cementitious materials. Nanotechnology in concrete has always been an area of focus for concrete technologists, and due to recent advances in nanomaterials, researchers have extended their investigations to include hybrid nanoparticles [30], nanoparticles with fibres [28], and nanoparticles with cementitious supplementary materials [21]. The incorporation of nanomaterials in concrete helped enhance its permeability properties by mitigating the material's water absorption and hindering the penetration and transport of deleterious ions [30]. This is attributed to nanoparticles' role in reducing pore size and connectivity by advancing the growth of C-S-H gel. Various studies investigated the influence of incorporating nanomaterials on the durability performance of concrete [11–14]. Hence, in this article, a comprehensive review was completed to highlight the effect of nanomaterials on cement gel in concrete and how their durability performance can be enhanced against heat, acid attack, sulphate attack, chloride attack, and surface abrasion.

This review article demonstrates the mechanism of action underlying how nanoparticles improve the durability properties of concrete by improving the hydration of binding gel and other hydration products. The improvement in concrete durability following nanomaterial incorporation can be summarised as enhanced residual mechanical strength, reduced concrete mass loss, reduced diffusion coefficients for thermal and chloride, improved performance against sulphates and acid attack, and increased surface resistance to abrasion. The technological progress in nanoconcrete is in line with the ninth United Nations Sustainable Development Goal, which is 'Industry, Innovation, and Infrastructure'. This review was performed by referring to the most up-to-date studies and highlighting their results to draw a robust conclusion. It provides useful ideas for future published research works related to the role of nanomaterials in improving concrete durability. Despite the literature having considerable review articles related to the effects of nanoparticles on the durability of cement-based composites, there is a lack of published reviews explaining the efficiency of nanomaterials in promoting C-S-H gel formation, which has a considerable impact on concrete's resistance to heat, acid, sulphate, chloride, and surface abrasion.

2. Microstructure of Concrete and C-S-H Gel Hydration with Nanomaterials

2.1. Development of C-S-H Gel in Concrete Induced with Nanomaterials

Several studies have shown that the addition of nanoparticles can greatly enhance the strength and durability of concrete by improving the hydration of tricalcium silicate (C_3S) found in cement. This is achieved as the C-S-H gel expands into the capillary pores, binding the solid products and sealing any microcracks that may impact the structural integrity of the concrete [13,30–32]. The exact mechanisms behind the improvement in hydration using nanomaterials have been studied extensively, and multiple theories have been proposed to explain their effectiveness [33]. An example of a material with a significant pozzolanic impact is silica nanoparticles. They interact with calcium hydroxide (CH), which encourages the expansion of the C-S-H gel, resulting in an improvement. Additionally, because of its sizeable specific area, nanosilica can act as a reactive siliceous surface, aiding in early C-S-H growth. Moreover, silica nanoparticles can serve as seeds that facilitate cement hydration and densify the gel structure by blocking the voids between C-S-H particles and thereby increasing its stiffness.

Incorporating C-S-H seeds into cementitious materials has proved to be effective in improving the early strength properties of concrete by accelerating the precipitation of silicates [34]. It was also discovered that the Ca/Si ratio in the C-S-H seeds had a significant impact on the effectiveness of the cement acceleration process, and a low Ca/Si ratio in C-S-H seeds further reduced the setting time of cement, shortening the induction period of the cement hydration process [35]. This could be attributed to the large surface areas provided by the low Ca/Si gel, which facilitated hydration. Land et al. [35] pointed out that the chemical composition and particle size of C-S-H seeds were the two primary factors that determined the effectiveness of the water–cement reaction. Moreover, incorporating nano-C-S-H particles with mineral admixtures hastened their secondary hydration process [36].

The introduction of nanoclay encouraged the production of C-S-H and carboaluminate (C-A-S-H) gels following its reaction with various cement components such as C_3S, C_3A, and C_4AF. In addition, a rise in the pore solution's Ca^{2+} ionic concentration was detected, and less consumption of silicate phases occurred in the hydration process. Microstructural observation of concrete [37] revealed that calcined nanoclay can refine the microstructure of concrete by filling in pores and exhibiting pozzolanic behaviour. This reaction occurs with Portlandite and results in the production of more C-S-H gel. Graphene oxide sheets promoted the formation of ettringite as flower-like crystals, while there was a dramatic consumption of silicates and an increase in Ca^{2+} ions with nanosilica incorporation [31]. Nanotitanium dioxide had a lower cement hydration acceleration compared to other nanomaterials and played an inert role in hydration to some extent [31]. The assumption was emphasised by the lower consumption of silicate phases. On the other hand, nanotitanium dioxide promoted the microstructure by stimulating more C-S-H gel and preventing CH growth, in addition to its filling and nucleation effects [38].

Nanometakaolin can generate additional quantities of C-S-H gel due to nanoparticles' role in catalysing the cement–water reaction. It was reported that nanometakaolin has the ability to increase the chemical bonding and cohesion forces in the C-S-H structure. Like nanosilica, nanometakaolin is a pozzolanic material that creates more C-S-H gel through a secondary reaction with CH crystals [39]. The inclusion of nano-ZrO_2 particles [40] can also improve the cement reaction with their nucleation effect in a similar manner to nanosilica. Devi et al. [41] discovered that C-S-H gel with graphene oxide was less porous because of its denser microstructure. This can be attributed to embedded CH crystals and ettringite needles while having interlocked and interweaved foil features. Observing the formation of massive and denser hydrated crystals revealed that graphene oxide can function as a concrete reinforcement agent. XRD results indicated that blending graphene oxide with cement assisted the hydration of unreacted cement particles at later ages, making the concrete stronger and more durable. Moreover, carbon nanotubes [42] can also facilitate C-S-H hydration by providing nucleation sites because of their higher surface area. Microstructure observation confirmed that the functional group (carboxyl) of carbon nanotubes can interact with CH, generating an additional amount of gel hydrates. On the other hand, it was reported that higher amounts of carbon nanotubes limited the hydration of C-S-H gel because they reduced the amount of free water, which is crucial for hydration to proceed. Furthermore, carbon nanotubes cover the unreacted cement particles, hindering water diffusion and preventing further reaction. Hence, the incorporation of carbon nanotubes can have a negative impact on cement hydration. Meanwhile, the size and shape of carbon nanotubes, in addition to their surface functional groups, are the major factors influencing their hydration effect. Wang et al. [43] used Si NMR spectroscopy to examine how nanofillers changed the structure of C-S-H gels.

The addition of nanofillers such as nanosilica has two major effects on the hydration of C-S-H gel. Firstly, they provide nucleation sites and undergo a pozzolanic reaction with CH hydrates. Secondly, their high water absorption capacity reduces the amount of proton water in the C-S-H gel, shortening the distance between the Ca, O, and Si atoms. Additionally, the nanofillers also break the Si-O-Si spatial structure of silicate tetrahedrons in the C-S-H gel and cause hydroxylation of the tetrahedron. These newly produced Si-OH monomers interact with one another and condense to create a higher polymerization silicate tetrahedron. Furthermore, one study found that incorporating fillers into a C-S-H gel reduced the amount of proton water in the gel, which, in turn, increased the strength of the chemical bond (ionic bond and covalent binding) between the gel's Ca, O, and Si atom structural groups. However, the nanofillers' excessive absorption of gel water could inhibit further hydration, and this could be indicated as an unfavourable phenomenon. Several comparison studies [12] confirmed that nanomaterials had a more effective role in improving cement hydration than supplementary cementitious materials (SCMs) when the pozzolanic reactivity of nanomaterials such as nanosilica outperformed common SCMs. Typically, the majority of nanoparticles aid in the process of cement gel hydration by creating more locations for nucleation and supporting the development of C-S-H gel through nucleation and growth processes. This was confirmed by Long et al. [44], who measured the heat released to different nanomaterial binders and found that reducing the period of the induction stage and the time for the main peak in heat flow by nanosilica and nano-C-S-H were indications of the nanoparticle effect on the rate of cement hydration. Higher heat release rate peak values were recorded with these types of nanoparticle binders. The study concluded that nanosilica and nano-C-S-H can remarkably increase the rate of hydrate nucleation and growth, while the effect was greater with increasing the rate of nucleation. Nano-$CaCO_3$ [45] had no effective role either in the nucleation rate or growth rate of hydration, and that was attributed to its low chemical activity. The effect of nanoparticles on cement hydration was dominated by the particle's size, affinity, and chemical activity. The C-S-H gels appear in a great number of clusters with a more uniform distribution in the presence of nanoparticles, binding various substances such as CH and unreacted cement particles with aggregates in concrete. This characteristic of C-S-H gel helps fill more voids at the microscopic level, making the microstructure of concrete more compact and homogeneous. By enhancing the concrete microstructure, durability will be significantly improved.

2.2. Microstructure of Concrete with Nanomaterials

Numerous microstructure observation techniques (e.g., SEM, XRD, and TGA) were used to investigate the kinetics of nanomaterials in cementitious materials and their mechanism in densifying and compacting the concrete microstructure [46,47]. The microstructure of concrete is primarily improved with nanoparticles through three effects, namely the nanofiller effect, pozzolanic reactivity, and the nucleation effect [48]. The nanofiller effect is attributed to nanoparticles' ultra-small size and higher surface area, which helps them to occupy a higher percentage of fine voids in concrete. Nanoparticles work as crystal nuclei, promoting cement hydration. In addition, nanoparticles generate more C-S-H gel through their reaction with CH. These effects remarkably enhance the microstructural packing efficiency of concrete and subsequently improve the material's durability and mechanical strength.

Figure 1 shows a diagram that summarises the mechanism by which nanoparticles improve the durability of concrete. Several scanning electron microscopic (SEM) images from different studies presented a denser, homogeneous, and more uniform microstructure of nano-modified concrete [11–48]. The interfacial transition zone (ITZ) was also improved with the addition of nanoparticles owing to their microaggregate filling effect [49,50]. This microstructure improvement implies that there was a significant increase in the concrete's resistance to acid and sulphate attack and a reduction in its permeability and water absorption. The SEM images demonstrated that the concrete microstructure containing nanoparticles exhibited denser textures and showed fewer destructive effects and microcracks due to the nanoparticle mechanism of action for advancing the growth of C-S-H gel [30–45]. Fewer CH crystals were observed in all types of nanomaterials incorporating concrete with higher amounts of gel-like hydrates. Even at higher temperatures, the microstructure of nanoconcrete was denser at 200 °C, more homogeneous at 400 °C, and had fewer cavities at 700 °C. Moreover, the concrete microstructure exhibited fewer microcracks at 400 and 600 °C.

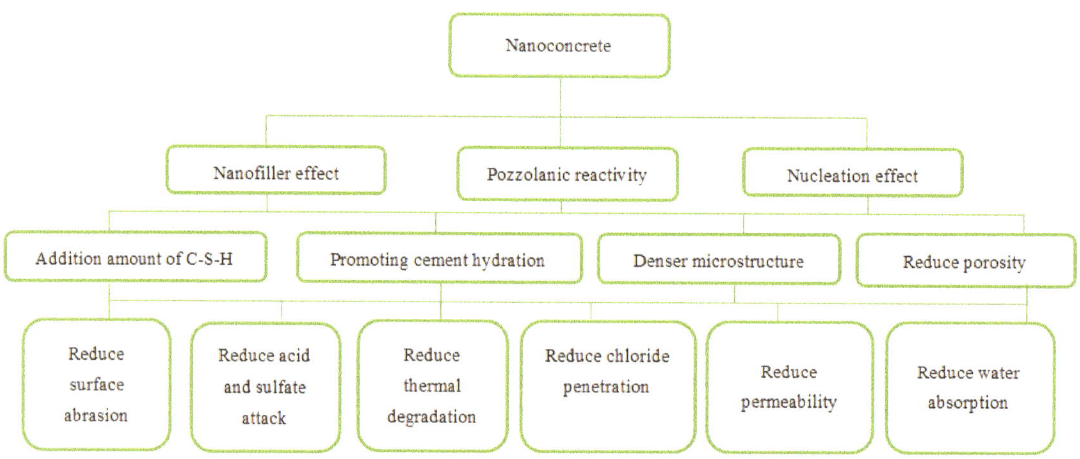

Figure 1. The action by which nanoparticles enhance the durability of concrete.

Figure 2 shows SEM images of concrete with nanoparticles [51]. The micrographs from several SEM studies indicated that the hydration products in the nanoconcrete microstructure were "distributed neatly and tidy, and the texture was denser with no obvious aciculate AFt or cubic CH to be traced" [52]. Furthermore, using the SEM images, it was observed that the microstructure of nanomodified concrete exhibited a lower amount of CH, gypsum, and ettringite needle crystals, in addition to a denser and more homogenous structure. X-ray diffraction (XRD) tests on nanoconcrete were performed and published in several research articles [11–48]. All the test results indicated a higher intensity of the C-S-H gel peak and a lower intensity of CH for concrete in the presence of nanomaterials.

Figure 2. SEM images showing concrete with nanoparticles. Reproduced with permission from [51] (*Journal of Building Engineering*); published by (Elsevier), (2021). (Note: milled nanometakaolin (NMK), nanowaste glass (NWG), nano rice husk ash (NRHA), nanosilica (NS), control concrete (C0)).

At a temperature of 200 °C, it was observed that nanoconcrete contained smaller CH crystals. The presence of such crystals indicated the pozzolanic reactivity of nanomaterials, while their smaller size and orientation were consistent with past research findings. Incorporating nanoparticles into concrete led to an upsurge in the volume of C-S-H gel, ascribed to their nucleation effect that expedited cement hydration. The high surface area of nanoparticles acted as nuclei for cement hydrates, hence enabling this effect. Additionally, the reaction between nanoparticles and CH crystals brought about significant enhancement in the interfacial transition zone (ITZ) between the hardened cement paste and aggregates. The substitution of CH crystals with C-S-H gel resulted in superior performance. When testing the concrete using XRD, good pozzolanic reactivity was demonstrated through the lower intensity peaks in CH, C_3S, and C_2S. However, there was an increasing trend in CH crystals as the inclusion of nanomaterials increased due to poor distribution. The EDX analysis revealed that the Ca/Si molar ratio increased with the incorporation of nanomaterials, indicating the positive impact of these materials on the microstructure of cementitious materials. The usage of DSC and TGA [51] in thermal analyses showed that concrete with varying nanoparticles had increased C-S-H gel and a significant decrease in CH contents. The XRD patterns for 28-day concrete with 3 wt.% nanowaste glass (NWG) and 3 wt.% nanowaste ceramic (NWC) [53] are illustrated in Figure 3.

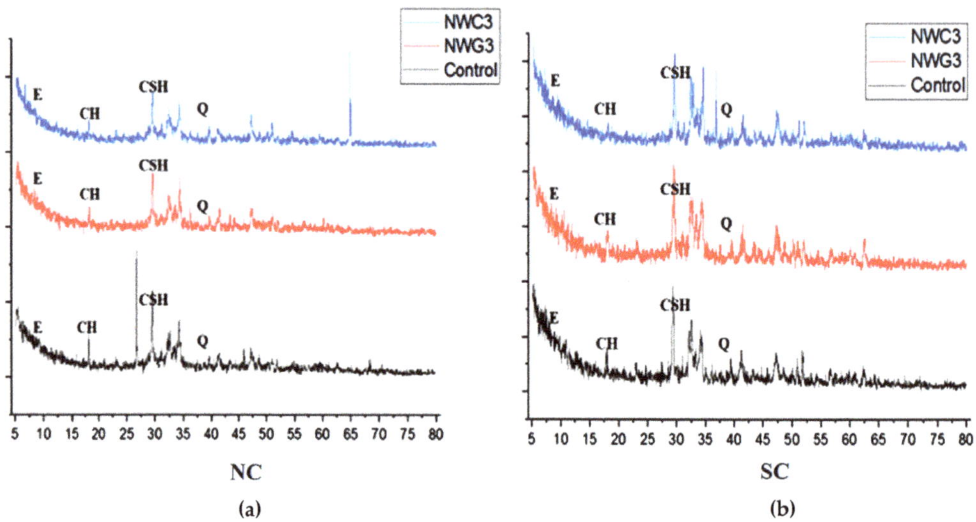

Figure 3. XRD patterns for 28-day self-compacted concrete under (**a**) normal curing (NC) and (**b**) self-curing (SC). Q = unhydrated SiO_2, E = ettringite, CSH = calcium silicate hydrate, and CH = calcium hydroxide. Reproduced with permission from [53] (*Journal of Building Engineering*); published by (Elsevier), (2022).

3. Improvement in the Heat Resistance of Concrete with Nanomaterials

Normal concrete as a building material has appreciable heat resistance with a very low value of thermal conductivity [54]. However, at elevated temperatures greater than 300 °C, the strength properties of concrete deteriorate, and at 600 °C, concrete loses its structural performance [55–57]. The deterioration mechanism for concrete exposed to elevated temperature results from a volume change with aggregate expansion (due to heat) versus cement paste shrinkage (due to water evaporation) [58]. The high temperature can cause damage to the concrete as a result of the drying out and breaking down of cement compounds. The C-S-H gel dehydrates and decomposes when concrete is exposed to elevated temperatures, while free CH loses its bound water, decomposing into calcium oxide. CH is rehydrated and expanded when exposed to moisture, generating built-in stresses and cracking [59,60]. The concrete mass loss is caused by heat due to capillary water evaporation in addition to hydrates and aggregate decomposition. When concrete is exposed to high temperatures, its mechanical features such as compressive, tensile, and flexural strengths are significantly decreased. Concrete spalling [61,62] is another great threat that can be induced by heat. The failure in sections closer to the concrete heated surface is a result of spalling caused by the interplay between the internal pore pressure and thermal stresses, which are characterised by the thermal gradient and thermal incompatibility [63]. For the purpose of improving the fire resistance of concrete, polypropylene fibres [64], light-weight aggregates [65], and supplementary cementitious materials (SCMs) [66] are used. Furthermore, nanomaterials are used in concrete to enhance the heat resistance of the material and reduce thermal degradation during exposure to elevated temperatures [67]. Table 1 summarises several studies conducted with regard to the improvement in concrete's resistance to heat with the addition of nanomaterials. From Table 1, it can be seen that the incorporation of different types of nanomaterials can help mitigate the thermal conductivity of concrete, decrease the mass loss caused by thermal degradation, and improve the mechanical properties of concrete.

Table 1. A summary of the reviews on the influence of various nanomaterials on concrete's resistance to heat.

Reference	Nanomaterial	Subjected Temperatures	Highlights
Brzozowski et al. [68]	Nanosilica	200, 400, 600, 800 °C	Nanosilica improved the residual strength and reduced the pore percentage, having a size of 0.3 to 300 μm.
Kumar et al. [69]	Nanosilica	200, 400, 600, 800 °C	Nanosilica reduced the thermal conductivity, mass loss, and the concrete elastic modules, and the residual strength increased.
Bastami et al. [70]	Nanosilica	400, 600, 800 °C	Nanosilica reduced the concrete mass loss, while residual strength increased.
Shah et al. [71]	Nanosilica	200, 500, 800 °C	Nanosilica concrete showed more significant cracks than microsilica concrete at 800 °C, but nanosilica concrete spalled less than micro- + nanosilica concrete.
Wang et al. [72]	Nanoclay	25–1000 °C	Nanoclay improved the residual strength; 0.1 wt.% nanoclay reduced the thermal conductivity coefficient of cement.
Chu et al. [73]	Graphene sulfonate nanosheets	Up to 1000 °C	Graphene sulfonate nanosheets increased compressive strength, splitting tensile strength, and thermal diffusivity. Porosity was reduced.
Nikbin et al. [74]	Nano bismuth oxide	200, 400, 600 °C	Nano-bismuth oxide particles with higher amounts reduce concrete weight loss.
Nikbin et al. [75]	Titanium dioxide	200, 400, 600 °C	Better performance regarding compressive strength was observed due to titanium dioxide incorporation.
Pachideh et al. [76]	Carbon nanotubes	100, 250, 500, 700 °C	Reduced the risk of fire-induced spalling.
Sakthirswaran et al. [77]	Nanoalumina	200, 400, 600, 800 °C	Nanoalumina particles strengthened the IZT zone and reduced porosity.
Mohammed et al. [78]	Graphene oxide	400, 600, 800 °C	The compressive strength of normal and high-strength concretes exposed to heat increased with Graphene oxide addition.

3.1. Improving Concrete Thermal Conductivity Using Nanomaterials

Several studies confirmed the ability of nanomaterials to improve concrete thermal conductivity. Reddy et al. [79] observed that adding ground-granulated blast furnace slag (GGBS) to concrete modified with 3 wt.% nanosilica resulted in lower thermal conductivity and greater thermal resistance properties. The steady-state box and transition method were used to assess concrete's thermal properties. The results of the steady-state box method demonstrated that blending 3 wt.% nanosilica with GGBS concrete reduced thermal conductivity and thermal diffusivity by 40% and 26%, respectively, as compared with a conventional concrete mix. However, incorporating a higher amount of nanosilica exhibited lower efficiency. The results of the transient method indicated that the average thermal conductivity values for concrete samples cured after 28 days were reduced by 12% with 3 wt.% nanosilica incorporated into GGBS concrete. Thus, 3 wt.% was the optimum dosage of nanosilica to improve the thermal resistance of concrete. Kumar et al. [70] observed that the incorporation of 3 wt.% nanosilica into high-strength concrete retarded heat transfer and reduced the rate of thermal degradation. The thermal conductivity for all concrete mixes exhibited a decreasing trend when subjected to 200, 400, 600, and 800 °C temperatures. However, the high-strength nanosilica concrete had lower thermal conductivity compared to the control high-strength concrete. Nanosilica incorporation helped reduce the rate of heat transfer in high-strength concrete by 11%, 18%, 22%, and 15% at 200, 400, 600, and 800 °C, respectively. High-temperature conditions and their influence on the thermal conductivity coefficient of concrete were investigated by Wang [73]. The concrete thermal conductivity coefficient was reduced with increasing temperatures. C-S-H gel decomposition, in addition to capillary water evaporation caused by heat, was the reason for the decrease. Furthermore, the inclusion of nanoclay in concrete at a weight percentage of 0.3 and 0.5 resulted in an improvement in the thermal conductivity coefficient when compared to control concrete. Conversely, the thermal conductivity coefficient decreased with the addition of 0.1 wt.% nanoclay. The study revealed that the higher amount of nanoclay increased the concrete's thermal conductivity coefficient when subjected to heat, and the cause was related to the concrete's proportion.

3.2. Residual Mechanical Strength Improvements with Nanomaterials

Figure 4 illustrates the collected values in the literature about concrete compressive strength under heat from [71,73–77,80–82]. The concrete compressive strength for all types, regardless of its mix proportion, exhibited a decreasing trend with increasing temperature. However, some concrete exhibited a higher compressive strength at 200 °C than at room temperature. This can be related to the influence of heat on the density of C-S-H gel in concrete, as it was reported that denser C-S-H with an increased chain length can be formed at 200 °C. The improvement in the compressive strength of concrete at high temperatures can be attributed to an increase in the van der Waals force during capillary water evaporation, thus increasing the bonding strength [70,75]. At temperatures greater than 200 °C, the compressive strength of concrete remarkably declined due to thermal degradation, and the values kept declining with increasing temperatures, reaching their lowest at 1000 °C. Furthermore, the figure emphasises the impact of nanomaterials on enhancing the residual mechanical strength of concrete, as higher values of compressive strength can be observed with nanomodified concrete than with control concrete.

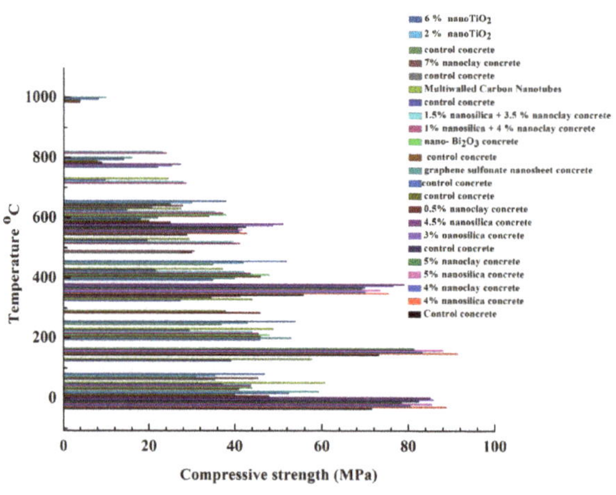

Figure 4. Compressive strength of concrete under different temperatures [71,73–77,80–82].

Figure 5 obviously highlights the impact of nanomaterials on enhancing compressive strength at elevated temperatures [80]. In Figure 5, the effect of nanomaterial enhancement on residual compressive strength is obvious when comparing the results to control concrete. It was reported that silica fume was added to the control sample [80], but the concrete with nanosilica and nanoclay exhibited higher compressive strength than the plain concrete under high temperatures, indicating that nanomaterials can provide concrete with better thermal properties than supplementary cementitious materials such as silica fume.

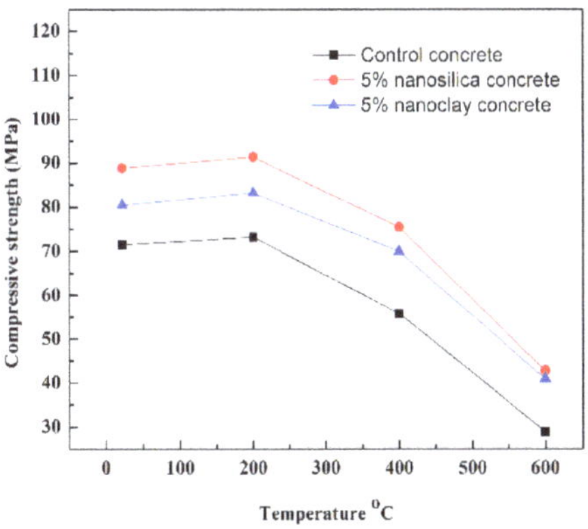

Figure 5. Compressive strength of control concrete, 5% nanosilica concrete, and 5% nanoclay concrete at 22, 200, 400, and 600 °C [80].

According to Figure 6, the results for concrete tensile strength at high temperatures [71,74,77] showed the same trend toward decreasing compressive strength, demonstrating the influence of temperature on tensile strength. The concrete's tensile strength was decreased after raising the temperature, which was identical to the findings for its compressive strength. However, the incorporation of nanomaterials such as nanosilica or multi-walled nanotubes reduced the thermal damage caused by heat, thus increasing the residual tensile strength. Figure 7 shows the compressive and tensile strength losses of concrete at 600 °C.

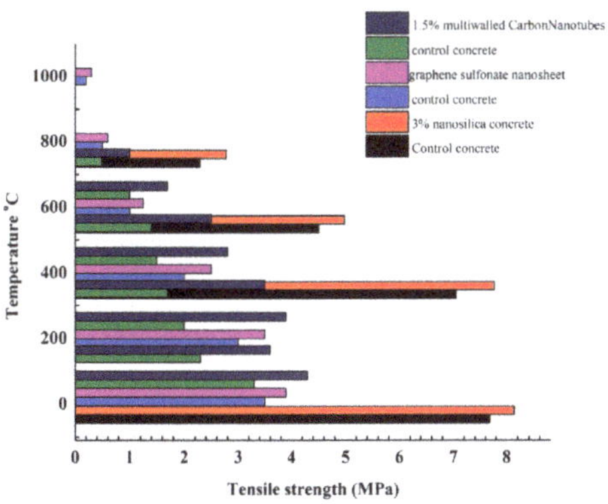

Figure 6. Tensile strength of concrete under different temperatures [71,74,77].

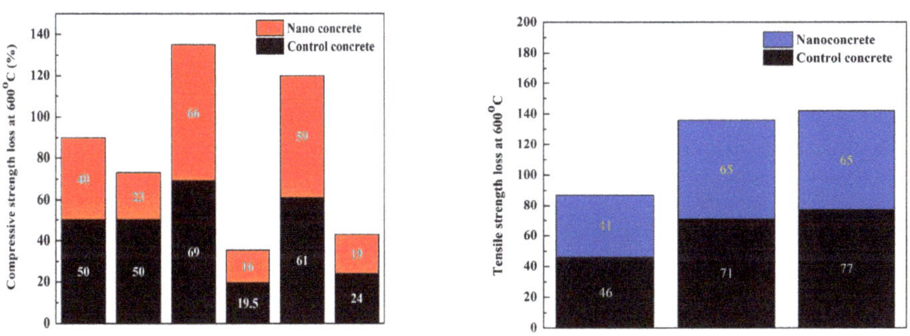

Figure 7. Compressive and tensile strength loss of concrete at 600 °C [71,73–77].

Brzozowski et al. [68] reported that the thermal resistance enhancement with nanosilica was attributed to its thermal characteristics as well as the cement paste porosity structure when they used a method that was good at dispersing the nanoparticles in concrete. It was reported that the 0.3–300 m diameter pores were reduced with nanosilica particles, which also limited the microcracks and strengthened the ITZ zone. It was shown that the utilization of nanosilica in concrete exhibited significant enhancements in its mechanical properties, even at elevated temperatures of 600 °C. The rationale behind this improvement was attributed to the dense formation of calcium silicate hydrate (C-S-H), which is characterised by high rigidity. In a study by Nikbin et al. [75], the application of titanium dioxide as a nanomaterial was also investigated for its impact on the shielding performance of heavy concrete. In order to measure the effect of high temperatures on concrete strength, compressive strength tests were conducted on samples exposed to various levels of heat, including 25, 200, 400, and 600 °C. According to the findings, the compressive strength of concrete at high temperatures was enhanced with the incorporation of titanium dioxide. Pachideh et al. [77] also investigated the influence of carbon nanotubes on post-heat-treated concrete by adding different amounts of carbon nanotubes (0.5, 1, and 1.5 wt.%) to concrete and exposing the samples to temperatures of 25 °C, 100 °C, 250 °C, 500 °C, and 700 °C using an electric furnace. The study showed that the compressive strength of concrete increased by 10% at 25 °C, 16% at 100 °C, 39% at 250 °C, 11% at 500 °C, and 76% at 700 °C when a 0.5% dosage of carbon nanotubes was used. Furthermore, with a 1.5% dosage of carbon nanotubes, the compressive strength increased by 152% at 700 °C. The results highlight that the strong bonding between the cement paste and aggregates is what makes carbon nanotubes particularly effective at increasing the compressive strength of concrete subjected to heat. Additionally, carbon nanotube-modified concrete also demonstrated an increase in tensile strength at elevated temperatures. According to the findings of the study, incorporating carbon nanotubes in concrete can significantly enhance its tensile strength, particularly at higher temperatures. The study examined the effect of carbon nanotubes on the modulus of elasticity of concrete at different elevated temperatures and found that it increased up to 1.33 times at 25 °C, 1.1 times at 100 °C, 1.12 times at 250 °C, 1.45 times at 500 °C, and 1.84 times at 700 °C compared to control concrete. This suggests that carbon nanotubes can effectively improve the performance of concrete under high-temperature conditions.

Another study conducted by Kumar et al. [70] demonstrated that nanosilica high-strength concrete experienced an increase in compressive and split-tensile strength when exposed to temperatures up to 400 °C for 2 h. In comparison with conventional high-strength concrete, nanosilica high-strength concrete displayed higher residual compressive and split-tensile strength during exposure to temperatures up to 800 °C for 2 h. Interestingly, nanosilica high-strength concrete showed brittle failure only up to 600 °C, whereas conventional high-strength concrete exhibited brittle failure only up to 400 °C. Therefore, incorporating nanosilica into concrete can provide superior resistance to high-temperature environments and prevent premature failure in the structure. It was noted that conventional high-strength concrete had better ductility behaviour than nanosilica high-strength concrete. Wang [73] discovered that nanoclay concrete had greater strength when the temperature was maintained below 300 °C. At temperatures varying from 440 to 580 °C, the compressive strength was significantly reduced. The nanoclay concrete showed a 10% decrease in strength at 1000 °C when compared to the original concrete. The study indicated that nanoclay increased the compressive strength at 0.3 wt.% and 0.5 wt.% by weight dosages.

In a study by Mohammed et al. [79], the impact of combining graphene oxide with normal and high-strength concrete was examined under high temperatures of 800 °C. The results showed that the mechanical characteristics of concrete were improved due to the addition of graphene oxide, as evidenced by the residual compressive strength of 70% for the graphene oxide concrete compared to only 35% for the reference concrete. Chu et al. [74] assessed the mechanical characteristics of ferro-siliceous sacrificial concrete embedded with graphene sulfonate nanosheets during high-temperature exposure. According to the research conducted, the inclusion of 0.1 wt.% graphene sulfonate nanosheets was found to enhance the compressive strength and splitting tensile strength of concrete that was subjected to temperatures of up to 1000 °C. The reason for this improvement was believed to be due to the strengthening and toughening impact on the concrete's microstructure, in addition to the decrease in overall porosity as the temperature increased. The conclusion drawn from this study was that graphene sulfonate nanosheets can significantly enhance the high-temperature integrity of concrete.

In a study by Nikbin et al. [75], the effect of different temperatures (25 °C, 200 °C, 400 °C, and 600 °C) on heavy-weight concrete incorporated with varying amounts of nano-bismuth oxide (Nano-Bi_2O_3) was analysed. Tests were conducted on the compressive strength of concrete exposed to high temperatures, and it was found that an increase in the content of nano-bismuth oxide led to a linear increase in the compressive strength. The study also revealed that there was an increase in the compressive strength for all samples as the temperature rose to 200 °C and 400 °C. However, at 600 °C, the compressive strength decreased, with the highest compressive strength seen in the samples containing 6 wt.% nano-bismuth oxide. In general, the study concluded that adding 6% of nano-bismuth oxide particles was the optimum content to enhance the mechanical properties of concrete exposed to high temperatures.

In their study, Sakthieswaran et al. [78] evaluated the behaviour of concrete modified with nanoalumina and containing zircon sand as fine aggregate when exposed to different elevated temperatures. To examine the performance of the concrete mixture, various tests were conducted on concrete samples exposed to temperatures ranging from 200 °C to 800 °C. The study found that the mechanical properties of the concrete improved due to the nanofiller effect of the nanoalumina, which reduced the pores and strengthened the interfacial transition zones in the concrete. The study also revealed that an increase in the percentage of nanoalumina in the concrete mixture resulted in higher compressive strength. In addition, the modified concrete exhibited greater flexural strength than the control concrete at all temperatures tested. These findings suggest that adding nanoalumina to the concrete mixture improves both its compressive and flexural strength, particularly at high temperatures. Furthermore, compared to the control concrete, the nanoalumina-modified concrete had considerably lower capillary porosity at normal and elevated temperatures. The decrease in porosity was attributable to the combined impact of zircon sand and the nano-alumina pore-filling capacity.

Overall, these findings suggest that incorporating nanoalumina into concrete is a promising approach for enhancing the mechanical properties of concrete, particularly when it is expected to be exposed to elevated temperatures.

3.3. *Mass Loss and Spalling Improvements with Nanomaterials*

Bastami et al. [71] conducted a study on the effects of heat on high-strength concrete that was blended with nanosilica, specifically looking at its mass loss and spalling tendencies. Concrete cracking and spalling were visibly observed at temperatures of 600 °C, and the decomposition of the aggregate occurred as the temperature reached 800 °C. The temperature at which nanosilica concrete experienced spalling and mass loss was greater than 400 °C, while the temperature at which ordinary concrete experienced spalling and mass loss was greater than 300 °C. At 600 and 800 °C temperatures, concrete spalling ranged from insignificant (a pitting surface) to severe (an explosive surface). The impact of nanosilica on spalling was not given much emphasis in the published work, as it mainly focused on its ability to decrease concrete permeability and enhance built-in tension stresses.

Shah et al. [72] performed a study showcasing how high-strength concrete's properties were influenced by the presence of both nano- and microsilica when subjected to high temperatures. The results showed better spalling resistance in nanosilica specimens but higher temperature-induced cracking. It was observed by Mohammed et al. [79] that graphene oxide concrete has better fire-induced spalling resistance due to the reinforcing mechanism of graphene oxide and its ability to create networks of microchannels, facilitating the release of vapour pressure.

Table 2 lists the results for mass loss in concrete modified with nanomaterials at high temperatures.

Table 2. Mass loss (%) of concrete modified with nanomaterials under different temperatures.

Reference	Concrete Admixture	200 °C	400 °C	500 °C	600 °C	800 °C	1000 °C
Bastami et al. [71]	0 wt.% nanosilica		4.13		9.26	18.56	
	1.5 wt.% nanosilica		4.03		7.16	14.26	
	3 wt.% nanosilica		3.38		7.24	10.82	
	4.5 wt.% nanosilica		3.20		8.48	11.76	
Shah et al. [72]	10 wt.% microsilica	2.50		5.00		15.00	
	5 wt.% nanosilica	3.00		7.00		10.00	
	10 wt.% microsilica + 5 wt.% nanosilica	3.00		10.00		20.00	
Chu et al. [74]	0 wt.% graphene sulfonate nanosheets	2.38	3.43		4.32	6.03	6.18
	0.1 wt.% graphene sulfonate nanosheets	2.12	3.08		4.04	5.42	5.61
Nikbin et al. [75]	0 wt.% Bi_2O_3	1.50	3.75		5.25		
	2 wt.% Bi_2O_3	1.00	3.10		5.30		
	4 wt.% Bi_2O_3	0.80	2.10		5.35		
	6 wt.% Bi_2O_3	0.8	2.00		5.25		
Nikbin et al. [76]	0 wt.% nanoTiO_2	1.20	3.30		5.20		
	2 wt.% nanoTiO_2	2.20	3.40		6.20		
	4 wt.% nanoTiO_2	2.50	4.50		6.50		
	6 wt.% nanoTiO_2	2.80	5.50		7.00		

3.4. *Microstructure Observation of C-S-H Gel with Nanomaterials at High Temperatures*

Increasing the temperature (i.e., less than 200 °C) can help make the concrete microstructure denser and more homogenous because it accelerates the cement hydration and increases the formed C-S-H gel stiffness. However, microcracks can be formed in plain concrete at temperatures higher than 400 °C, while temperatures above 800 °C cause the decomposition of cement gel. The incorporation of nanomaterials assists in providing a denser and more compacted microstructure even at higher temperatures, such as 400 °C. The improvement can be attributed to the nanoparticle effect on forming high-density C-S-H gel [70,75]. High-density C-S-H gel has higher thermal stability, which helps mitigate heat damage. Nanomaterials help create more arranged clusters of C-S-H gel in concrete when subjected to high temperatures. Several SEM images revealed that the concrete microstructure incorporated with nanomaterials at high temperatures had fewer cracks and cavities as compared to the control sample. Nanomaterials enhance the bond strength between cement paste and aggregates, which helps improve concrete's resistance to heat. XRD images showed that nanoconcrete had a lower CH peak and a higher intensity of C-S-H at 200–400 °C, indicating the formation of an additional content of cement gel in the presence of nanoparticles.

4. Acid Resistance of Concrete Enhanced with Nanomaterials

The sustained attack of acid on concrete can severely harm and degrade its microstructure, ultimately resulting in material destruction [83]. The concrete comprising industrial construction such as sewer pipes, wastewater treatment plants, silos, dairies, and cooling towers is prone to acid attack, and the damage is extremely severe when the concrete surface is not protected. The acidity level affects how intense an acid attack is, and concrete's significant damage occurs when the pH value of the acid solution is lower than 4.5 [84]. The three deterioration mechanisms of acidolysis, complexolysis, and decalcification were used to describe how acid solutions cause concrete to deteriorate [85]. The severity of an acid attack [86,87] is conditioned by several variables, such as concrete permeability, cement alkalinity, the content of calcium hydroxide in the cement matrix, the pH values or concentrations of acid solutions, and the solubility of formed salts. Based on their aggressiveness level, acid solutions can be classified into high-aggressive acids (e.g., hydrochloric, acetic, nitric, and sulfuric acids) and low-aggressive acids (e.g., phosphoric and humic acids). The most common acids encountered by concrete are sulfuric acid, nitric acid, hydrochloric acid, and carbonic acid [88]. The degradation mechanism underlying an acid attack on cementitious materials is mainly associated with the decalcification process of cement hydrates. Decalcification is the process of removing calcium and hydroxide ions from cement hydrates during exposure to acid solutions. The decalcification process causes the dissolution of the hydrated phases of cement paste. Based on hydrate solubility, CH dissolves first, followed by the hydrated phases with a greater Ca/Si ratio. At slightly lower pH values, the AFm and AFt phases typically dissolve. When calcium hydroxide is completely consumed, the dissolution of the C-S-H gel begins. The Ca/Si ratio affects the rate of C-S-H dissolution as well. In the most severe cases of acid attack, the dissolution of calcium–silicate–hydrate gel can seriously harm concrete's structural integrity.

Certain techniques and strategies can be used to modify concrete's resistance to an acid attack, such as concrete surface coating, altering the environment to make it less hostile to the concrete, adding fibres, and enhancing the impermeability of concrete. Incorporating nanomaterials helps increase the mechanical properties of concrete and its durability. Therefore, nanomaterials offer a wide range of applications for improving concrete's resistance to an acid attack.

The effect of nanosilica and nanometakaolin on the resistance of high-strength and high-performance concrete against an acid attack was studied by Diab et al. [89]. The study focused on the analysis of the effect of various nanomaterials on concrete's porosity, compressive strength, and water capillary absorption, as well as the impact of an acid attack on the material's expansion strain, mass, and ultrasonic pulse velocity (UPV). The findings demonstrated that nanomaterials significantly reduced concrete's porosity and improved its microstructure, ultimately enhancing its acid resistance. Moreover, nanomaterials reduced the water absorption percentage due to the nanofiller effect and porosity refinement mechanism. The bonding strength of the cement paste and aggregate was improved with nanomaterial incorporation. The higher the added content of nanomaterials, the lower the water absorption of concrete. The study revealed that nanomaterials can efficiently reduce the compressive strength loss of concrete caused by acid attacks. Furthermore, the results demonstrated that nanomaterials considerably reduced the concrete mass loss and UPV loss. Using scanning electron microscopic images, the microstructures of nanomaterial concrete were found to exhibit a lower amount of CH, gypsum, and ettringite needle crystals, in addition to a denser and more homogenous structure. The expansion strain that was caused by a nitric acid attack was greatly reduced with the incorporation of nanomaterials in concrete. In general, the stronger concrete exhibited less expansion after an acid attack; therefore, the compressive strength enhancement provided by nanomaterials proved to have better concrete expansion resistance. Due to nanomaterials' pozzolanic activity that reduces the hydroxyl ion content and forms a higher amount of calcium silicate gel, which seals the capillary pores and increases concrete's impermeability, less permeable concrete exhibits lower expansion under an acid attack.

In a study by Mahdikhani et al. [90], concrete's mechanical properties and durability were investigated using the inclusion of nanosilica at varying concentrations of 0–6% and exposure to acid rain environments. The study evaluated mass loss, compressive strength, electrical resistivity, and water absorption of concrete under conditions of acid rain with different pH levels. To create acid rain, the researchers utilised sulfuric acid salt at a pre-determined pH level. The findings suggested that incorporating nanosilica improved concrete's ability to resist acidic environments. Additionally, increasing the pH level of the acid rain resulted in enhanced mechanical properties and durability of the concrete. Various pH values were used to test the compressive strength of concrete containing nanosilica. With increasing the nanosilica amount in concrete at a higher pH value, the compressive strength was obviously increased. The destruction of concrete was observed in solutions with a lower pH value of 2.5, while at higher pH values of 7, 5.5, and 4.4, the damage caused by the acid attack was negligible. Sujay et al. [91] conducted experimental research to assess the durability of composite fibre-reinforced high-performance concrete incorporating nanosilica and ultra-fine fly ash. The concrete was immersed in solutions containing 5% concentrations of hydrochloric acid, sulfuric acid, and magnesium sulphate. Weight loss, residual compressive strength, and compressive strength were tested for concrete subjected to acid attack. With increasing the acid immersion period, the weight loss of concrete was increased. Sulfuric (H_2SO_4) acid immersion had the greatest weight loss effect on concrete, while hydrochloric (HCl) acid immersion had the least effect when only minimal weight loss was observed. This can be attributed to sulfuric acid's combined damage from sulphate attack and acid attack. Compared to control concrete, nanomodified concrete had a lower weight loss. According to the study, the decline in the concrete's residual compressive strength increased with the duration of acid soaking. Acids reduced the residual compressive strength of concrete by increasing the formation of ettringite in the concrete's microscopic structure as the duration of acid soaking increased. However, adding a higher amount of nanosilica and ultra-fine fly ash worked to reduce the loss in residual compressive strength of concrete during its exposure to HCl and H_2SO_4. Decreasing residual strength loss was a result of the packing effect of nanomaterials, resulting in a dense microstructure and a reduction in capillary pores due to the higher amount of cement gel that improves the concrete's resistance against acid attacks.

Praveenkumar et al. [92] studied the effects of a HCl attack on the strength and durability properties of concrete with TiO_2 nanoparticles and rice husk ash. The study revealed that the longer the duration of acid immersion, the greater the weight loss of the concrete. Furthermore, the control concrete exhibited more damage than the concrete with nanomaterials due to acid exposure. The combination of 10% rice husk ash and different nano-TiO_2 contents improved the concrete's resistance to the HCl acid deterioration effect. The results indicated that 3 wt.% nano-TiO_2 was the best dosage to improve acid resistance in concrete. Cao et al. [93] evaluated the effect of multi-walled carbon nanotubes on cementitious material resistance to sulfuric acid. The study used the diffusion–dissolution–precipitation mechanism in its explanation for acid degradation. Using their observation of the microstructure, it was concluded that gypsum formation was the main degradation effect through the sulfuric acid attack, owing to gypsum's ability to provide pathways for ions to invade the materials, thus creating a damage cycle. The effect of multi-walled carbon nanotubes on enhancing sulfuric acid corrosion resistance was attributed to their role in mitigating ion diffusion by compacting the microstructure, promoting the hydration process, and forming a protective barrier in acid conditions. The inclusion of multi-walled carbon nanotubes was found to regulate the onset of microcracks and impede the spread of flaws such as cracks and detrimental pores. Furthermore, after incorporating multi-walled carbon nanotubes, the permeability and resistance to cracking of the cement matrix were enhanced. At the same time, the microstructure improvement induced with nanoparticles mitigated the concrete volume change that caused by gypsum formation and weakened expansion pressure. Table 3 summarises the results of studies that examined the impact of utilising nanomaterials on the acid resistance of concrete.

Figure 8 is a schematic drawing demonstrating the damage zones caused by an acid attack between plain concrete and nanomodified concrete. It can be seen from the figure that the nanoparticles have a positive effect by compacting the microstructure, thereby decreasing the effect of acid invasion and reducing the depth of the damage zone compared to plain concrete.

Table 3. Summary of several studies that investigated the effect of using nanomaterials on the acid resistance of concrete.

Reference	Nanomaterials	Level Content (wt.%)	Acid Solutions	Key Findings	Discussion
Diab et al. [89]	Nanosilica	0.5%, 1%, 1.5%, 2%	Nitric acid, sulfuric acid	The effect of nitric acid was decreased with nanomaterial incorporation when the water capillary absorption, porosity, compressive strength loss caused by the acid attack, and expansion strains caused by the acid attack were reduced. The nanomaterial enhancement of concrete's resistance to sulfuric acid was better than concrete's resistance to nitric acid.	Microstructure improvement due to the filling effect of nanomaterials and pozzolanic activity.
	Nanometakaolin	1%, 3%, 6%, 9%			
Mahdikhani et al. [90]	Nanosilica	0%, 2%, 4%, 6%	Acid rain using sulfuric acid salt	Nanosilica improved the compressive strength of concrete in sulfuric conditions and the impermeability and durability of concrete.	Reduced porosity with the precipitation of higher hydrates due to changes in pore solution composition as well as an improvement in the bond strength between the matrix substances, owing to the filling impact of nanosilica.
Sujay et al. [91]	Nanosilica	15%	Hydrochloric acid, sulfuric acid	Lower weight loss with the addition of nanomaterials. In addition, the higher the content of nanosilica, the higher the residual compressive strength of concrete exposed to acids.	Microstructure improvement due to the filling effect of nanomaterials and pozzolanic activity.
	Ultra-fine fly ash	1.5%, 3%, 4.5%			
Praveenkumar et al. [92]	NanoTiO_2	1%, 2%, 3%, 4%, 5%	Hydrochloric acid	Overall, 3% nanoTiO_2 was the optimum amount for a lower mass loss of concrete exposed to hydrochloric acid. The combination of 10% rice husk ash and different nano-TiO_2 amounts had greater resistance to deterioration when subjected to acidic conditions.	Improved the durability and strength of concrete.

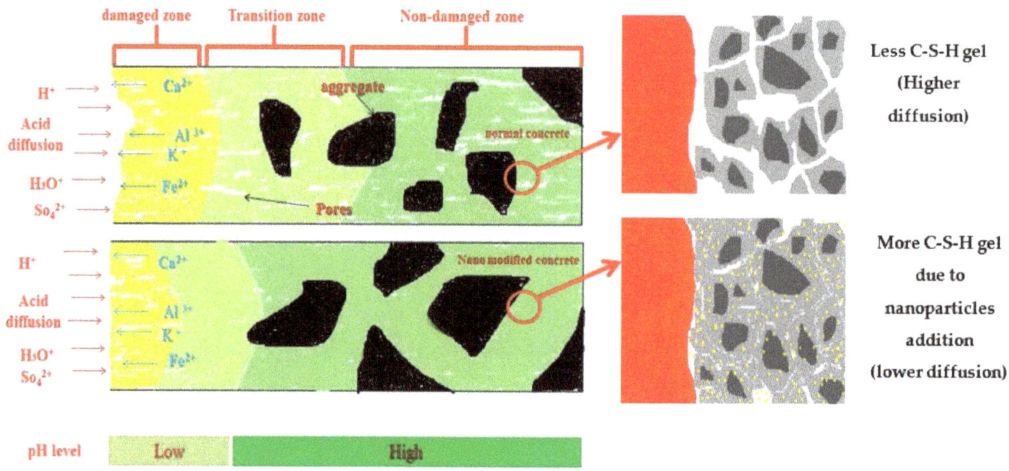

Figure 8. Schematic illustration showing the damage zones caused by an acid attack between plain concrete and nanomodified concrete.

5. Sulphate Resistance of Concrete Enhanced with Nanomaterials

A sulphate attack [94,95] has a dramatic deterioration effect on concrete through the sulphate reaction with cement hydrates, producing new solid phases at a relatively large volume. These new reaction products in the hardened cementitious matrix exert tensile stresses due to crystallisation pressure. Once these tensile stresses locally suppress the tensile strength of the material, cracks would develop, compromising the structure of the concrete. A sulphate attack can occur either as an external attack or as an internal attack. An external sulphate attack is characterised by sulphate contained in the surrounding environment (e.g., soils, groundwater, seawater, etc.) that infiltrates through the pores in the concrete and reacts with compounds in the cement paste such as CH, C-S-H gel, and monosulphate. An internal sulphate attack occurs when there is an excessive amount of sulphate in the raw materials of concrete, such as cement, aggregates, water, mineral and chemical admixtures, etc. The attack of sulphate on concrete can be characterised by chemical or physical degradation. An external chemical attack happens with the penetration of sulphate-containing solutions into the pores in the concrete, which increases the concentration of sulphate ions ($SO4^{2-}$) that react with the source of calcium (Ca^{2+}) and alumina to form ettringite ($CaO{\cdot}Al_2O_3{\cdot}3CaSO_4{\cdot}32H_2O$) and with the source of calcium to form gypsum ($CaSO_4{\cdot}2H_2O$). Ettringite is an expansive component, and its formation in hardened concrete could result in expansion, cracking, and mass loss, particularly when restrained. The formation of gypsum results in the softening of the concrete microstructure and loss of cohesion. A chemical attack by sulphate can involve the destabilisation of C-S-H gel and CH, generating microcracks within the concrete structure. A physical salt attack by sulphate can happen when concrete is in contact with soil. The groundwater contains deleterious ions, mostly sulphate, that infiltrate through concrete pores either by capillary sorption or diffusion. When the water inside the pores is evaporated by the dry air, salt will crystallise in the pores near the surface of the concrete. Salt precipitation and crystal growth exert crystallisation pressure, and whenever that pressure is greater than the tensile strength of the pore wall in which the crystals grow, surface scaling can occur, such as that caused by cycles of freezing and thawing. The most commonly occurring and naturally occurring sulphates that attack concrete are calcium ($CaSO_4{\cdot}2H_2O$), sodium ($NaSO_4$), and magnesium sulphate ($MgSO_4$). The main three forms of damage caused by these salts are (1) the formation of ettringite and gypsum that cause expansion and strength loss; (2) the salt's physical attack; and (3) thaumasite formation, which causes C-S-H destabilisation.

The denser microstructure and refined pore structure due to nanomaterial incorporation improve the permeability-related properties of concrete. Many studies revealed that nanomaterial addition to concrete had a huge impact on reducing concrete's water absorption, providing denser microstructure in the interfacial transition zone (ITZ), and reducing pore connectivity [27,96,97]. Thus, the incorporation of nanomaterials can significantly improve concrete's resistance to sulphate attack and reduce the deterioration effect of sulphate on its mechanical properties.

Figure 9 demonstrates the effect of nanomaterials on the compressive strength loss of concrete due to a sulphate attack, as found in different studies. In Figure 9a,b, it is shown that the addition of nanosilica and nanometakaolin to high-strength concrete (55 MPa) subjected to 360 days of immersion in a 10% magnesium sulphate solution can linearly increase the enhancement in residual compressive strength [66]. The higher the added dosage, the higher the reduction in compressive strength loss. Magnesium sulphate has a considerable level of aggressiveness due to its combined effect of forming an expansive compound such as gypsum and destabilising cement hydrates such as CH and C-S-H gel by producing magnesium hydroxide. At a later stage of attack, magnesium sulphate reacts with C-S-H gel to form gypsum, magnesium hydroxide ($MgOH_2$), and silica gel. The $MgOH_2$ lowers the pH of the pore solution and causes decalcification of the C-S-H gel by exchanging calcium ions with magnesium ions, forming a weak non-cementitious substance known as magnesium silicate hydrate. Therefore, the destructive effect of a magnesium sulphate attack is associated with both cracking and softening. The longer the immersion duration of concrete in a magnesium solution, the higher the loss in compressive strength. However, the study showed that nanomaterials (nanosilica and nanometakaolin) enhanced the strength of concrete against sulphate attack due to their filling action and pozzolanic reactivity, which contribute to porosity reduction and compacting the microstructure. Multi-walled carbon nanotubes (MWCNTs) [98] improved the compressive strength loss of concrete due to a sulphate attack after 90 days of exposure. In the study, the concrete specimens with MWCNTs were immersed in a sodium solution for 28, 50, and 90 days. Like the previous study, the results indicated that the higher the exposure to sodium sulphate attack, the greater the reduction that can be induced in concrete's compressive strength. It was observed that the reduction in concrete's compressive strength loss increased with the addition of MWCNTs in all immersion periods until a certain limit of addition. The results of the study revealed that the specimens with a 0.25% addition of MWCNTs can be regarded as having the optimum content for minimising the compressive strength loss of the concrete. In a study by Reshma et al. [99], the impact of zinc oxide (ZnO) and titanium dioxide (TiO_2) on the mechanical properties of concrete was assessed both with and without polypropylene fibres. The results presented in Figure 9d [99] illustrate the positive effects of incorporating ZnO and TiO_2 on the residual compressive strength of non-fibrous concrete. The data showed a linear relationship between the amount of nanomaterial added and the reduction in compressive strength loss of the concrete. The improvements in compressive strength loss were due to the ability of ZnO and TiO_2 to enhance C-S-H gel formation and increase the bond between aggregate and cement in the concrete.

Faried et al. [51] performed a study to investigate how the mechanical strength and durability of ultra-high-performance concrete were affected by varying curing conditions. To achieve the study's aim, four different nanowaste materials—milled nanometakaolin, nanowaste glass, nanorice husk ash, and chemically prepared nanosilica—were added to the ultra-high-performance concrete. The study measured the residual compressive strength of several samples of the modified concrete after exposing them to a concentrated solution of sodium sulphate and magnesium sulphate for a period of 28 days. The research findings indicated that all the concrete specimens experienced reduced compressive strength due to a sulphate attack. However, the residual compressive strength of the nanomaterial-modified concrete was considerably higher compared to the unmodified concrete. Figure 10 presents the results for the residual compressive strength of the air-cured, nanomodified concrete that was exposed to the sulphate solution [51].

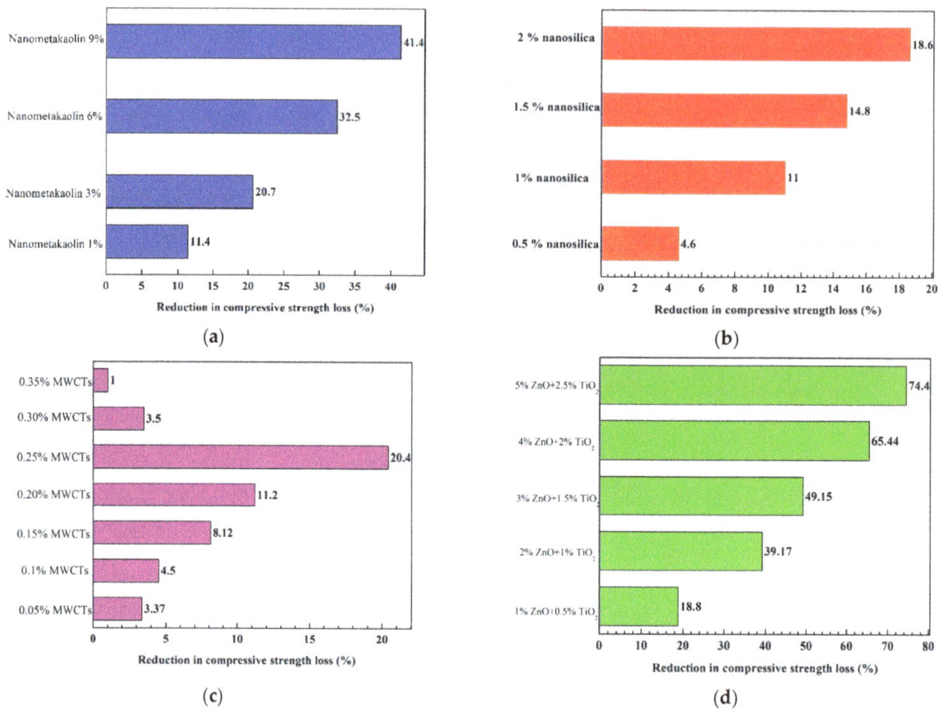

Figure 9. The reduction in concrete compressive strength loss (%) due to a sulphate attack after the addition of nanomaterials (**a**) The effect of different contents of nanometakaolin on reducing compressive strength loss of concrete subjected to 360 days of magnesium sulphate [89]. (**b**) The effect of different contents of nanosilica on reducing compressive strength loss of concrete subjected to 360 days of magnesium sulphate [89]. (**c**) The effect of different contents of multi-walled carbon nanotubes (MWCNTs) on reducing compressive strength loss of concrete subjected to 90 days of sodium sulphate [98]. (**d**) The effect of different contents of ZnO + TiO_2 on reducing compressive strength loss of concrete subjected to 28 days of a sulphate solution [99].

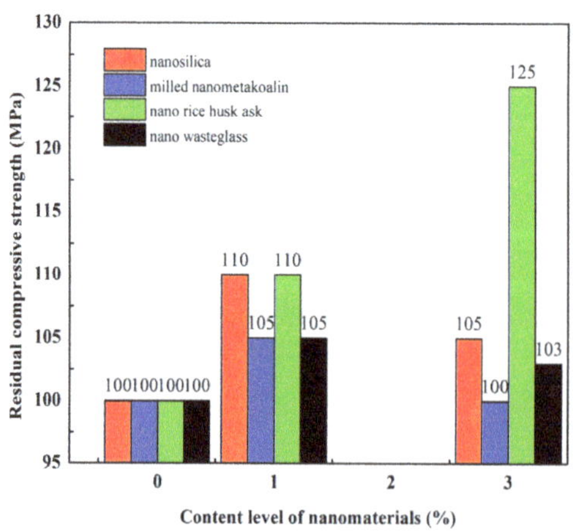

Figure 10. Residual compressive strength of air-cured, nanomodified concrete subjected to a sulphate solution [51].

Several studies were also conducted to assess the weight loss of nanomaterial-modified concrete as a result of a sulphate attack [89,98–101]. Table 4 shows the data collected from various studies, which indicated that prolonged exposure to a sulphate solution increased the weight loss for all concrete samples. However, the use of nanomaterials significantly contributed to reducing the weight loss of the modified concrete.

Figure 11 depicts how nanomaterials enhanced the performance of concrete by reducing weight loss and compressive strength loss using a relationship that was developed between the mass loss and residual compressive strength improvement based on the data from several studies [89,98–101]. The fitting curve shows a good correlation between the two parameters. The higher the residual compressive strength of concrete exposed to sulphate attack, the lower the weight loss.

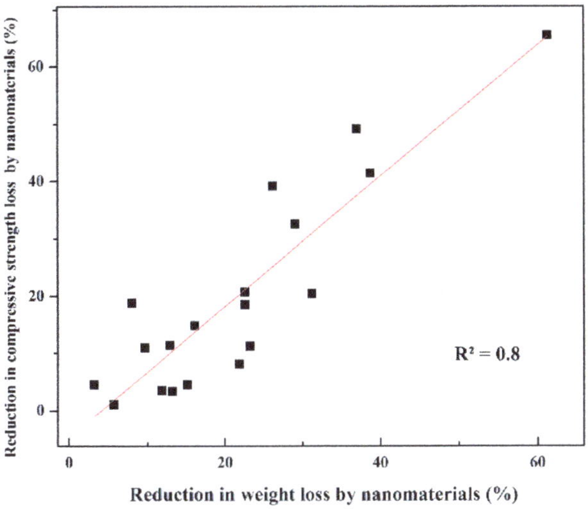

Figure 11. Correlation between the reduction in compressive strength loss and reduction in weight loss for concrete with nanomaterials [89,98–101].

However, a study by Vishwakarma et al. [102] reported that when subjected to ammonium and sodium sulphate solutions, concrete structures containing nanoparticles were shown to be less resistant to sulphate attack than fly ash-modified concrete. Three different types of concrete were subjected to 3% ammonium and sodium sulphate for 90 days. The test results for these three types of concrete indicated a significant weight loss for specimens incorporated with fly ash, 2 wt.% nano-TiO_2, and nano-$CaCO_3$ compared to fly ash concrete without nanomaterial incorporation when the concrete specimens were subjected to 3% ammonium sulphate. Nano-TiO_2-fly ash concrete exhibited higher weight loss than fly ash concrete. However, concrete with fly ash and nano-$CaCO_3$ had the lowest weight loss percentage among other types of concrete. For concrete specimens subjected to a 3% sodium sulphate solution, the nano-TiO_2-fly ash concrete had the highest percentage of weight loss, followed by fly ash-nano-$CaCO_3$ concrete. Table 5 lists the weight loss for concrete with different types of nanomaterials. The fly ash concrete type had the lowest weight loss percentage among the other types. The researchers reached the conclusion that fly ash concrete treated with nanoparticles is not suitable for a sulphate-rich environment and that fly ash is better for sulphate attack resistance. This conclusion was attributed to the fact that fly ash decreases the percentage of C_3A components, reducing the risk of sulphate attack by reducing the potential for formatting ettringite and gypsum.

Table 4. The weight loss for nanomaterial-modified concrete exposed to a sulphate attack.

Reference	Nanomaterials	Level of Content (%)	Sulphate Solution	Exposure Duration (Days)	Weight Loss (%)
Vijayabhaskar et al. [98]	MWCNTs	0	Na_2SO_4	90	2.11
	MWCNTs	0.05	Na_2SO_4	90	1.83
	MWCNTs	0.10	Na_2SO_4	90	1.79
	MWCNTs	0.15	Na_2SO_4	90	1.65
	MWCNTs	0.2	Na_2SO_4	90	1.62
	MWCNTs	0.25	Na_2SO_4	90	1.45
	MWCNTs	0.3	Na_2SO_4	90	1.86
	MWCNTs	0.35	Na_2SO_4	90	1.99
	MWCNTs	0.4	Na_2SO_4	90	3.01
Reshma et al. [99]	ZnO + TiO_2	0	Sulphate solution	28	6.12
	ZnO + TiO_2	1 + 0.5	Sulphate solution	28	5.63
	ZnO + TiO_2	2 + 1	Sulphate solution	28	4.52
	ZnO +TiO_2	3 + 1.5	Sulphate solution	28	3.86
	ZnO + TiO_2	4 + 2	Sulphate solution	28	2.37
	ZnO + TiO_2	5 + 2.5	Sulphate solution	28	2.21
Moslemi et al. [100]	NanoSiO_2	0	Na_2SO_4	180	3.51
	NanoSiO_2	2	Na_2SO_4	180	2.40
	NanoSiO_2	4	Na_2SO_4	180	2.23
	NanoSiO_2	6	Na_2SO_4	180	1.13
	NanoSiO_2	8	Na_2SO_4	180	1.00
Diab et al. [89]	Nanometakaolin	0	$MgSO_4$	360	3.10
	Nanometakaolin	1	$MgSO_4$	360	2.70
	Nanometakaolin	3	$MgSO_4$	360	2.40
	Nanometakaolin	6	$MgSO_4$	360	2.20
	Nanometakaolin	9	$MgSO_4$	360	1.90
	NanoSiO_2	0	$MgSO_4$	360	3.10
	NanoSiO_2	0.5	$MgSO_4$	360	3.00
	NanoSiO_2	1	$MgSO_4$	360	2.80
	NanoSiO_2	1.5	$MgSO_4$	360	2.60
	NanoSiO_2	2	$MgSO_4$	360	2.40
Sathe et al. [101]	NanoAl_2O_3	0	$MgSO_4$	28	3.86
	NanoAl_2O_3	2.3	$MgSO_4$	28	1.52

Table 5. Weight loss for concrete with different types of nanomaterials [102].

Concrete Type	Sulphate Solution	Weight Loss (%)
Fly ash concrete	Ammonium sulphate	1.8
Fly ash + Nano-TiO_2 concrete	Ammonium sulphate	2.5
Fly ash + Nano-$CaCO_3$ concrete	Ammonium sulphate	1.75
Fly ash + Nano-TiO_2 + Nano-$CaCO_3$ concrete	Ammonium sulphate	2.70
Fly ash concrete	Sodium sulphate	1.30
Fly ash + Nano-TiO_2 concrete	Sodium sulphate	2.80
Fly ash + Nano-$CaCO_3$ concrete	Sodium sulphate	2.20
Fly ash + Nano-TiO_2 + Nano-$CaCO_3$ concrete	Sodium sulphate	2.10

The promotion of C-S-H gel formation with the pore refinement effect has been found to have a noteworthy correlation with the improvement in concrete's mechanical properties and durability when added with nanomaterials. Solid evidence and explanation revealed that the total capillary porosity of concrete is decreased with nanomaterial incorporation due to the nanoparticle filler mechanism and high surface area; moreover, nanoparticles can act as nucleation sites for additional hydrate formation, increasing the C-S-H gel content in the cement matrix. External sulphate ingress through the concrete is controlled by permeability (porosity-dependent) when the phase composition and pore network determine the rate of sulphate ion penetration. In addition, the amount of water absorbed is directly related to the level of porosity in the concrete. Due to their super fineness, nanoparticles, such as nanosilica, have proven to be highly efficient in decreasing the water absorption into concrete. The nanosilica's ability to densify and improve the microstructure of concrete by filling the porous hydration products and reacting with CH to produce more gel hydrates would block water transport channels and inhibit the diffusion of aggressive ions. On the other hand, Wang et al. [103] found that the content level of nanosilica is a crucial factor for water absorption; when the content level exceeded 5%, it had an adverse impact on material resistance to sulphate penetration and concentration. The reason for this adverse impact was attributed to the nanosilica particle agglomeration effect, which has a negative impact on hydration phases and material structure compactness. The research also showed that adding nanosilica to concrete decreased the SO_4^{2-} apparent diffusion coefficients; however, an excessive nanosilica incorporation had a larger deviation rate of the apparent diffusion coefficients. Reshma et al. [99] investigated the impact of nanomaterials such as ZnO and TiO_2 on the water absorption and sorptivity of concrete and reported a beneficial outcome. It is clear that the addition of nanomaterials to a cement mixture can enhance the durability of concrete, as it improves its microstructure by filling gaps and increasing pozzolanic reactivity, thus resisting sulphate attacks.

6. Chloride Ion Penetration of Concrete Enhanced with Nanomaterials

Reinforced concrete structures are very vulnerable to chloride ion ingress, which can result in steel corrosion [104]. When steel bars corrode due to chloride-induced corrosion, it weakens the structure's strength and safety by causing spalling and reducing the steel cross-sectional area. Therefore, it is critical and urgent to safeguard concrete against chloride ion corrosion [105]. The permeability of materials, the capacity of chloride binding, and the ion exchange capacity all greatly impact the diffusion of chloride ions [106] into concrete.

The addition of nanoparticles to concrete helps to enhance its mechanical characteristics, including compressive strength, by elevating the quantity of C-S-H gel and optimising the structure of the concrete's pores. This is achieved when the particles react with CH to form a denser microstructure, thus enhancing material durability. Multiple experiments and explanations have confirmed that nanomaterials can improve concrete's resistance to chloride attacks.

Singh et al. [107] reported that the incorporation of nanosilica significantly improved resistance to chloride ion diffusion. Their study comparing the effectiveness of nanosilica and silica fume in mitigating chlorine attacks found that adding 3% nanosilica increased resistance to chloride ions by as much as 43%, whereas silica fume only improved it by 15%. The enhancement in nanosilica's ion penetration resistance was attributed to its ability to promote the cement hydration process and create a denser microstructure as compared to silica fume. This assumption was supported by monitoring the cement microstructure development of both nanosilica and silica fume using SEM, TGA, and compressive strength. An examination was carried out by Rezakhani et al. [108] to investigate the effects of water pressure and rebar corrosion on chloride ion (Cl^-) diffusion in graphene oxide-modified concrete. Various types of concrete were subjected to constant water pressures of 0.3, 0.5, and 0.7 MPa for a duration of 144 h to infiltrate chloride ions. The outcome of the research showed that the rate of chloride penetration was increased by both the water pressure and the depth of immersion in concrete, regardless of the type of concrete. However, the graphene oxide–slag-based, cement-modified concrete displayed improved resistance against ion penetration. The positive impact of nanomaterials on this type of concrete was observed at higher pressure, which caused a significant reduction in the measured diffusion coefficients by a factor of 3. The nanofiller effect of graphene oxide was responsible for blocking the pores and cutting the ions' diffusion paths in concrete. However, the higher content level of graphene oxide might be detrimental due to the agglomeration effect of concrete, highlighting the importance of caution regarding the use of nanomaterials.

Liu et al. [109] investigated the chloride ion migration coefficient for steel fibre-reinforced concrete treated with graphene oxide. Their findings revealed that resistance to chloride ion ingress into concrete was significantly enhanced by increasing the curing time and incorporating graphene oxide. The experimental results demonstrated that 0.03 wt.% and 0.05 wt.% of graphene oxide significantly reduced the chloride ion penetration depth and lowered the chloride migration coefficient. Moreover, 0.03 wt.% of graphene oxide was found to be the optimum amount for improving the concrete's resistance to chloride. The chloride ion resistance of graphene oxide-modified, steel fibre-reinforced concrete is illustrated in Figure 12 [109].

Figure 12. The chloride ion resistance of graphene oxide-modified, steel fibre-reinforced concrete. Reproduced with permission from [109], (*Cement and Concrete Composites*); published by (Elsevier), (2022) (Note: SFRC: 0% graphene oxide, GOSFR-1: 0.01%wt. graphene oxide, GOSFRC-3: 0.03 wt.%, GOSFRC-5: 0.05 wt.%, GOSFRC: 0.07% graphene oxide).

Zhang et al. [110] evaluated the durability of marine concrete doped with nanoparticles. The concentration of free chlorides and carbonation depth of the concrete were regarded as the durability index. Nanosilica and nano-ZnO were used, and the test was performed by measuring the chloride content and carbonation depth after immersing the concrete in a 5% NaCl solution for a total of 56 cycles. The concentration of free chlorides was higher at lower depths and decreased with increasing depth, indicating that the Cl^- ions accumulated closer to the concrete surface. Moreover, it can be observed from Figure 13 [110] that the free concentration of Cl^- ions in nanoconcrete was reduced compared to that of plain concrete. This suggests that the addition of nanomaterials enhanced the concrete's ability to resist penetration by chloride ions. However, the 2% addition of nanosilica was more effective at increasing the concrete's resistance to chlorides compared to the 1% nano-ZnO concrete. The carbonation depth of nanoconcrete is lower compared to plain concrete. The optimal performance of concrete regarding anti-carbonation and chloride resistance was achieved by incorporating the proper content of nanoparticles into the concrete, and adding 1% of nano-ZnO and 2% of nanosilica was noted to be the optimum content levels. In contrast, the concrete's chloride resistance and anti-carbonation performance were reduced by increasing the amount of nanomaterials. The adverse impact of exceeding the optimal content was due to the agglomeration effect and insufficient local cement hydration. Based on XRD test results, it was affirmed that nanosilica is more effective in consuming calcium hydroxide than nano-ZnO because nanosilica has smaller nanoparticles and thereby greater surface activity. By consuming the $Ca(OH)_2$ phase, nanomaterials can drastically improve concrete's resistance to chlorides due to the phase negative impact by reacting with Cl^- to form complexes that facilitate chloride diffusion and reacting with CO_2 to form $CaCO_3$, which reduces the concrete pH and accelerates steel corrosion. The incorporation of nanoparticles lowers the $CaCO_3$ amount in concrete and produces a higher C-S-H gel content, thus enhancing the anti-carbonation performance of concrete under chlorine attack. On the other hand, nanosilica can enhance durability in a more functional way than nano-ZnO, due to the former's advanced effect in promoting the formation of Friedel's salts and improving the concrete microstructure.

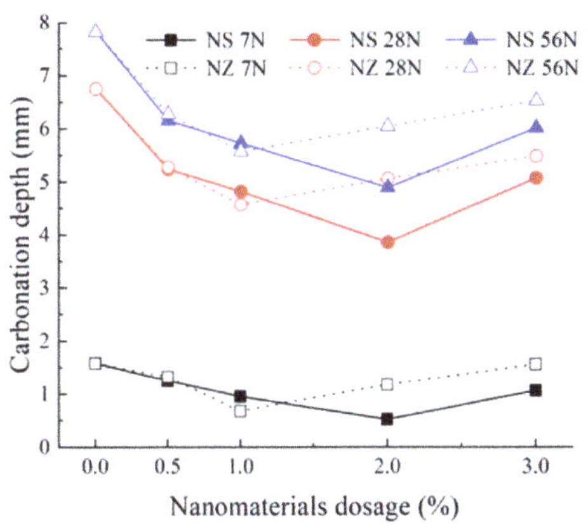

Figure 13. Free Cl^- concentration in nanoconcrete at different nanomaterial dosages (0–5 mm). Reproduced with permission from [110], (*Case Studies in Construction Materials*); published by (Elsevier), (2023) (Note: NS: $NanoSiO_2$, NZ: Nano-ZnO, N = number of cycles).

Joshaghani et al. [111] conducted a study on concrete that had been modified using different types of nanomaterials such as nano-TiO_2, nano-Al_2O_3, and nano-Fe_2O_3. Their research covered areas such as water penetration and electrical resistivity, as well as the rapid chloride permeability test and the rapid chloride migration test. The study emphasised that nanomaterial-modified concrete had lower permeability and lower water penetration than plain concrete due to the nanoparticles' role in reducing the pore volume and connectivity. A lower permeability indicates lower chemical attacks due to the inhibition of the ingress of aggressive ions such as chlorides into concrete. The 91-day water penetration depths of 0.32 w/c ratio concrete incorporated with 5% TiO_2, Al_2O_3, and Fe_2O_3 were found to be 51%, 59%, and 64% lower than the control concrete, respectively. The improvement in water penetration with nanomaterial incorporation was interpreted as the cause of nanoparticle dilution and heterogeneous nucleation effects. Promoting cement hydration through pozzolanic reactivity with nanoparticle addition was expected to result in a higher amount of cement gel, which would reduce the pores connectivity and hinder the water transport within concrete. The electrical resistivity of concrete surfaces increased over time as the hydration process continued. This increase in the solid phase blocks electrical currents. The nanomodified concrete had a greater electrical resistivity than control concrete due to the porosity-refining effect of nanomaterials. The electrical response of concrete is a useful indicator of permeability. A linear correlation between water penetration and surface resistivity tests was established. An increase in electrical resistivity led to a decrease in water penetration depth. Incorporating nanoparticles reduced the concrete's penetrability, as confirmed using the rapid chloride migration tests. The chloride migration coefficient decreased significantly with the addition of nanoparticles. Electrical resistivity and chloride migration coefficients depended on the water-to-binder ratio and nanomaterial incorporation. These parameters were controlled by concrete porosity, pore size, and pore connectivity. The rapid chloride permeability tests showed decreased charge passing through concrete specimens when the w/b ratio was lowered and when nanoparticles were added. Nanoparticles improved concrete's resistance to chloride penetration, especially with a higher w/b ratio, due to their porosity-refining effect. Nanotechnology helped to create durable concrete with low permeability and low chloride penetration rates. A linear relationship was found between electrical resistivity and the rapid chloride permeability of the concrete. Figure 14 demonstrates correlations between the electrical resistivity and durability parameters of nanoconcrete, which indicate that durability was strongly correlated to permeability. Thus, nanoparticle addition can enhance the microstructure and promote C-S-H gel content to improve concrete durability properties.

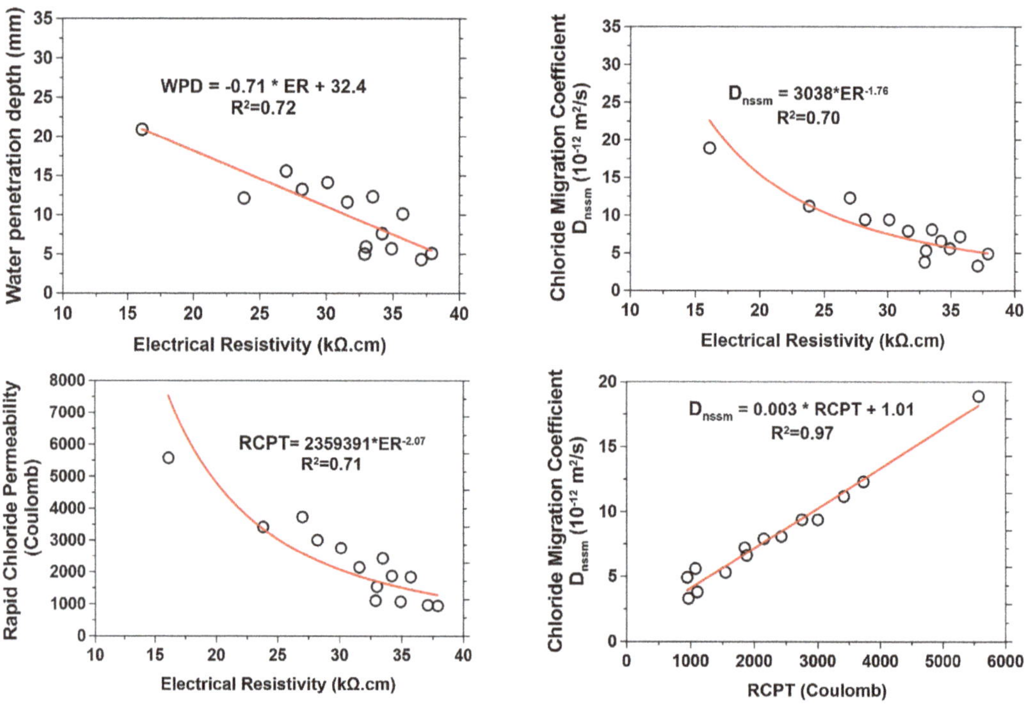

Figure 14. The relationship between electrical resistivity and durability parameters of nanoconcrete. Reproduced with permission from [111], (*Construction and Building Materials*); published by (Elsevier), (2020).

A rapid chloride permeability test was carried out by Sujay et al. [91] on high-performance concrete reinforced with composite fibres and treated with nanosilica and ultra-fine fly ash. It was deduced that the chloride ion permeability for composite fibre-reinforced, high-performance concrete with nanomaterials was significantly lower than plain concrete. Nanosilica acted by reducing the amount of charge passed within the concrete, and the effect was more clearly demonstrated with increasing the amount of nanosilica. It was found that the ideal replacement for concrete with nanosilica was 3%. The research by Wang et al. [112] indicated that the incorporation of carbon nanofiber into concrete, with a volume fraction ranging from 0.1 to 0.5%, resulted in a significant reduction in water seepage height and the relative permeability coefficient by 14.1%, 26.6%, 39.1%, 37.5%, and 32.8%, respectively, when compared to control concrete. In addition, the carbon nanofiber-modified concrete with a volume fraction of 0.3% had the lowest permeability. The improvement in permeability was mainly related to the pore structure in the concrete. Carbon nanofiber addition dramatically reduced the average pore size, median pore diameter, largest pore size, and total pore volume in the concrete. Figure 15 shows the porosity percentage for the carbon nanofiber-modified concrete [112], and it was observed that the percentage of macropores within carbon nanofiber-modified concrete was extremely lower than plain concrete. Nanomaterial incorporation increased the nanopores and micropores due to their high surface area and filler effect on reducing the size of macropores. Furthermore, the concrete containing 0.3% carbon nanofiber possessed the finest pore structure. When compared to ordinary concrete, adding 0.1 to 0.5% carbon nanofiber decreased the average size, median diameter, total volume of pores, and percentage of macropores by 58.9%, 81.9%, 99.6%, 16.4%, and 45.6%, respectively. However, the volume fraction of 0.5% had higher macropores than the optimum content of 3%, indicating that the pore structure of carbon nanofiber concrete can provide a better improvement. In all other respects, the maximum proportion of 0.5% had a better improvement over plain concrete. It was determined from the findings that the correct amount of carbon nanofibre should be added to the concrete to reduce capillary porosity and improve its microstructure.

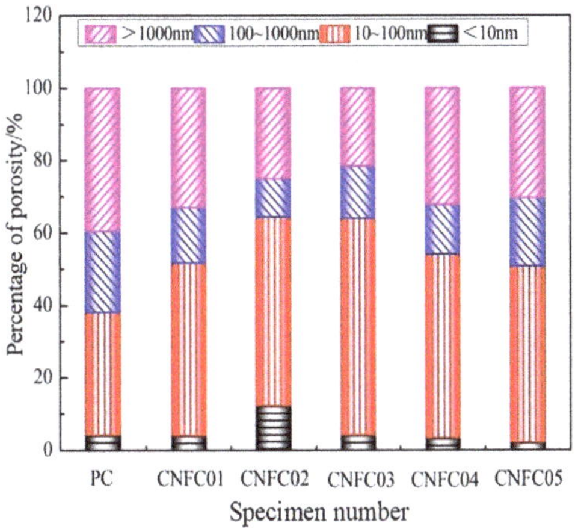

Figure 15. Percentage of porosity for carbon nanofiber concrete (CNFC) in different dosages (0.1–0.5%). Reproduced with permission from [112], (*Construction and Building Materials*); published by (Elsevier), (2020).

Tests were carried out by Li [113] to evaluate the effects of nanosilica, nano-$CaCO_3$, and multi-walled carbon nanotubes on the chloride permeability and chloride binding capacity of concrete. The study applied rapid chloride migration and equilibrium tests to quantify these properties. The results indicated that the inclusion of 2 wt.% nanosilica proved to be extremely effective at decreasing the chloride diffusion coefficient of the concrete, with the coefficient of nanosilica-enhanced concrete being 85.3% lower than that of plain concrete. The improvement using nanosilica was attributed to the nano-filler effect and its role in blocking the harmful pores and microcracking inside the concrete matrix, intensifying the microstructure, and hindering chloride transport. Furthermore, nanosilica promoted the hydration process with a nucleation effect that leads to the formation of higher amounts of gel-like hydrates, which increase the strength of cement paste and their bond to aggregate. Reducing the CH crystals with the pozzolanic activity of nanosilica was associated with improving the concrete's strength by forming a denser C-S-H gel and strengthening the interfacial transition zone. These three effects of nanosilica (the nanofiller effect, nucleation effect, and pozzolanic effect) decreased the pore volume and subsequently lowered chloride diffusion. However, at higher incorporated amounts of nanosilica, such as 4 wt.%, the concrete resistance was less compared to 2 wt.%. The adverse impact was attributed to agglomeration and poor dispersion of nanoparticles owing to the van der Waals force, which causes weak zones and defects in the cement matrix. Regarding the nano-$CaCO_3$ incorporation, the effect was also significant in improving the concrete resistance to chlorides by reducing the diffusion coefficient with a low-dosage addition, due to the nanofiller effect and nucleation effect as well as the chemical effect. The addition of 4 wt.% of nano-$CaCO_3$ had a negative impact on chloride diffusion as compared to plain concrete. The negative impact was attributed to the agglomeration effect of nanoparticles. For the multi-walled carbon nanotube addition, the results showed remarkable improvement in regard to chloride resistance by reducing the chloride diffusion coefficient. The improvement resulted from the chemical effect of absorbing the Ca^{2+} from the pore solution, promoting the crystallisation and production of hydration products, in addition to the nucleation and bridging effects of bonding the C-S-H and improving the microstructure. The agglomeration effect was also present with a 4 wt.% addition, which reduced the effectiveness of nanomaterial enhancement of chloride resistance. Mercury intrusion porosimetry (MIP) was conducted to link the pore characteristics with concrete resistance to chlorides. The results showed that nanomaterial incorporation reduced the total pore volume of concrete. The content of smaller-diameter pores was increased with nanomaterial incorporation, which was considered harmless. The results indicated a refinement in pore structure with nanomaterials. Small pores are characterised as isolated local pores, while large pores are often connected as interconnected networks in concrete. Nanomaterials, due to their filler effect, turn the large pores into small pores, increasing the tortuosity of the system and hindering chloride transport in concrete. The diffusion coefficient was increased by increasing the capillary pores in the concrete. The increased concrete diffusion caused by increasing porosity was attributed to an increase in connected pores. Chloride binding can mitigate the chloride attack and reduce rebar corrosion in concrete by removing the chloride ions. However, the incorporation of nanomaterials compromises the chloride binding in concrete due to reducing the amount of CH by the pozzolanic reaction and reducing the Ca/Si ratio of C-S-H. By reducing the amount of CH, the pH in concrete would be reduced, the alkalinity would be lowered, and this would lower the chloride binding capacity.

Recent studies by Lu et al. [49] and Liu et al. [50] focused on the effect of nanomaterial incorporation on the durability of recycled aggregate concrete. The investigation also included testing the water sorptivity and chloride ion permeability. A graphene oxide–cement slurry was used to coat the recycled aggregates with the purpose of strengthening the interface transition zone (ITZ) microstructure [49]. The results revealed that the graphene oxide coating helped reduce the water absorption coefficient by 15.1%, improved the microstructure, reduced the microcracks, reduced the ITZ by 44.4%, lowered the total charge passed by 63.4%, and significantly mitigated the chloride ion penetration of concrete. The impermeability, chloride ion resistance, and freezing–thawing resistance were investigated for recycled coarse aggregate concrete modified by coating the coarse aggregate with nanosilica and sodium silicate [50]. Regarding the concrete, the coating technique utilizing nanoparticles and coarse aggregates yielded increased mechanical strength, superior durability characteristics, and reduced pore content. Nanosilica has widespread use not only because of its lower cost compared to other nanomaterials but also because of its filling, nucleation, and pozzolanic properties.

7. Surface Abrasion Resistance of Concrete Optimised with Nanomaterials

Concrete durability mainly depends on the abrasion resistance property. Abrasion resistance defines how surfaces resist rubbing and friction [114]. This characteristic is crucial for pavements subjected to continuous dynamic loads from traffic. The strength, porosity, and aggregate properties of concrete greatly influence its ability to resist abrasion [115]. Concrete with low compressive strength is susceptible to rapid abrasion, as it affects its surface integrity. A previous published work showed a linear correlation between compressive strength and abrasion resistance [116]. Incorporating nanomaterials into concrete significantly improved its mechanical properties, and therefore its abrasion resistance. Li et al. [117] demonstrated that the addition of nano-TiO_2 and nanosilica into concrete enhanced its abrasion resistance compared to concrete containing polypropylene fibres. The effect of nanomaterial incorporation was best observed with 1 wt.% nanosilica or nano-TiO_2, which had an index increment of up to 157% and 180%, respectively. These enhancements were due to the nanomaterials' ability to control crystallisation in cement paste, prompting the formation of more C-S-H gel, and leading to a compacted and homogenous cement matrix. However, the amount of nanoparticles added should be optimum, as higher amounts result in less improvement. A linear relationship between abrasion resistance and compressive strength was identified for plain concrete, concrete with fibres, and concrete with nanomaterials as shown in Figure 16. Compressive strength is a primary factor influencing the abrasion resistance of concrete.

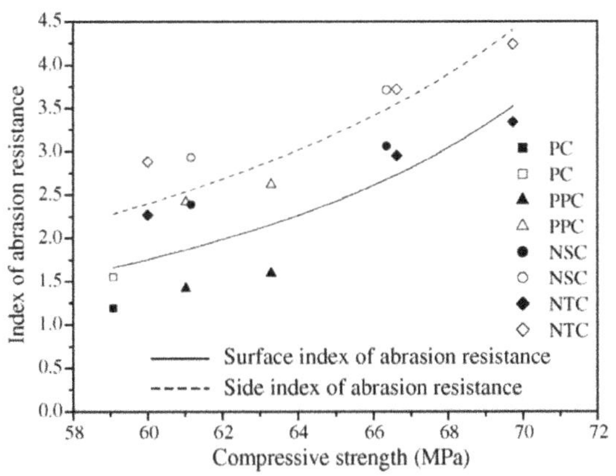

Figure 16. The relationship between the index of abrasion resistance and compressive strength for all mixtures of concrete. Reproduced with permission from [117], (*Wear*); published by (Elsevier), (2006) (Note: PC: control concrete, PPC: polypropylene fibre concrete, NSC: nanosilica concrete, NTC: nano-TiO_2 concrete).

Gao et al. [52] investigated the wear resistance of fly ash concrete that underwent modification using nano-silica or nanoSiC. The results of the study revealed that the incorporation of nanomaterials significantly enhanced the wear resistance of fly ash concrete when compared to reference concrete. Furthermore, the analysis showed that the wear loss in the presence of nanosilica and nano-SiC was remarkably reduced. However, it was discovered that increasing the concentration of nanoparticles led to a linear decrease in the wear loss, and consequently, the optimal content of nanomaterial was determined to be 2 wt.%. To visually represent the wear loss in concrete incorporated with the two types of nanomaterials, Figure 17 was presented. The outcomes of the investigation indicated that the effect of wear resistance was better on nano-double-doped concrete as opposed to that of nano-single-doped concrete. Of all the concrete types tested, H3 concrete displayed the best effect. The findings also revealed that H1 had superior performance concerning wear loss compared to H2, suggesting that nano-SiC had a more notable impact on wear resistance. In summary, the study highlighted the significant improvement of wear resistance in fly ash concrete with the addition of nanomaterials, most notably nano-silica and nanoSiC. Wang et al. [118] examined the enhancement caused by nanoparticles on the wear resistance of reactive powder concrete. Three types of nanomaterials were used for the investigations: nanosilica, nano-TiO_2, and nano-ZrO_2. Figure 18 shows the abrasion loss for concrete with different nanomaterial content using two curing approaches: ambient curing and heat curing.

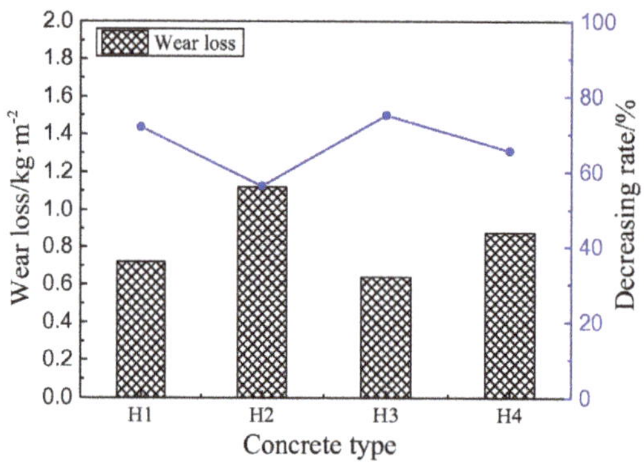

Figure 17. The results of wear resistance for nano-double-doped concrete. Reproduced with permission from [52], (*Construction and Building Materials*); published by (Elsevier), (2017) (Note: H1: 1% nanoSiO_2 + 3% nanoSIC, H2: 3% nanosilica + 1% nanoSIC, H3: 2% nanoSiO_2 + 2% nanoSIC, H4: 2% nanoSiO_2 + 1% nanoSIC).

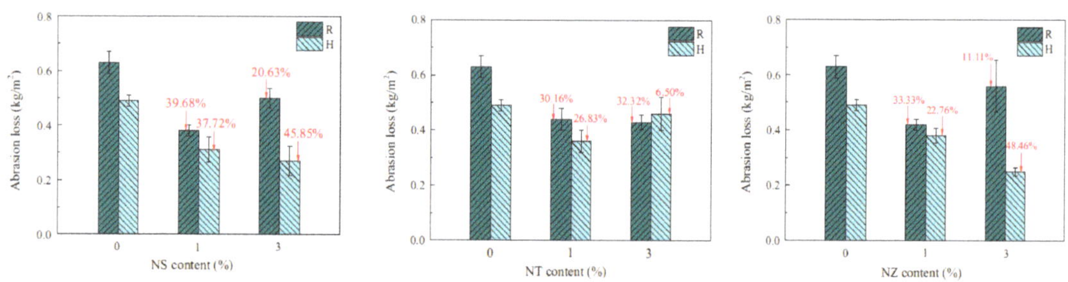

Figure 18. The abrasion loss of reactive powder concrete containing nanomaterials. Reproduced with permission from [118], (*Construction and Building Materials*); published by (Elsevier), (2018). (Notes: R: cured at room temperature, H: heat curing, NS: nano-SiO_2, NT: nano-TiO_2, NZ: nano-ZrO_2).

From Figure 18, it can be observed that all types of nanomaterials improved the abrasion resistance of reactive powder concrete. However, better abrasion resistance was observed for reactive powder concrete containing nanomaterials cured at elevated temperatures, except for 3 wt.% nano-TiO_2. The nanomaterial content level for the three concretes was between 1 and 3 wt.%. For room temperature curing, 1 wt.% content was the optimum dosage for the three types of nanomaterials regarding abrasion loss. Nanosilica reactive powder concrete had the best enhancement effect on abrasion loss, whether using room temperature curing or at heat curing, when the enhancement rate reached 39.68% for a 1 wt.% dosage and 37.72% for a 3 wt.% dosage. In ascending order, the wear resistance of room temperature-cured reactive powder concrete was improved by nanoparticles, namely nanoZrO_2-3%, nanosilica-3%, nanoTiO_2-1%, nanoTiO_2-3%, nanoZrO_2-1%, and nanosilica-1%. For heat-cured reactive powder concrete, the improvement increased in the order of nanoTiO_2-3%, nanoZrO_2-1%, nanoTiO_2-1%, nanosilica-1%, nanosilica-3%, and nanoZrO_2-3%. The improvement in wear resistance with nanomaterials was attributed to their effect on enhancing the interfacial transit zone and structure of cement. The results of the study elucidated that increasing the content of nanoparticles decreased the concrete's resistance to abrasion due to an increasing trend in CH crystal orientation. Microstructural analysis and surface characteristics of nanoparticle-containing concrete pavements were studied by Ghoddousi et al. [119]. The abrasion resistance was tested for concrete containing nanomontmorillonite, nanosilica, and nanohalloysite. Similar to the above-mentioned results, the study revealed that nanomaterials profoundly improved the concrete pavement wear resistance when wear loss was remarkably lower than that of the reference concrete. Figure 19 depicts the loss of concrete volume per surface area after 16 wear cycles. The results showed that nanomaterial-modified concrete had less volume loss as compared to silica fume concrete and reference concrete. The improvement of 3 wt.% nano-montmorillonite, 2 wt.% nano-silica, and 3 wt.% nano-halloysite to the wear resistance of concrete was 23.2%, 21.2%, and 10.3%, respectively. For concrete with 10 wt.% silica fume, the improvement was 15%. The reason for the lower improvement was due to the higher replacement level of silica fume, increasing the content of cementitious composite and reducing the presence of aggregate, which greatly contributed to concrete abrasion resistance. However, the blending silica fume and nanomaterials together into concrete can profoundly increase the material's wear resistance. The greatest improvement was observed in the result for the silica fume-nanomontmorillonite-modified concrete sample. It was believed that the enhancement brought about using silica fume and nanoparticles had a more important effect on strengthening the abrasion resistance of the concrete samples than the reduction in the presence of aggregates that occurred as a result of SCM addition.

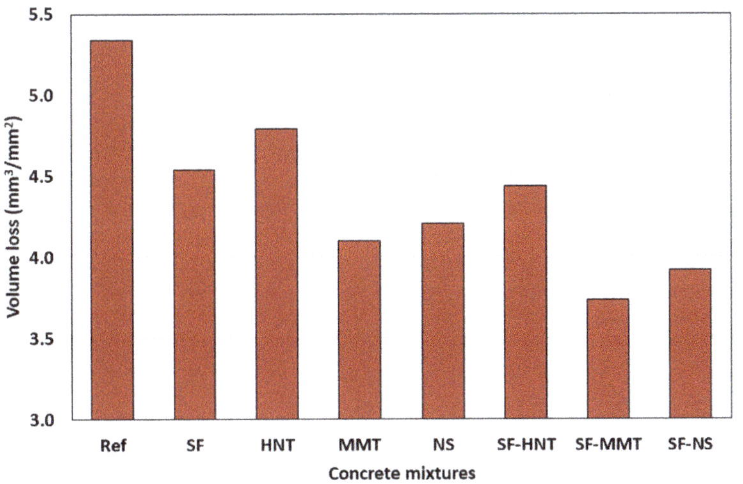

Figure 19. The loss of concrete volume per surface area after 16 periods of wearing. Ref: control concrete, SF: silica fume concrete, HNT: nano-halloysite MMT: montmorillonite, NS: nanosilica. Reproduced with permission from [119], (*Construction and Building Materials*); published by (Elsevier), (2020).

To evaluate how porosity and the pore size distribution impact abrasion resistance, concrete samples underwent Mercury intrusion porosimetry (MIP) testing. Figure 20 presents the volume of small and large capillary and gel pores across all concrete types. Among these, the silica fume and nanomaterials concrete sample had a reduced content of large capillary pores, resulting in higher resistance against abrasion. Comparatively, the nano-halloysite concrete exhibited greater volume loss and a higher content of large capillary porosity. The study found that nano-halloysite had less effect on converting large porosity into small capillary pores and gel pores when compared to nano-montmorillonite and nanosilica. As a result, the total porosity of nano-halloysite increased by 21.1% and was 14.8% less than that of nano-montmorillonite and nanosilica, respectively. Silica fume concrete showed only a 3.5% reduction in pore volume when compared with reference concrete. Therefore, the impact of silica fume on concrete porosity was considered insignificant compared to the effects of nanomaterials. The weight loss was correlated with the compressive strength and average pore diameter of the concrete. The study revealed that a higher compressive strength leads to lower volume loss due to abrasion, while a larger average pore diameter results in higher weight loss.

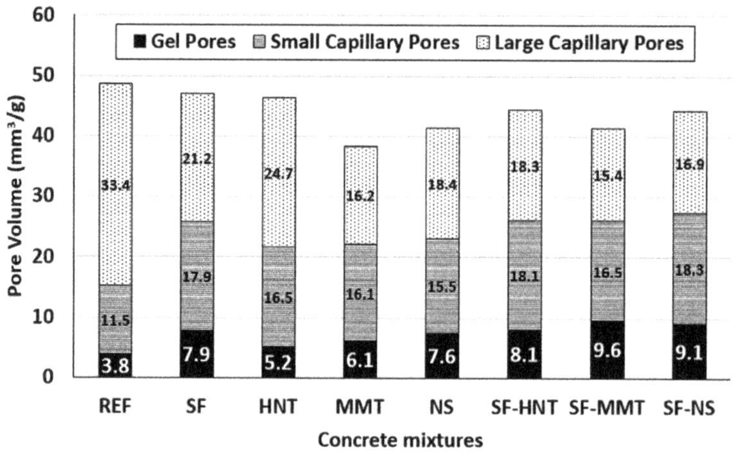

Figure 20. The pore volume of nanoconcrete. REF: control concrete, SF: silica fume concrete, HNT: nano-halloysite MMT: montmorillonite, NS: nanosilica. Reproduced with permission from [119], (*Construction and Building Materials*); published by (Elsevier), (2020).

Wang et al. [120] performed pore structural and fractal analysis on the abrasion resistance and cracking resistance of concrete. It was discovered that concrete porosity and pore surface fractal dimensions measured using MIP were linearly correlated to the concrete abrasion resistance. In addition, the dimensional intricacy or unevenness of a material's pores, known as pore surface fractal dimensions, has a greater impact on the abrasion resistance of concrete than its porosity. The study revealed that both concrete strength and abrasion resistance increased with decreasing porosity and increasing pore surface fractal dimensions. As such, using ultrafine materials such as nanoparticles to refine the pore structure and diameter can greatly increase concrete's surface abrasion resistance and mechanical strength.

NanoMgO is another nanomaterial used to reduce the surface cracking of concrete caused by autogenous shrinkage and thermal stresses [121]. On the other hand, microMgO [122] functions as an agent that limits concrete cracking caused by thermal shrinkage in mass concrete and self-desiccation from utilizing a low w/c ratio and fine material. However, Wang et al. [123] reported that excessive amounts of microMgO can deteriorate compressive strength, tensile strength, and abrasion resistance in concrete. Therefore, nanoMgO is a better alternative for reducing shrinkage and improving compressive strength of concrete simultaneously. A few studies [124–126] indicated that nanoMgO is more effective at reducing autogenous shrinkage compared to microMgO while also enhancing the mechanical properties and permeability of cementitious materials by developing a more uniform and compact microstructure. Adding 1% nanoMgO to concrete reduced permeability by up to 63% and improved other mechanical properties such as compressive and tensile strength, according to Yazdchi et al. [124]. Ye et al. [127] conducted SEM and XRD investigations to examine the hydration and microstructure effects of nanoMgO on cementitious material and found that it improved its microstructure.

8. Conclusions

This article concentrates on enhancing the hydration of C-S-H gel while examining the robustness of nanoconcrete. Incorporating nanomaterials into concrete reduces porosity and enhances its microstructure by increasing the content of cement gel. The mechanical strength and permeability of concrete are improved by the nanofiller, pozzolanic, and nucleation effects of nanoparticles. In addition, the nanoparticle effect reduces concrete capillary porosity and total pore volume by promoting the hydration of unhydrated cement particles and pozzolanic reactions with CH crystals. By increasing the amount of C-S-H gel produced with the nanomaterial reaction, the material's resistance to external attack by sulphates or acids is significantly increased, and the permeability and water absorption of concrete are reduced. The addition of nanomaterials such as nanosilica, nanoclay, and graphene oxide was proven to enhance concrete's durability against high temperatures as well as acid or sulphate degradation. Furthermore, nanoconcrete exhibits better performance against surface abrasion with the reduction in chloride penetration and wear load sustained. While an optimum amount of nanoparticles provides more durable and robust concrete, a higher amount could cause an adverse impact due to the agglomeration effect. To enhance the performance of nanoconcrete, recommendations include conducting more studies on its durability and resistance to other types of threats, broadening the types of nanomaterials added, and exploring better techniques and approaches for better nanoparticle dispersion in concrete.

Author Contributions: Conceptualization, F.Y.A.-s.; validation, L.S.W.; formal analysis, F.Y.A.-s. and L.S.W.; investigation, F.Y.A.-s.; writing—original draft preparation, F.Y.A.-s. and L.S.W.; writing—review and editing, S.C.P.; supervision, L.S.W.; funding acquisition, L.S.W. All authors have read and agreed to the published version of the manuscript.

Funding: The funding related to this review article was provided by the MINISTRY OF HIGHER EDUCATION MALAYSIA via the Fundamental Research Grant Scheme (FRGS), grant number FRGS/1/2022/TK01/UNITEN/02/1. The article processing charge (APC) for this review article was sourced from the UNIVERSITI TENAGA NASIONAL through the Internal Research Grant OPEX; grant type J510050002—IC-6 BOLDREFRESH2025—CENTRE OF EXCELLENCE.

Data Availability Statement: No new data were created or analysed in this study. Data sharing is not applicable to this article.

Acknowledgments: The authors would like to express gratitude for the administrative assistance provided by the staffs of the Innovation & Research Management Centre (iRMC) (Universiti Tenaga Nasional), which greatly contributed to the publication of this review article.

Conflicts of Interest: The authors declare no conflict of interest.

References

1. Nazeer, M.; Kapoor, K.; Singh, S.P. Strength, durability and microstructural investigations on pervious concrete made with fly ash and silica fume as supplementary cementitious materials. *J. Build. Eng.* **2023**, *69*, 106275. [CrossRef]
2. Mehta, P.K.; Monteiro, P.J.M. *Durability in Concrete: Microstructure, Properties, and Materials*; McGraw-Hill Companies, Inc.: New York, NY, USA, 2005; pp. 121–199.
3. Al-Jabari, M. Concrete durability problems: Physicochemical and transport mechanisms. In *Integral Waterproofing of Concrete Structures*; Woodhead Publishing Series in Civil and Structural Engineering; Woodhead Publishing: Cambridge, UK, 2022; pp. 69–107. [CrossRef]
4. Li, X.; Xu, Q.; Chen, S. An experimental and numerical study on water permeability of concrete. *Constr. Build. Mater.* **2016**, *105*, 503–510. [CrossRef]
5. Cyr, M. Influence of supplementary cementitious materials (SCMs) on concrete durability. In *Eco-Efficient Concrete*; Woodhead Publishing: Cambridge, UK, 2016; pp. 153–197. [CrossRef]
6. Pan, X.; Shi, Z.; Shi, C.; Ling, T.C.; Li, N. A review on concrete surface treatment Part I: Types and mechanisms. *Constr. Build. Mater.* **2017**, *132*, 578–590. [CrossRef]

7. Isayed, S.H.; Amjad, M.A. Strength, water absorption and porosity of concrete incorporating natural and crushed aggregate. *J. King Saud Univ. Eng. Sci.* **1996**, *8*, 109–119. [CrossRef]
8. Claisse, P. *Measurement of Porosity as a Predictor of the Transport Properties of Concrete in Transport Properties of Concrete: Modelling the Durability of Structures*; Transport Properties of Concrete; Woodhead Publishing: Cambridge, UK, 2021; pp. 115–147. [CrossRef]
9. Baghabra Al-Amoudi, O.S.; Al-Kutti, W.A.; Ahmad, S.; Maslehuddin, M. Correlation between compressive strength and certain durability indices of plain and blended cement concretes. *Cem. Concr. Compos.* **2014**, *31*, 672–676. [CrossRef]
10. Peter, C. *Hewlett and Martin Liska. Lea's Chemistry of Cement and Concrete*, 5th ed.; Elsevier Ltd.: Amsterdam, The Netherlands, 2017. [CrossRef]
11. Goel, G.; Sachdeva, P.; Chaudhary, A.K.; Singh, Y. The use of nanomaterials in concrete: A review. *Mater. Today Proc.* **2022**, *69*, 365–371. [CrossRef]
12. Reches, Y. Nanoparticles as concrete additives: Review and perspectives. *Constr. Build. Mater.* **2018**, *175*, 483–495. [CrossRef]
13. Yoo, D.Y.; Oh, T.; Banthia, N. Nanomaterials in ultra-high-performance concrete (UHPC)—A review. *Cem. Concr. Compos.* **2022**, *134*, 104730. [CrossRef]
14. Silvestre, J.; Silvestre, N.; de Brito, J. Review on concrete nanotechnology. *Eur. J. Environ. Civ. Eng.* **2015**, *20*, 455–485. [CrossRef]
15. Li, L.; Wang, B.; Hubler, M.H. Carbon nanofibers (CNFs) dispersed in ultra-high performance concrete (UHPC): Mechanical property, workability and permeability investigation. *Cem. Concr. Compos.* **2022**, *131*, 104592. [CrossRef]
16. Piro, N.S.; Salih, A.; Hamad, S.M.; Kurda, R. Comprehensive multiscale techniques to estimate the compressive strength of concrete incorporated with carbon nanotubes at various curing times and mix proportions. *J. Mater. Res. Technol.* **2021**, *15*, 6506–6527. [CrossRef]
17. Abhilash, P.P.; Nayak, D.K.; Sangoju, B.; Kumar, R.; Kumar, V. Effect of nano-silica in concrete; a review. *Constr. Build. Mater.* **2021**, *278*, 122347. [CrossRef]
18. Liu, R.; Xiao, H.; Geng, J.; Du, J.; Liu, M. Effect of nano-$CaCO_3$ and nano-SiO_2 on improving the properties of carbon fibre-reinforced concrete and their pore-structure models. *Constr. Build. Mater.* **2020**, *244*, 118297. [CrossRef]
19. Ansari rad, T.; Tanzadeh, J.; Pourdada, A. Laboratory evaluation of self-compacting fiber-reinforced concrete modified with hybrid of nanomaterials. *Constr. Build. Mater.* **2020**, *232*, 117211. [CrossRef]
20. Ruan, Y.; Han, B.; Yu, X.; Zhang, W.; Wang, D. Carbon nanotubes reinforced reactive powder concrete. *Compos. Part A Appl. Sci. Manuf.* **2018**, *112*, 371–382. [CrossRef]
21. Khalaf, M.A.; Cheah, C.B.; Ramli, M.; Ahmed, N.M.; Al-Shwaiter, A. Effect of nano zinc oxide and silica on mechanical, fluid transport and radiation attenuation properties of steel furnace slag heavyweight concrete. *Constr. Build. Mater.* **2021**, *274*, 121785. [CrossRef]
22. Maohua, Z.; Zhengyi, L.; Jiyin, C.; Zenong, T.; Zhiyi, L. Durability of marine concretes with nanoparticles under combined action of bending load and salt spray erosion. *Adv. Mater. Sci. Eng.* **2022**, *2022*, 1968770. [CrossRef]
23. Khooshechin, M.; Tanzadeh, J. Experimental and mechanical performance of shotcrete made with nanomaterials and fiber reinforcement. *Constr. Build. Mater.* **2018**, *165*, 199–205. [CrossRef]
24. Ahmed, T.I.; El-Shafai, N.M.; El-Mehasseb, I.M.; Sharshir, S.W.; Tobbala, D.E. Recent advances in the heating resistance, thermal gravimetric analysis, and microstructure of green concrete incorporating palm-leaf and cotton-stalk nanoparticles. *J. Build. Eng.* **2022**, *61*, 105252. [CrossRef]
25. Yan, J.; Liu, X.; Wang, X.; Wang, L.; Weng, W.; Yu, X.; Xing, G.; Xie, J.; Lu, C.; Luo, Y.; et al. Influence of nano-attapulgite on compressive strength and microstructure of recycled aggregate concrete. *Cem. Concr. Compos.* **2022**, *134*, 104788. [CrossRef]
26. Narasimman, K.; Jassam, T.M.; Velayutham, T.; Yaseer, M.; Ruzaimah, R. The synergic influence of carbon nanotube and nanosilica on the compressive strength of lightweight concrete. *J. Build. Eng.* **2020**, *32*, 101719. [CrossRef]
27. Du, S.; Ge, Y.; Shi, X. A targeted approach of employing nano-materials in high-volume fly ash concrete. *Cem. Concr. Compos.* **2019**, *104*, 103390. [CrossRef]
28. Wang, X.; Ding, S.; Qiu, L.; Ashour, A.; Wang, Y.; Han, B.; Ou, J. Improving bond of fiber-reinforced polymer bars with concrete through incorporating nanomaterials. *Compos. B Eng.* **2022**, *239*, 109960. [CrossRef]
29. Jia, H.; Cui, B.; Niu, G.; Chen, J.; Yang, Y.; Wang, Q.; Tang, C. Experimental and mechanism study on the impermeability and thermal insulation of foam concrete regulated by nano-silica and fluorine-free foam. *J. Build. Eng.* **2023**, *64*, 105675. [CrossRef]
30. Shahpari, M.; Khaloo, A.; Rashidi, A.; Saberian, M.; Li, J. Synergetic effects of hybrid nano-blended cement on mechanical properties of conventional concrete: Experimental and analytical evaluation. *Structures* **2023**, *48*, 1519–1536. [CrossRef]
31. Meng, T.; Ying, K.; Yang, X.; Hong, Y. Comparative study on mechanisms for improving mechanical properties and microstructure of cement paste modified by different types of nanomaterials. *Nanotechnol. Rev.* **2021**, *10*, 370–384. [CrossRef]
32. Wang, Y.; Lu, H.; Wang, J.; He, H. Effects of Highly Crystalized Nano C-S-H Particles on Performances of Portland Cement Paste and Its Mechanism. *Crystals* **2020**, *10*, 816. [CrossRef]
33. Liu, C.; He, X.; Deng, X.; Wu, Y.; Zheng, Z.; Liu, J.; Hui, D. Application of nanomaterials in ultra-high performance concrete: A review. *Nanotechnol. Rev.* **2020**, *9*, 1427–1444. [CrossRef]
34. Wang, B.; Yao, W.; Stephan, D. Preparation of calcium silicate hydrate seeds by means of mechanochemical method and its effect on the early hydration of cement. *Adv. Mech. Eng.* **2019**, *11*, 168781401984058. [CrossRef]
35. Land, G.; Stephan, D. Controlling cement hydration with nanoparticles. *Cem. Concr. Compos.* **2015**, *57*, 64–67. [CrossRef]

36. Xue, Q.; Ni, C.; Wu, Q.; Yu, Z.; Shen, X. Effects of Nano-CSH on the hydration process and mechanical property of cementitious materials. *J. Sustain. Cem.-Based Mater.* **2021**, *11*, 378–388. [CrossRef]
37. Hakamy, A.; Shaikh, F.; Low, I. Characteristics of nanoclay and calcined nanoclay-cement nanocomposites. *Compos. Part B Eng.* **2015**, *78*, 174–184. [CrossRef]
38. Yuenyongsuwan, J.; Sinthupinyo, S.; O'Rear, E.A.; Pongprayoon, T. Hydration accelerator and photocatalyst of nanotitanium dioxide synthesized via surfactant-assisted method in cement mortar. *Cem. Concr. Compos.* **2019**, *96*, 182–193. [CrossRef]
39. Khamchin, F.; Rasiah, S.; Sirivivatnanon, V. Properties of Metakaolin Concrete—A Review. In Proceedings of the International Conference on Sustainable Structural Concrete, La Plata, Argentina, 15–18 September 2015.
40. Nazari, A.; Riahi, S. The effects of ZrO_2 nanoparticles on physical and mechanical properties of high strength self-compacting concrete. *Mater. Res.* **2010**, *13*, 551–556. [CrossRef]
41. Devi, S.C.; Khan, R.A. Compressive strength and durability behavior of graphene oxide reinforced concrete composites containing recycled concrete aggregate. *J. Build. Eng.* **2020**, *32*, 101800. [CrossRef]
42. Zhang, P.; Su, J.; Guo, J.; Hu, S. Influence of carbon nanotube on properties of concrete: A review. *Constr. Build. Mater.* **2023**, *369*, 130388. [CrossRef]
43. Wang, J.; Han, B.; Li, Z.; Yu, X.; Dong, X. Effect Investigation of Nanofillers on C-S-H Gel Structure with Si NMR. *J. Mater. Civ. Eng.* **2019**, *31*, 04018352. [CrossRef]
44. Long, G.; Li, Y.; Ma, C.; Xie, Y.; Shi, Y. Hydration kinetics of cement incorporating different nanoparticles at elevated temperatures. *Thermochim. Acta* **2018**, *664*, 108–117. [CrossRef]
45. Sun, Y.; Zhang, P.; Guo, W.; Bao, J.; Qu, C. Effect of Nano-$CaCO_3$ on the mechanical properties and durability of concrete incorporating Fly Ash. *Adv. Mater. Sci. Eng.* **2020**, *2020*, 7365862. [CrossRef]
46. Hussien, R.M.; Abd el-Hafez, L.; Mohamed, R.; Faried, A.S.; Fahmy, N.G. Influence of nano waste materials on the mechanical properties, microstructure, and corrosion resistance of self-compacted concrete. *Case Stud. Constr. Mater.* **2022**, *16*, e00859. [CrossRef]
47. Li, G.; Chen, X.; Zhang, Y.; Zhuang, Z.; Lv, Y. Studies of nano-SiO_2 and subsequent water curing on enhancing the frost resistance of autoclaved PHC pipe pile concrete. *J. Build. Eng.* **2023**, *69*, 106209. [CrossRef]
48. Ren, J.; Lai, Y.; Gao, J. Exploring the influence of SiO_2 and TiO_2 nanoparticles on the mechanical properties of concrete. *Constr. Build. Mater.* **2018**, *175*, 277–285. [CrossRef]
49. Lu, D.; Wang, D.; Wang, Y.; Zhong, J. Nano-engineering the interfacial transition zone between recycled concrete aggregates and fresh paste with graphene oxide. *Constr. Build. Mater.* **2023**, *384*, 131244. [CrossRef]
50. Liu, X.; Xie, X.; Liu, R.; Lyu, K.; Zuo, J.; Li, S.; Liu, L.; Shah, S.P. Research on the durability of nano-SiO_2 and sodium silicate co-modified recycled coarse aggregate (RCA) concrete. *Constr. Build. Mater.* **2023**, *378*, 131185. [CrossRef]
51. Faried, A.S.; Mostafa, S.A.; Tayeh, B.A.; Tawfik, T.A. Mechanical and durability properties of ultra-high performance concrete incorporated with various nano waste materials under different curing conditions. *J. Build. Eng.* **2021**, *43*, 102569. [CrossRef]
52. Gao, Y.; He, B.; Li, Y.; Tang, J.; Qu, L. Effects of nano-particles on improvement in wear resistance and drying shrinkage of road fly ash concrete. *Constr. Build. Mater.* **2017**, *151*, 228–235. [CrossRef]
53. Fahmy, N.G.; Hussien, R.M.; el-Hafez, L.A.; Mohamed, R.; Faried, A.S. Comparative study on fresh, mechanical, microstructures properties and corrosion resistance of self compacted concrete incorporating nanoparticles extracted from industrial wastes under various curing conditions. *J. Build. Eng.* **2022**, *57*, 104874. [CrossRef]
54. Asadi, I.; Shafigh, P.; Hassan, Z.F.B.A.; Mahyuddin, N.B. Thermal conductivity of concrete—A review. *J. Build. Eng.* **2018**, *20*, 81–93. [CrossRef]
55. Saif, M.S.; Shanour, A.S.; Abdelaziz, G.E.; Elsayad, H.I.; Shaaban, I.G.; Tayeh, B.A.; Hammad, M.S. Influence of blended powders on properties of ultra-high strength fibre reinforced self compacting concrete subjected to elevated temperatures. *Case Stud. Constr. Mater.* **2023**, *18*, e01793. [CrossRef]
56. Ma, Q.; Guo, R.; Zhao, Z.; Lin, Z.; He, K. Mechanical properties of concrete at high temperature—A review. *Constr. Build. Mater.* **2015**, *93*, 371–383. [CrossRef]
57. Almasaeid, H.H.; Suleiman, A.; Alawneh, R. Assessment of high-temperature damaged concrete using non-destructive tests and artificial neural network modelling. *Case Stud. Constr. Mater.* **2022**, *16*, e01080. [CrossRef]
58. Jiao, Y.; Liu, H.; Wang, X.; Zhang, Y.; Luo, G.; Gong, Y. Temperature effect on mechanical properties and damage identification of concrete structure. *Adv. Mater. Sci. Eng.* **2014**, *2014*, 191360. [CrossRef]
59. Lin, S.; Wang, Y.; Suzuki, Y. High-temperature CaO hydration/$Ca(OH)_2$ decomposition over a multitude of cycles. *Energy Fuels* **2009**, *23*, 2855–2861. [CrossRef]
60. Janotka, I.; Nürnbergerová, T. Effect of temperature on structural quality of the cement paste and high-strength concrete with silica fume. *Nucl. Eng. Des.* **2005**, *235*, 2019–2032. [CrossRef]
61. Ali, M.H.; Dinkha, Y.Z.; Haido, J.H. Mechanical properties and spalling at elevated temperature of high performance concrete made with reactive and waste inert powders. *Eng. Sci. Technol.* **2017**, *20*, 536–541. [CrossRef]
62. Mohammed, H.; Ahmed, H.; Kurda, R.; Alyousef, R.; Deifalla, A.F. Heat-induced spalling of concrete: A review of the influencing factors and their importance to the phenomenon. *Materials* **2022**, *15*, 1693. [CrossRef]
63. Kannangara, T.; Joseph, P.; Fragomeni, S.; Guerrieri, M. Existing theories of concrete spalling and test methods relating to moisture migration patterns upon exposure to elevated temperatures—A review. *Case Stud. Constr. Mater.* **2022**, *16*, e01111. [CrossRef]

64. Zeiml, M.; Leithner, D.; Lackner, R.; Mang, H.A. How do polypropylene fibers improve the spalling behavior of in-situ concrete? *Cem. Concr. Res.* **2006**, *36*, 929–942. [CrossRef]
65. Bao, J.; Zheng, R.; Sun, Y.; Zhang, P.; Cui, Y.; Xue, S.; Song, Q.A. state-of-the-art review on high temperature resistance of lightweight aggregate high-strength concrete. *J. Build. Eng.* **2023**, *69*, 106267. [CrossRef]
66. Lu, D.; Tang, Z.; Zhang, L.; Zhou, J.; Gong, Y.; Tian, Y.; Zhong, J. Effects of combined usage of supplementary cementitious materials on the thermal properties and microstructure of high-performance concrete at high temperatures. *Materials* **2020**, *13*, 1833. [CrossRef] [PubMed]
67. Sikora, P.; Abd Elrahman, M.; Stephan, D. The influence of nanomaterials on the thermal resistance of cement-based composites—A review. *Nanomaterials* **2018**, *8*, 465. [CrossRef] [PubMed]
68. Brzozowski, P.; Strzałkowski, J.; Rychtowski, P.; Wróbel, R.; Tryba, B.; Horszczaruk, E. Effect of nano-SiO_2 on the microstructure and mechanical properties of concrete under high temperature conditions. *Materials* **2022**, *15*, 166. [CrossRef] [PubMed]
69. Kumar, R.; Singh, S.; Singh, L.P. Studies on enhanced thermally stable high strength concrete incorporating silica nanoparticles. *Constr. Build. Mater.* **2017**, *153*, 506–513. [CrossRef]
70. Bastami, M.; Baghbadrani, M.; Aslani, F. Performance of nano-silica modified high strength concrete at elevated temperatures. *Constr. Build. Mater.* **2014**, *68*, 402–408. [CrossRef]
71. Shah, A.H.; Sharma, U.K.; Roy, D.A.B.; Bhargava, P. Spalling behaviour of nano SiO_2 high strength concrete at elevated temperatures. In *MATEC Web of Conferences*; EDP Sciences: Les Ulis, France, 2013; p. 01009.
72. Wang, W.-C. Compressive strength and thermal conductivity of concrete with nanoclay under various high-temperatures. *Constr. Build. Mater.* **2017**, *147*, 305–311. [CrossRef]
73. Chu, H.; Jiang, J.; Sun, W.; Zhang, M. Effects of graphene sulfonate nanosheets on mechanical and thermal properties of sacrificial concrete during high temperature exposure. *Cem. Concr. Compos.* **2017**, *82*, 252–264. [CrossRef]
74. Nikbin, I.M.; Rafiee, A.; Dezhampanah, S.; Mehdipour, S.S.; Mohebbi, R.; Moghadam, H.H.; Sadrmomtazi, A. Effect of high temperature on the radiation shielding properties of cementitious composites containing nano-Bi_2O_3. *J. Mater. Res. Technol.* **2020**, *9*, 11135–11153. [CrossRef]
75. Nikbin, I.M.; Mehdipour, S.; Dezhampanah, S.; Mohammadi, R.; Mohebbi, R.; Moghadam, H.H.; Sadrmomtazi, A. Effect of high temperature on mechanical and gamma ray shielding properties of concrete containing nano-TiO_2. *Radiat. Phys. Chem.* **2020**, *174*, 108967. [CrossRef]
76. Pachideh, G.; Gholhaki, M.; Moshtagh, A.; Felaverjani, M.K. An Investigation on the effect of high temperatures on the mechanical properties and microstructure of concrete containing multiwalled carbon nanotubes. *Mater. Perform. Charact.* **2018**, *8*, 503–517. [CrossRef]
77. Sakthirswaran, N.; Jeyamurugan, A.; Sophia, M.; Suresh, P. Effect of elevated temperatures on the properties of nano alumina modified concrete containing zircon sand as fine aggregate. *Rom. J. Mater.* **2020**, *50*, 175–182.
78. Mohammed, A.; Sanjayan, J.G.; Nazari, A.; Al-Saadi, N.T.K. Effects of graphene oxide in enhancing the performance of concrete exposed to high-temperature. *Aust. J. Civ. Eng.* **2017**, *15*, 61–71. [CrossRef]
79. Reddy, L.S.I.; Vijayalakshmi, M.M.; Praveenkumar, T.R. Thermal conductivity and strength properties of nanosilica and GGBS incorporated concrete specimens. *Silicon* **2022**, *14*, 145–151. [CrossRef]
80. Sherif, M.A. Effect of elevated temperature on mechanical properties of nano materials concrete. *Int. J. Innov. Res. Sci. Eng Technol.* **2017**, *7*, 1–9.
81. Shalby, O.B.; Elkady, H.M.; Nasr, E.A.R.; Kohail, M. Assessment of mechanical and fire resistance for hybrid nano-clay and steel fibres concrete at different curing ages. *J. Struct. Fire Eng.* **2019**, *11*, 189–203. [CrossRef]
82. Dahish, H.A.; Almutairi, A.D. Effect of elevated temperatures on the compressive strength of nano-silica and nano-clay modified concretes using response surface methodology. *Case Stud. Constr. Mater.* **2023**, *18*, e02032. [CrossRef]
83. Nijland, T.; Larbi, J. *Microscopic Examination of Deteriorated Concrete in Non-Destructive Evaluation of Reinforced Concrete Structures. Deterioration Processes and Standard Test Methods*; Woodhead Publishing Series in Civil and Structural Engineering; Woodhead Publishing: Cambridge, UK, 2010; Volume 1, pp. 137–179. [CrossRef]
84. Bensted, J.; Rbrough, A.; Page, M. *Chemical Degradation of Concrete in Durability of Concrete and Cement Composites*; Woodhead Publishing Series in Civil and Structural Engineering; Woodhead Publishing: Cambridge, UK, 2007; pp. 86–135. [CrossRef]
85. Dyer, T. Influence of cement type on resistance to organic acids. *Mag. Concr. Res.* **2017**, *69*, 175–200. [CrossRef]
86. Dong, Z.; Wan, Y.; Wang, P.; Chen, Z.; He, X.; Hui, X. Effect of long-term acid attack on impermeability and microstructure of compacted cement-bound soils. *Environ. Technol.* **2022**, 1–9. [CrossRef]
87. Zivica, V.; Bajza, A. Acidic attack of cement based materials—A review. *Constr. Build. Mater.* **2001**, *15*, 331–340. [CrossRef]
88. Mariam Ninan, C.; Ajay, A.; Ramaswamy, K.P.; Thomas, A.V.; Bertron, A. A critical review on the effect of organic acids on cement-based materials. *IOP Conf. Ser. Earth Environ. Sci.* **2020**, *491*, 012045. [CrossRef]
89. Diab, A.M.; Elyamany, H.E.; Abd Elmoaty, A.E.M.; Sreh, M.M. Effect of nanomaterials additives on performance of concrete resistance against magnesium sulfate and acids. *Constr. Build. Mater.* **2019**, *210*, 210–231. [CrossRef]
90. Mahdikhani, M.; Bamshad; Shirvani, F. Mechanical properties and durability of concrete specimens containing nano silica in sulfuric acid rain condition. *Constr. Build. Mater.* **2018**, *167*, 929–935. [CrossRef]
91. Sujay, H.; Nair, N.A.; Sudarsana Rao, H.; Sairam, V. Experimental study on durability characteristics of composite fiber reinforced high-performance concrete incorporating nanosilica and ultra fine fly ash. *Constr. Build. Mater.* **2020**, *262*, 120738. [CrossRef]

92. Praveenkumar, T.; Vijayalakshmi, M.; Meddah, M. Strengths and durability performances of blended cement concrete with TiO_2 nanoparticles and rice husk ash. *Constr. Build. Mater.* **2019**, *217*, 343–351. [CrossRef]
93. Cao, R.; Yang, J.; Li, G.; Zhou, Q.; Niu, M. Durability performance of multi-walled carbon nanotube reinforced ordinary Portland/calcium sulfoaluminate cement composites to sulfuric acid attack at early stage. *Mater. Today Commun.* **2023**, *35*, 105748. [CrossRef]
94. Piasta, W. Analysis of carbonate and sulphate attack on concrete structures. *Eng. Fail. Anal.* **2017**, *79*, 606–614. [CrossRef]
95. Elahi, M.M.A.; Shearer, C.R.; Naser Rashid Reza, A.; Saha, A.K.; Khan, M.N.N.; Hossain, M.M.; Sarker, P.K. Improving the sulfate attack resistance of concrete by using supplementary cementitious materials (SCMs): A review. *Constr. Build. Mater.* **2021**, *281*, 122628. [CrossRef]
96. Singh, G.; Saini, B. Nanomaterial in cement industry: A brief review. *Innov. Infrastruct. Solut.* **2022**, *7*, 45. [CrossRef]
97. Wang, Y.; Zeng, D.; Ueda, T.; Fan, Y.; Li, C.; Li, J. Beneficial effect of nanomaterials on the interfacial transition zone (ITZ) of non-dispersible underwater concrete. *Constr. Build. Mater.* **2021**, *293*, 123472. [CrossRef]
98. Vijayabhaskar, A.; Shanmugasundaram, M. Enhancing the durability properties of concrete prepared with multiwalled carbon nanotubes. *Electron. J. Struct. Eng.* **2018**, *18*, 117–127. [CrossRef]
99. Reshma, T.V.; Manjunatha, M.; Bharath, A.; Tangadagi, R.B.; Vengala, J.; Manjunatha, L. Influence of ZnO and TiO_2 on mechanical and durability properties of concrete prepared with and without polypropylene fibers. *Materialia* **2021**, *18*, 101138. [CrossRef]
100. Moslemi, A.M.; Khosravi, A.; Izadinia, M.; Heydari, M. Application of nano silica in concrete for enhanced resistance against sulfate attack. *Adv. Mater. Res.* **2013**, *829*, 874–878. [CrossRef]
101. Sathe, S.; Zain Kangda, M.; Amaranatha, G. Resistance against sulphate attack in concrete by addition of nano alumina. *Mater. Today Proc.* **2022**, *60*, 294–298. [CrossRef]
102. Vishwakarma, V.; Uthaman, S.; Dasnamoorthy, R.; Kanagasabai, V. Investigation on surface sulfate attack of nanoparticle-modified fly ash concrete. *Environ. Sci. Pollut. Res.* **2020**, *27*, 41372–41380. [CrossRef]
103. Wang, T.; Cao, L.; Zhang, F.; Luo, J.; Jiang, S.; Chu, H.; Jiang, L. Reduction of SO_4^{2-} and Cl^- migration rates and degradation of silica nanoparticles incorporated cement pastes exposed to co-existence of sulfate, chloride and electric fields. *Constr. Build. Mater.* **2022**, *344*, 128234. [CrossRef]
104. Zhu, F.; Ma, Z.; Zhang, M. Chloride penetration into concrete under the coupling effects of internal and external relative humidity. *Adv. Civ. Eng.* **2020**, *2020*, 1468717. [CrossRef]
105. Hu, J.; Zhang, S.; Chen, E.; Li, W. A review on corrosion detection and protection of existing reinforced concrete (RC) structures. *Constr. Build. Mater.* **2022**, *325*, 126718. [CrossRef]
106. Li, D.; Li, L.Y.; Wang, X. Chloride diffusion model for concrete in marine environment with considering binding effect. *Mar. Struct.* **2019**, *66*, 44–51. [CrossRef]
107. Singh, L.; Bhattacharyya, S.; Sharma, U.; Mishra, G.; Ahalawat, S. Microstructure improvement of cementitious systems using nanomaterials: A key for enhancing the durability of concrete. In Proceedings of the Ninth International Conference on Creep, Shrinkage, and Durability Mechanics (CONCREEP-9), Cambridge, MA, USA, 22–25 September 2013. [CrossRef]
108. Rezakhani, D.; Jafari, A.; Hajabassi, M. Durability, mechanical properties and rebar corrosion of slag-based cement concrete modified with graphene oxide. *Structures* **2023**, *49*, 678–697. [CrossRef]
109. Liu, C.; Hunag, X.; Wu, Y.Y.; Deng, X.; Zheng, Z.; Yang, B. Studies on mechanical properties and durability of steel fiber reinforced concrete incorporating graphene oxide. *Cem. Concr. Compos.* **2022**, *130*, 104508. [CrossRef]
110. Zhang, M.; Du, L.; Li, Z.; Xu, R. Durability of marine concrete doped with nanoparticles under joint action of Cl^- erosion and carbonation. *Case Stud. Constr. Mater.* **2023**, *18*, e01982. [CrossRef]
111. Joshaghani, A.; Balapour, M.; Mashhadian, M.; Ozbakkaloglu, T. Effects of nano-TiO_2, nano-Al_2O_3, and nano-Fe_2O_3 on rheology, mechanical and durability properties of self-consolidating concrete (SCC): An experimental study. *Constr. Build. Mater.* **2020**, *245*, 118444. [CrossRef]
112. Wang, T.; Xu, J.; Meng, B.; Peng, G. Experimental study on the effect of carbon nanofiber content on the durability of concrete. *Constr. Build. Mater.* **2020**, *250*, 118891. [CrossRef]
113. Li, C. Chloride permeability and chloride binding capacity of nano-modified concrete. *J. Build. Eng.* **2021**, *41*, 102419. [CrossRef]
114. Li, T.; Liu, X.; Wei, Z.; Zhao, Y.; Yan, D. Study on the wear-resistant mechanism of concrete based on wear theory. *Constr. Build. Mater.* **2021**, *271*, 121594. [CrossRef]
115. Yazıcı; İnan, G. An investigation on the wear resistance of high strength concretes. *Wear* **2006**, *260*, 615–618. [CrossRef]
116. Adewuyi, A.P.; Sulaiman, I.A.; Akinyele, J.O. Compressive strength and abrasion resistance of concretes under varying exposure conditions. *Open J. Civ. Eng.* **2017**, *07*, 82–99. [CrossRef]
117. Li, H.; Zhang, M.H.; Ou, J.P. Abrasion resistance of concrete containing nano-particles for pavement. *Wear* **2006**, *260*, 1262–1266. [CrossRef]
118. Wang, D.; Zhang, W.; Ruan, Y.; Yu, X.; Han, B. Enhancements and mechanisms of nanoparticles on wear resistance and chloride penetration resistance of reactive powder concrete. *Constr. Build. Mater.* **2018**, *189*, 487–497. [CrossRef]
119. Ghoddousi, P.; Zareechian, M.; Shirzadi Javid, A.A.; Habibnejad Korayem, A. Microstructural study and surface properties of concrete pavements containing nanoparticles. *Constr. Build. Mater.* **2020**, *262*, 120103. [CrossRef]
120. Wang, L.; Jin, M.; Guo, F.; Wang, Y.; Tang, S. Pore structural and fractal analysis of the influence of fly ash and silica fume on the mechanical property and abrasion resistance of concrete. *Fractals* **2021**, *29*, 2140003. [CrossRef]

121. Polat, R.; Demirboğa, R.; Karagöl, F. The effect of nano-MgO on the setting time, autogenous shrinkage, microstructure and mechanical properties of high performance cement paste and mortar. *Constr. Build. Mater.* **2017**, *156*, 208–218. [CrossRef]
122. Zhang, J. Recent advance of MgO expansive agent in cement and concrete. *J. Build. Eng.* **2022**, *45*, 103633. [CrossRef]
123. Wang, L.; Zeng, X.; Li, Y.; Yang, H.; Tang, S. Influences of MgO and PVA fiber on the abrasion and cracking resistance, pore structure and fractal features of hydraulic concrete. *Fractal Fract.* **2022**, *6*, 674. [CrossRef]
124. Yazdchi, M.; Foroughi Asl, A.; Talatahari, S.; Gandomi, A.H. Evaluation of the mechanical properties of normal concrete containing nano-MgO under freeze–thaw conditions by evolutionary intelligence. *Appl. Sci.* **2021**, *11*, 2529. [CrossRef]
125. Ye, Q.; Kaikai Yu, K.; Zenan Zhang, Z. Expansion of ordinary Portland cement paste varied with nano-MgO. *Constr. Build. Mater.* **2015**, *78*, 189–193. [CrossRef]
126. Moradpour, R.; Taheri-Nassaj, E.; Parhizkar, T.; Ghodsian, M. The effects of nanoscale expansive agents on the mechanical properties of non-shrink cement-based composites: The influence of nano-MgO addition. *Compos. Part B Eng.* **2014**, *55*, 193–202. [CrossRef]
127. Ye, Y.; Liu, Y.; Shi, T.; Hu, Z.; Zhong, L.; Wang, H.; Chen, Y. Effect of nano-magnesium oxide on the expansion performance and hydration process of cement-based materials. *Materials* **2021**, *14*, 3766. [CrossRef] [PubMed]

gels

Article

Layered Sol–Gel Deposition of a Sn, Ti, Zn, and Pr Mixed Oxide Thin Film with Electrical Properties for Gas Sensing

Izabella Dascalu, Cristian Hornoiu *, Jose Maria Calderon Moreno *, Petre Osiceanu and Simona Somacescu

"Ilie Murgulescu" Institute of Physical Chemistry—Romanian Academy, Splaiul Independentei 202, 060021 Bucharest, Romania; idascalu@icf.ro (I.D.); posic@icf.ro (P.O.); ssimona@icf.ro (S.S.)
* Correspondence: chornoiu@icf.ro (C.H.); calderon@icf.ro (J.M.C.M.)

Abstract: This article presents a layered mixed oxide thin film composed of Sn, Ti, Zn, and Pr obtained by sol–gel deposition for gas sensing applications. The film was characterized by X-ray diffraction (XRD), X-ray photoelectron spectroscopy (XPS), Raman spectroscopy, UV-Vis spectroscopy, Scanning electron microscopy with energy dispersive X-ray spectroscopy (SEM-EDX), and Electrochemical impedance spectroscopy (EIS). X-ray diffraction results showed the presence of a single crystalline phase with a cassiterite-like structure. Raman spectroscopy revealed characteristic bands of oxygen-deficient SnO_2-based nanocrystallites. The band gap energy calculated from UV-Vis spectroscopy is Eg = 3.83 eV. The XPS proved the presence on the surface of all elements introduced by the inorganic precursors as well as their oxidation states. Thus, Sn^{4+}, Ti^{4+}, Zn^{2+}, and Pr^{3+} were detected on the surface. Moreover, by XPS, we highlighted the presence of OH groups and water adsorbed on the surface. SEM showed the five-layer morphology of the film after five successive depositions. Electrochemical properties were determined by EIS-impedance spectroscopy. The selectivity for gas sensing was also investigated for methane, propane, and formaldehyde and the gas sensing mechanism was explained. The results indicated that the mixed oxide thin film exhibited high sensitivity and selectivity towards specific gases.

Keywords: sol gel; SnO_2; ZnO; TiO_2; praseodynium; gas sensing

1. Introduction

Semiconductor oxides have great potential as basic materials in high-performance and modern technologies. Research in this field has led to the development of new materials with structures and characteristics of great interest that open up wide areas of applicability such as chemical engineering (nanopolymers, nanowires/nanotubes), optics (TCO materials, solar cells), optoelectronics (photodetectors, LEDs, lasers), electronics (varistors, transistors, transparent displays), medicine (diagnostic techniques, prostheses and implants), chemical sensors, piezoelectric devices, and other functional devices. These semiconductors have also attracted considerable attention because of their low cost, ease of use, and ability to detect a large number of gases.

Nanocrystalline semiconductor metal oxides with controlled composition are of great interest in the field of gas sensors. Recently, many papers have been reported regarding the combination of SnO_2, SiO_2, In_2O_3, CeO_2, TiO_2, and ZnO to obtain materials with structural, optical, electrical, photocatalytic, and sensing properties [1–5].

Rare earth oxides, such as Ce_2O_3, Pr_2O_3, and Tb_2O_3 do not have stable valence and therefore are rarely studied as gas sensors only, but for other applications in optics and photonics, electronics, and catalysis [6–10].

They can be used in small amounts as a doping material for various n-type semiconductors or as mixed oxides with n-type semiconductors. It could be that in combination with other oxides, the sensing properties of n-type oxide semiconductors mixed or doped with Pr_2O_3 are improved [11,12].

The powder of the mixed oxide of Sn, Ti, Zn, and Pr was shown to improve electrical properties by doping, and consequently Pr can increase the sensory performances of materials [1].

Materials based on metal oxides can be obtained by using the sol–gel method [13–15], one of the most widely used methods that offers the possibility of producing various materials with new, predefined properties in a simple process and at a relatively low cost.

The sol–gel method is versatile and allows control of material composition, morphology, and properties, making it very attractive for various applications, such as catalysts, sensors, optics, and ceramics. Recent studies show that this is a promising method for the deposition of gas sensors. Gas sensors obtained by the sol–gel method are used in a wide range of applications, including to monitor and detect hazardous gases such as CO, CO_2, NO_2, and CH_4 [16–18] to ensure safety and air quality.

In this article, we attempted to obtain a sensor based on the previously studied composition (Sn:Ti:Zn:Pr molar ratios of 60:20:15:5) that exhibits a very good sensory response to CO. Considering the influence of the calcination temperature on the obtained electrical properties (the sensitivity to CO decreases with increasing calcination temperature), it was expected that a more sensitive material would be obtained at the treatment temperature of the film (450 °C). In this paper, we present the results of XRD, SEM-EDX, Raman, UV-Vis, XPS, and IS investigations used to study our film.

2. Results and Discussion

2.1. Structural and Morphological Properties

2.1.1. XRD

The diffraction pattern for the Sn, Ti, Zn, and Pr mixed oxide thin film after heat treatment is shown in Figure 1a. The PDXL software from the Rigaku equipment, connected to the ICCD database, identified the cassiterite phase, characteristic of SnO_2, with maximum diffraction peaks at 27.15°, 34.72°, and 52.42° matching with diffracting lines (110), (101), and (211) of SnO_2 as tetragonal cassiterite (PDF 01-072-1147 [19]) as the single-component crystalline phase of the deposited STZP film. The crystal lattice planes appeared to shift towards higher 2-theta values, indicating a reduction in the lattice cell parameters (Table 1) due to the introduction of dopants in the cassiterite lattice, compared to standard lattice parameters of tetragonal SnO_2 a = b = 4.737 Å, c = 3.185 Å.

Table 1. Structural properties of the STZP film.

Phase	a (Å)	b (Å)	c (Å)	$\alpha = \beta = \gamma$ (°)	Crystallite Size (nm)
Cassiterite	4.70 (6)	4.70 (6)	3.11 (7)	90	6.4

The crystallite size was obtained using the Debye–Scherrer formula by direct measurement of the full width at half peak (2.85 deg) of the (211) peak at ~52° (Table 1).

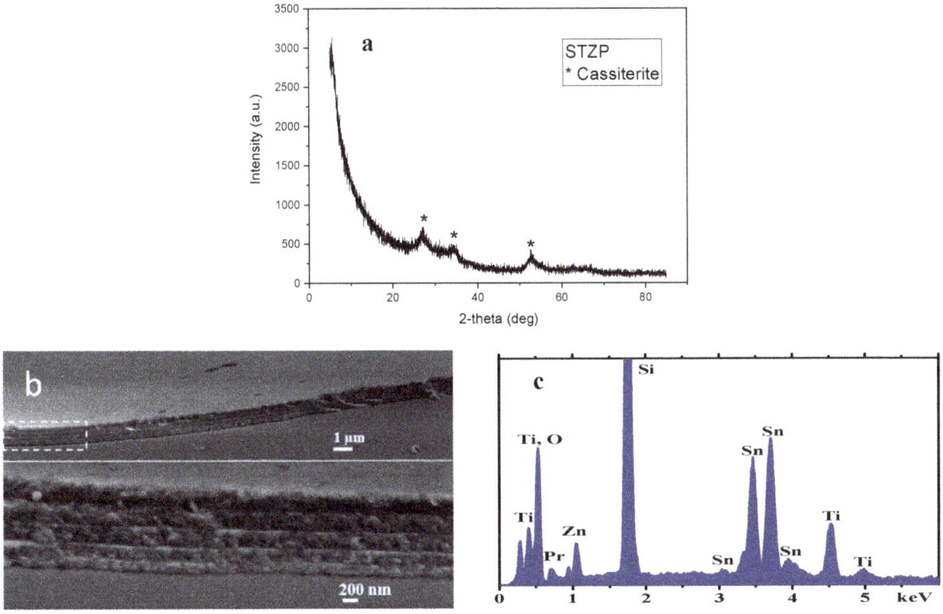

Figure 1. XRD pattern (**a**) SEM images at different magnification (**b**) and EDX spectrum (**c**) for STZP film.

2.1.2. SEM-EDX

The cross-section SEM images of the STZP film using various magnifications reveal the film morphology. At low magnification, the film appears in the cross-section image with a flat surface with a uniform thickness of about one micron. The absence of any major defects or irregularities can be observed. Increasing the magnification level reveals details in the thin film's cross-section. Distinct layers can be observed within the film, composed of five distinguishable layers with boundaries clearly visible at this magnification and thicknesses of the individual layers around 200 nm, increasing slightly from the first inner layer, close to the substrate, to the fifth, outer layer. EDX elemental analysis detects the film elements O, Pr, Sn, Ti, Zn, and Si element from the substrate. Measurements at different areas revealed a homogeneous composition of the film, with Sn:Ti:Zn:Pr cationic ratios of 60:20:15:5.

2.1.3. Raman

The Raman spectra of the STZP film based on the cassiterite lattice with significant substitution by lower valence cations (Zn^{2+}, $Pr^{2+/3+}$) can provide valuable information about the structural properties of the material. Cassiterite has several phonon modes that are typically observed in its Raman spectra. In the Raman spectrum of a cassiterite film with oxygen vacancies, these phonon modes will be present but may exhibit variations due to the structural changes caused by the vacancies. Group theory predicts four Raman-active modes for tin oxide with cassiterite structure. Several Raman bands are observed (Figure 2), out of which we can assign the one at 617 cm^{-1} to the A1g mode, the one at 450 cm^{-1} to the Eg mode, and the one at 780 cm^{-1} to the B2g mode [20–23].

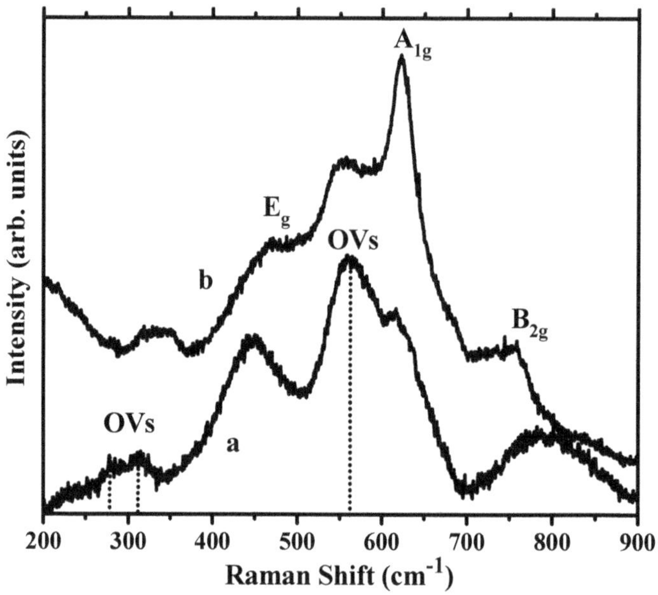

Figure 2. Raman spectra of the STZP film (**a**) and undoped SnO_2 (**b**), showing intrinsic Raman modes of cassiterite and OVs induced modes (marked by dotted lines).

However, additional Raman bands, marked with dotted lines in Figure 2, are not intrinsic Raman modes because they do not appear in bulk SnO_2 according to symmetry analysis. The presence of oxygen vacancies (OVs) gives rise to these additional Raman modes that are characteristic of defect-related vibrations [24]. These additional modes appear as distinct peaks in the Raman spectrum and can provide specific information about the presence and nature of the vacancies. The most intense additional band observed at ~560 cm^{-1} in Figure 2 has been reported by several studies of nanosized SnO_2 and assigned its origin to the activation of the Raman inactive A2g mode, related to the presence of OVs and local lattice disorder [24–27]. We attribute, therefore, the strongest observed mode at 560 cm^{-1} to surface in-plane OVs in cassiterite nanocrystallites, in agreement with previous studies. The presence of OVs can induce disorder and lattice strain into the cassiterite film, resulting in peak broadening in the Raman spectra. The very broad peaks observed in Figure 2 can be explained by the presence of oxygen vacancies induced by lower valence cations. We also observe a significant red shift in the characteristic A1g mode, from ~640 to 617 cm^{-1}. Shifts in the peak positions of the Raman bands compared to a purely stoichiometric tin dioxide film also arise due to local lattice distortions or changes in the bonding environment caused by the vacancies [24–26].

Furthermore, an additional OVs related band is observed at ~315 cm^{-1}. This band has been rarely observed in Raman studies of SnO_2 nanocrystals. Density functional theory (DFT) calculations [26] reported that this band appears only in non-stoichiometric cassiterite with a certain density of OVs (SnO_{2-x}) when the value of x is at least 0.06. We observe a broad split band extended towards lower frequencies, with a shoulder at 278 cm^{-1}, a position of the OVs-induced peak at low frequencies that according to Li et al. appears at values of x over 0.20. Our results confirm Li's calculations; considering the elemental composition of the film from EDX measurements ($Sn_{0.6}Ti_{0.2}Zn_{0.15}\ Pr_{0.05}O_{2-\delta}$) and the cation valencies determined by XPS measurements, we estimated an average value of $\delta = 0.19$ in our non-stoichiometric STZP film. Oxygen deficiency can cause a decrease in the ionic character of the Sn^{4+} state in the Sn-O bond and increases the density of the s-electron surrounding the Sn atom. This results in a decreased binding energy observed in the XPS spectra.

2.1.4. UV-Vis Spectroscopy

Room temperature UV-Vis spectra of STZP film is shown in Figure 2.

In order to determine the band gap energy, the absorption coefficient (α) was calculated according to Lambert's law which is indicated by the following relation:

$\alpha = (1/d) \times \ln(1/T)$, where T is the measured transmittance (%) and d—the thickness of the samples.

The band gap energy was calculated using the Tauc method for direct band gap semiconductors. The linear region of the rising part of the curve $[F(R) \times h\nu]^2$ vs. $h\nu$ was extrapolated to zero to obtain the band gap (Figure 3 inset), where $h\nu$ is the energy of the photons. Tauc's plot shows an indirect band semiconductor with $E_g = 3.83$ eV value similar to the literature value of the energy for SnO_2 [28].

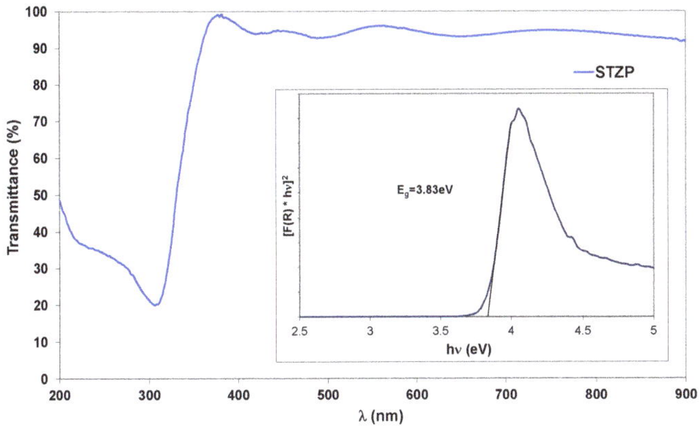

Figure 3. Room temperature UV-Vis spectra of STZP film annealed at 450 °C. The inset shows a corresponding optical band-gap plot.

2.1.5. XPS

X-ray Photoelectron Spectroscopy (XPS) analysis was performed to determine the elements present on the outermost surface layer (<10 nm) used as inorganic precursors, together with their corresponding chemical species. Thus, the high-resolution spectra for the most prominent transitions: Sn3d, Ti2p, Zn2p3/2, Pr3d5/2, and O1s were recorded.

Sn was detected on the surface as Sn^{4+} at the binding energy (BE) ~486.1 eV (Figure 4a) It is worth mentioning that this value was shifted towards the lower BEs side as compared to the standard photoelecton line at ~487 eV. The chemical shift could be attributed to the lattice distortions and/or changes in the bonding environment caused by the presence of the lower valence cations which generate oxygen vacancies, as was highlighted by Raman Spectroscopy. Moreover, this finding is supported by our previous works [22]. Thus, we found that $Sn3d_{5/2}$ is shifted towards lower BEs when cations such as In^{3+} and Cu^{1+} were incorporated in the SnO_2 lattice. Ti was detected as Ti^{4+} (BE~458.3 eV), (Figure 4b).

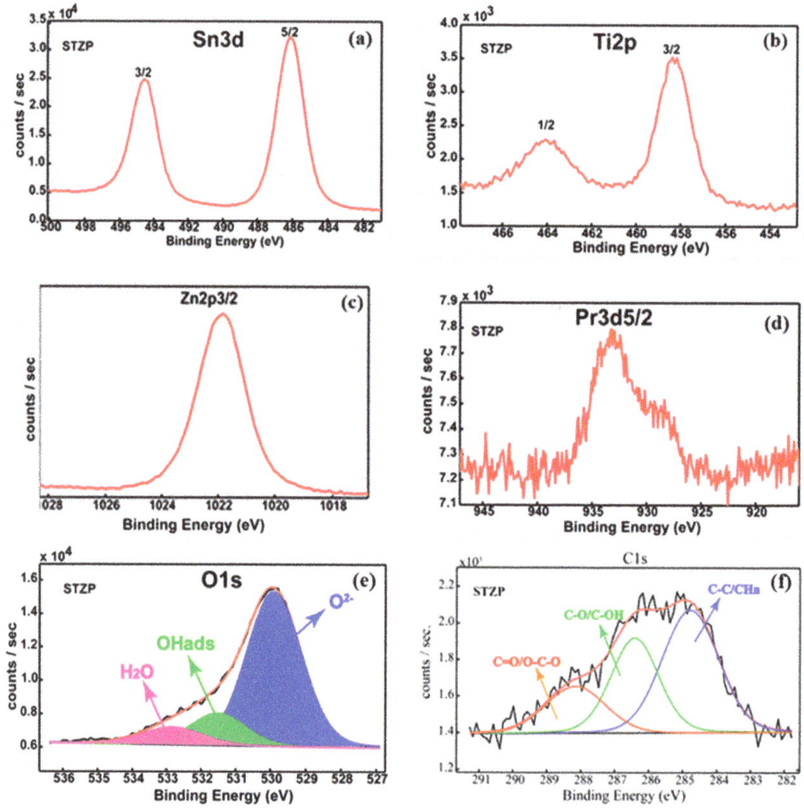

Figure 4. High-resolution spectra for the most prominent XPS transitions of the SZTP film Sn3d (**a**), Ti2p (**b**), Zn2p3/2 (**c**), Pr 3d5/2 (**d**), O1s (**e**), and C1s (**f**).

In line with the aforementioned statement, the value of the Ti2p3/2 BE is lower in comparison with the standard one (~458.8 eV), beyond our experimental errors (±0.2 eV), suggesting that an interaction should take place inside the SnO_2 lattice. We emphasize that the structural analysis does not reveal the formation of a secondary TiO_2 phase; therefore, Ti is incorporated into the cassiterite structure as shown by XRD results. Zn and Pr were detected on the surface as Zn^{2+} (BE~1021.8 eV) (Figure 4c) and Pr^{3+} (BE~933.0 eV) (Figure 4d), according to NIST XPS Databases and Handbook of X-ray Photoelectron spectroscopy [29].

The oxygen chemistry (Figure 4e) displays the presence of the lattice oxygen (BE~530.0 eV—~71%), OH groups adsorbed on the top of the surface layer (BE~531.5 eV—~19%), and adsorbed H_2O (BE~533.0 eV—~10%). The O1s spectrum clearly reveals a hydroxylated surface sample. One can notice that the hydroxylation degree of the surface is strongly affected by the synthesis protocols. Thus, our previous work highlighted that SnO_2 and Pd/SnO_2 sensors obtained by different chemical routes exhibit percentages of OH groups and H_2O in the range of 20–40% [30,31]. Figure 4f shows the C1s deconvoluted spectrum, which exhibits three components attributed as follows: (1) C-C/CHn at 284.8 eV—used for callibration; (2) C–O/C–OH at 286.4 eV and (3) C=O, O=C–O at 288.3 eV [29]. We emphasize that cation relative concentrations after quantification from HR XPS spectra reveal the following atomic ratio: Sn:Ti:Zn:Pr—70.4:20.0:8.1:1.5 in agreement with EDX elemental analysis.

2.2. Electrical Properties Sensitivity Measurements of STZP Film

The bulk, the grain boundaries, and the electrode contributions to the overall resistance of the films can be resolved by exploiting the differences in their responses to an applied alternating current (AC) signal.

The software Z-view was used to fit the impedances of specific electrical circuits to the Nyquist experimental data (Nyquist representations) [1,13,14,32].

The electrical responses of the films were observed for different concentrations of CO (between 0 and 12,000 ppm). The gas sensing properties are characterized in terms of resistance and gas-sensing response. The resistance of the STZP film was decreased upon exposure to reducing gas (CO) and increased upon exposure to air, indicating that they are n-type semiconductors.

The impedance spectroscopy measurements used to identify the electrical behavior of the STZP film were made in CO (in air) reducing environment, in the frequency range 100 Hz–1 MHz. Figure 5 shows the impedance spectra for samples at 350 °C in air and CO in air (500–2000 ppm). The observed two overlapping semicircles can be interpreted with the classical series circuit of two parallel resistance-constant phase elements (CPE).

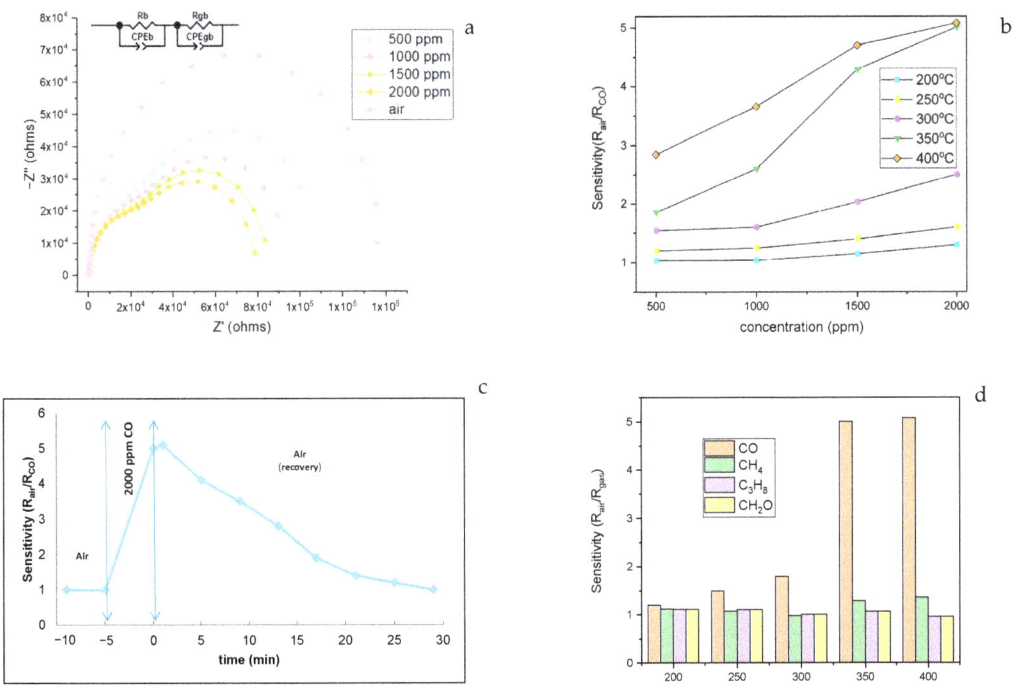

Figure 5. Impedance spectrum (Nyquist representation) frequency range 100 Hz to 1 MHz at 350 °C (**a**); sensitivity of STZP film at different CO concentrations (**b**); recovery time of STZP film at 350 °C for 2000 ppm CO in air (**c**); selectivity of STZP film to 2000 ppm (CO, CH_4, C_3H_8) and 20 ppm (CH_2O) (**d**).

In order to evaluate the sensitive properties as a sensor for the detection of CO, film tests were performed in the temperature range of 200–400 °C at different CO concentrations between 500 and 2000 ppm. The bulk resistance of the sample was determined from the interceptions of the first high-frequency arc with the real axis at low frequencies. However, with the increase in temperature, the grain boundary contribution decreased simultaneously.

For STZP film, the highest CO sensitivity (~5) was obtained for CO 2000 ppm concentrations at 350 and 400 °C, as shown in Figure 4b. The higher value of the sensitivity to CO of the film compared to the powder [1] is due to the lower treatment temperature of the mixed oxides.

Figure 4c shows the recovery time of the film at 2000 ppm CO and 350 °C. The response of the sensing film is the time needed to reach 90% of the resistance saturation value under exposure to the analyzed gas. The recovery time of the sensing film is the time needed to recover 10% of the original resistance value in air after exposure to the gas. The response for STZP film was found to be approximately 4 min and the recovery time was found to be approximately 20 min.

The correspondence of the semicircles obtained with the different regions of the sample, in this case the Nyquist representations, showed two semicircles for CO atmosphere [1]. The impedance parameters obtained by fitting experimental impedance data are listed in Table 2 (where Rgb is grain boundary resistance, Rb is bulk resistance, and CPE-T and CPE-P are constant phase parameters).

Table 2. The fitted parameters related to impedance measurement at 350 °C.

CO (ppm)	R_{gb}	CPE1-T	CPE1-P	R_b	CPE2-T	CPE2-P
500	41,923	5.68×10^{-10}	0.96204	57,931	1.38×10^{-10}	1.272
1000	33,000	5.20×10^{-10}	0.96767	44,720	1.29×10^{-10}	1.298
1500	21,052	5.49×10^{-10}	0.96591	42,905	1.26×10^{-10}	1.319
2000	18,181	5.41×10^{-10}	0.96777	38,829	1.30×10^{-10}	1.307
air	80,123	4.71×10^{-11}	1.253	44,574	7.79×10^{-10}	0.948

The sensor response strongly depends on the nature of the predominant defects involved in the conduction mechanism (mainly oxygen defects in oxide semiconductors).

2.2.1. Selectivity Tests of STZP Film

To demonstrate selectivity (Figure 4d), the film was also tested with other pollutant gases, such as methane (CH_4 2000 ppm), propane (C_3H_8 2000 ppm), and formaldehyde (CH_2O 20 ppm).

The selectivity coefficients (SCO/Sgas the sensor response ratio between the target gas—CO and other gas) are presented in Table 3.

Table 3. Selectivity coefficients.

Temperature (°C)	S_{CO}/S_{CH4}	S_{CO}/S_{C3H8}	S_{CO}/S_{CH2O}
200	1.07	1.15	1.08
250	1.40	1.40	1.36
300	1.83	2.23	1.79
350	3.12	4.40	3.76
400	3.55	5.48	4.99

2.2.2. Sensing Mechanism

Considering that the obtained film contains the highest proportion of SnO_2, the mechanism is similar to this one, and is shown in Figure 6.

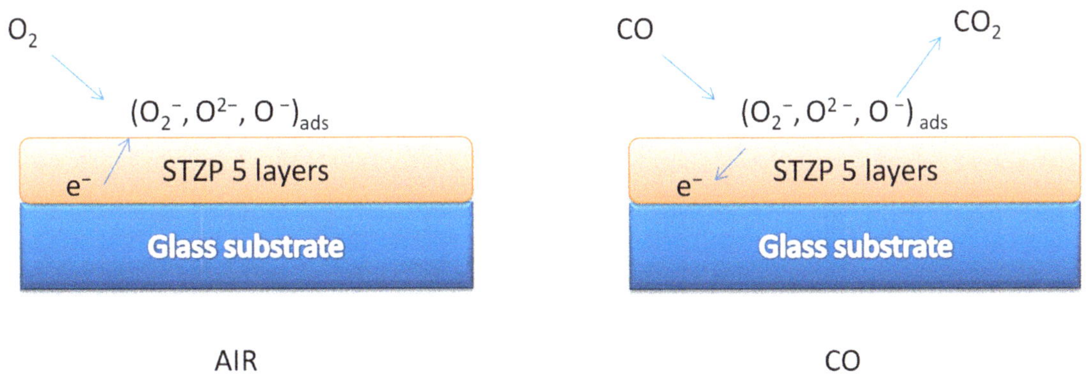

Figure 6. Intuitive schematic CO-sensing mechanism for STZP film.

The adsorbed oxygen on the surface can be of several forms: O_2, O_2^-, O^-, and O^{2-}. Of these species, O_2 is quite inactive because its activation energy is high; its concentration is also very low. The sensing mechanism can be explained as follows. Stable oxygen ions species were O_2^- below 200 °C, O^- between 200 and 300 °C, and O_2^- above 300 °C [33].

The reactions of the oxygen species with CO molecules at different operating temperatures can be described using the following equations:

$$2CO + O_2^- \rightarrow 2CO_2 + e^-$$

$$CO + O^- \rightarrow CO_2 + e^-$$

$$CO + O^{2-} \rightarrow CO_2 + 2e^-$$

The carbon monoxide which is the reducing agent, reacts rapidly with O^- presented on the surface, but very slowly with O_2^-.

The n-type gas sensor changes its resistance (R) as the concentration of adsorbed oxygen changes. Adsorbed oxygen from clean air will be consumed in contact with carbon monoxide, with the resulting decrease in R being used to estimate the carbon monoxide concentration. The sensor recovers its original resistance level when the carbon monoxide is turned off.

The sensor response strongly depends on the nature of the predominant defects involved in the conduction mechanism of STZP film (mainly oxygen defects in oxide semiconductors).

3. Conclusions

In summary, by a sol–gel deposition method, we successfully synthesized a mixed oxide thin film based on SnO_2 with Ti, Zn, and Pr incorporated in the lattice for enhanced CO sensing. The film was deposited by a spin-coating procedure on the glass substrate. Five layers were deposited and thermally treated at 280 °C after each step for consolidation. Finally, the thermal treatment was performed at 450 °C to obtain a crystalline film, as X-ray diffraction results confirmed.

The structural properties of the thin film obtained by XRD showed only the presence of SnO_2-cassiterite, with a crystallite size of 6.4 nm.

SEM analysis showed five distinct layers with visible boundaries with individual layer thicknesses around 200 nm, and elemental EDX analysis detected the elements present in the film O, Pr, Sn, Ti, Zn, and the element Si in the substrate. The Raman spectra of the STZP film revealed the cassiterite lattice with significant substitution with lower valence cations

(Ti^{4+}, Zn^{2+}, $Pr^{2+/3+}$), agreeing with the structural properties of the material shown by XRD analysis. The band gap energy of the STZP film calculated from UV-Vis spectroscopy was Eg = 3.83 eV, which is similar to the value of SnO_2.

The surface chemistry assessment, by XPS, highlighted the presence on the surface of all elements introduced from inorganic precursors. Sn^{4+}, Ti^{4+}, Zn^{2+}, and Pr^{3+} were detected on the surface after processing the Sn3d, Zn2p3/2, and Pr3d5/2 HR spectra. Surface hydroxylation was also proved by XPS; OH groups adsorbed and water were revealed by deconvolution of the O1s singlet.

Electrical measurements showed that the best sensitivity of STZP film to carbon monoxide (CO) was ~5 at 350 °C, the response was ~4 min, and the recovery time was ~20 min. STZP film had a good selectivity for CO in the presence of other pollutant gases (CH_4, C_3H_8, and CH_2O).

The results of this study provide insight into the potential of the mixed oxide thin film as a gas sensor material for various applications.

4. Materials and Methods

In this work, we obtained a film based on Sn, Ti, Zn, Pr, through a modified sol–gel method in accordance with the synthesis presented in the article [1]. In this sol–gel synthesis route, the inorganic precursors Tin chloride (IV) anhydrous, Titanium propoxide (IV), Zinc acetate dehydrated, and Praseodymium acetylacetonate (III) hydrate were dispersed and stabilized by using Polyvinylpyrolidone (PVP360000). As template we used a nonionic surfactant Triton X100 and Tripropylamine (TPA) was added as a chelating agent.

The films based on Sn, Ti, Zn, and Pr mixed oxide were obtained by the spin-coating deposition method. Five layers were deposited on a glass substrate, with a consolidation heat treatment at 280 °C for 30 min for each deposited layer followed by a final treatment at 450 °C for 4 h. The films are called STZP.

X-ray diffraction (XRD) measurements were performed using a Rigaku Ultima IV-type diffractometer with Cu Kα radiation. The equipment operated at parameters of 40 kV and 30 mA, in the range of $15° < 2\theta < 90°$, being equipped with a device for measurements at low incidence angles $\omega = 0.5°$.

The morphological analysis of the obtained films by scanning electron microscopy (SEM) was performed using a 3D FEI Quanta microscope, coupled with an energy-dispersive X-ray (EDX) spectrometer, at operating voltages of 5 kV, for secondary electron SEM images and 20 kV for elemental analysis EDX spectra.

Raman spectra were acquired in a Horiba Jobin–Yvon LabRam HR spectrometer with a He-Cd laser, using 325 nm as excitation wavelength. The power on the sample was kept below 0.5 mW to avoid sample damage due to irradiation. UV Raman spectra were collected using a 40× objective, in a UV-enhanced CCD camera and corrected by subtracting the background spectrum. The range of Raman shift recorded was between 200 and 700 cm^{-1} and the integration time of the measurements was 60 s. Ultraviolet-Visible (UV-VIS) Spectroscopy Diffuse reflectance spectroscopy was used to evaluate the optical properties and the band gap of STZP film. The spectra were recorded using a Perkin Elmer Lambda 35 spectrophotometer, in the range 200–1100 nm, with spectralon as a reference.

X-ray photoelectron spectroscopy (XPS) Surface analysis performed by X-ray photoelectron spectroscopy (XPS) was carried out on PHI Quantera equipment with a base pressure in the analysis chamber of 10-9 Torr. The X-ray source was monochromatized Al Kα radiation (1486.6 eV), while the unavoidable charging effect was minimized by using a dual beam (electrons and Ar ions) as a neutralizer. The Binding Energy (BE) scale was calibrated by C1s = 284.8 eV photoelectron line from the adsorbed environmental hydrocarbon on the surface (C–C or (CH)n bondings).

The electrical measurements were realized using a four-probe method AC impedance spectrometer with an electrochemical interface (Solartron 1260). The AC bias amplitude applied was 500 mV. Electrochemical Impedance spectra (EIS) were recorded in the frequency range of 1 MHz to 100 Hz from room temperature to 400 °C with a ProboStat cell (NorECs, Oslo, Norway).

The films were placed in a controlled atmosphere with a continuous gas flow of 177 mL/min (containing air and CO in He). Gas flows were provided by a calibrated system of mass flow controllers. Complex impedance measurements were used for the understanding of the mechanism of gas/solid interactions and the determination of the active regions in the films (surface, grain, and grain boundaries) that are involved in the detection of analyzed gas. The impedances $Z = Z' + j Z''$ (Z' and Z'' being, respectively, the real and imaginary components) were represented using Nyquist plots (Z'' vs. Z').

Author Contributions: Conceptualization, I.D., C.H., J.M.C.M., P.O. and S.S.; methodology, I.D. and S.S.; software, C.H.; validation, I.D., C.H., J.M.C.M., P.O. and S.S.; formal analysis, C.H. and J.M.C.M.; investigation, I.D., C.H., J.M.C.M., P.O. and S.S.; data curation, I.D., C.H., J.M.C.M., P.O. and S.S.; writing—original draft preparation, I.D., C.H., J.M.C.M., P.O. and S.S.; writing—review and editing C.H. and J.M.C.M.; visualization, I.D., C.H., J.M.C.M. and S.S.; supervision, C.H. and S.S. All authors have read and agreed to the published version of the manuscript.

Funding: This work was supported by the Romanian National Authority for Scientific Research and Innovation, UEFISCDI, project number PN-III-P2-2.1-PTE-2021-0592 (BioenergCell).

Institutional Review Board Statement: Not applicable.

Informed Consent Statement: Not applicable.

Data Availability Statement: The datasets generated during and/or analyzed during the current study are available from the corresponding author on reasonable request.

Acknowledgments: The paper was carried out within the research program "Science of Surfaces and Thin Layers" of the "Ilie Murgulescu" Institute of Physical Chemistry of the Romanian Academy.

Conflicts of Interest: The authors declare no conflict of interest.

References

1. Dascalu, I.; Culita, D.; Calderon-Moreno, J.M.; Osiceanu, P.; Hornoiu, C.; Anastasescu, M.; Somacescu, S.; Gartner, M. Structural, Textural, Surface Chemistry and Sensing Properties of Mesoporous Pr, Zn Modified SnO_2–TiO_2 Powder Composites. *Ceram. Int.* **2016**, *42*, 14992–14998. [CrossRef]
2. Mahajan, S.; Jagtap, S. Metal-Oxide Semiconductors for Carbon Monoxide (CO) Gas Sensing: A Review. *Appl. Mater. Today* **2020**, *18*, 100483. [CrossRef]
3. Tonezzer, M.; Thi Thanh Le, D.; Van Duy, L.; Hoa, N.D.; Gasperi, F.; Van Duy, N.; Biasioli, F. Electronic Noses Based on Metal Oxide Nanowires: A Review. *Nanotechnol. Rev.* **2022**, *11*, 897–925. [CrossRef]
4. Kononova, I.; Moshnikov, V.; Kononov, P. SnO_2-Based Porous Nanomaterials: Sol-Gel Formation and Gas-Sensing Application. *Gels* **2023**, *9*, 283. [CrossRef]
5. Mitran, R.-A.; Culita, D.C.; Atkinson, I. Thermal Stability Enhancement of Mesoporous SBA-15 Silica through Nanoconfinement of Ceria Nanoparticles. *Microporous Mesoporous Mater.* **2020**, *306*, 110484. [CrossRef]
6. Chakrabarty, N.; Dey, A.; Krishnamurthy, S.; Chakraborty, A.K. CeO_2/Ce_2O_3 Quantum Dot Decorated Reduced Graphene Oxide Nanohybrid as Electrode for Supercapacitor. *Appl. Surf. Sci.* **2021**, *536*, 147960. [CrossRef]
7. Wang, C.-M.; Wang, J.-F.; Chen, H.-C.; Su, W.-B.; Zang, G.-Z.; Qi, P.; Ming, B.-Q. Effects of Pr_2O_3 on the Nonlinear Electrical Characteristics and Dielectric Properties of $SnO_2{\cdot}Co_2O_3{\cdot}Ta_2O_5$ Varistor Systems. *Solid State Commun.* **2004**, *132*, 163–167. [CrossRef]
8. Zhang, G.; Liu, M.; Fan, G.; Zheng, L.; Li, F. Efficient Role of Nanosheet-Like Pr_2O_3 Induced Surface-Interface Synergistic Structures over Cu-Based Catalysts for Enhanced Methanol Production from CO_2 Hydrogenation. *ACS Appl. Mater. Interfaces* **2022**, *14*, 2768–2781. [CrossRef]
9. Hu, D.; Li, X.; Zhang, L.; Snetkov, I.; Chen, P.; Dai, Z.; Balabanov, S.; Palashov, O.; Li, J. Terbium (III) Oxide (Tb_2O_3) Transparent Ceramics by Two-Step Sintering from Precipitated Powder. *Magnetochemistry* **2022**, *8*, 73. [CrossRef]
10. Ibáñez, J.; Sans, J.Á.; Cuenca-Gotor, V.; Oliva, R.; Gomis, Ó.; Rodríguez-Hernández, P.; Muñoz, A.; Rodríguez-Mendoza, U.; Velázquez, M.; Veber, P.; et al. Structural and Lattice-Dynamical Properties of Tb_2O_3 under Compression: A Comparative Study with Rare Earth and Related Sesquioxides. *Inorg. Chem.* **2020**, *59*, 9648–9666. [CrossRef]
11. Cristina de Oliveira, R.; Cabral, L.; Cabral, A.C.; Almeida, P.B.; Tibaldi, N.; Sambrano, J.R.; Simões, A.Z.; Macchi, C.E.; Moura, F.; Marques, G.E.; et al. Charge Transfer in Pr-Doped Cerium Oxide: Experimental and Theoretical Investigations. *Mater. Chem. Phys.* **2020**, *249*, 122967. [CrossRef]

12. Kim, J.-S.; Na, C.W.; Kwak, C.-H.; Li, H.-Y.; Yoon, J.W.; Kim, J.-H.; Jeong, S.-Y.; Lee, J.-H. Humidity-Independent Gas Sensors Using Pr-Doped In_2O_3 Macroporous Spheres: Role of Cyclic Pr^{3+}/Pr^{4+} Redox Reactions in Suppression of Water-Poisoning Effect. *ACS Appl. Mater. Interfaces* **2019**, *11*, 25322–25329. [CrossRef]
13. Dascalu, I.; Calderon-Moreno, J.M.; Osiceanu, P.; Bratan, V.; Hornoiu, C.; Somacescu, S. Amorphous ZnO Modified Anatase TiO_2 Thin Films Templated by Tripropylamine and Their Electrical Properties. *Thin Solid Films* **2021**, *729*, 138697. [CrossRef]
14. Dascalu, I.; Hornoiu, C.; Calderon-Moreno, J.M.; Enache, M.; Culita, D.; Somacescu, S. Sol-Gel Synthesis of ZnO/Zn_2-$XFexTiO_4$ Powders: Structural Properties, Electrical Conductivity and Dielectric Behavior. *J. Sol-Gel Sci. Technol.* **2018**, *86*, 151–161. [CrossRef]
15. Nguyen, C.M.; Rao, S.; Yang, X.; Dubey, S.; Mays, J.; Cao, H.; Chiao, J.-C. Sol-Gel Deposition of Iridium Oxide for Biomedical Micro-Devices. *Sensors* **2015**, *15*, 4212–4228. [CrossRef] [PubMed]
16. Belaid, W.; Houimi, A.; Zaki, S.E.; Basyooni, M.A. Sol-Gel Production of Semiconductor Metal Oxides for Gas Sensor Applications. In *Sol-Gel Method—Recent Advances*; Singh, D.J.P., Acharya, D.S.S., Kumar, D.S., Dixit, D.S.K., Eds.; IntechOpen: Rijeka, Croatia, 2023; ISBN 978-1-80355-415-0.
17. Kim, J.-H.; Mirzaei, A.; Kim, H.W.; Kim, S.S. Low-Voltage-Driven Sensors Based on ZnO Nanowires for Room-Temperature Detection of NO_2 and CO Gases. *ACS Appl. Mater. Interfaces* **2019**, *11*, 24172–24183. [CrossRef] [PubMed]
18. Vatandoust, L.; Habibi, A.; Naghshara, H.; Aref, S.M. Fabrication and Investigation of $TiO_{1.5}$/ZnO Nanocomposite Nanosensor for Detection of CO and CH_4 Gases. *Surf. Interfaces* **2022**, *31*, 102001. [CrossRef]
19. Gates-Rector, S.; Blanton, T. The Powder Diffraction File: A quality materials characterization database. *Powder Diffr.* **2019**, *34*, 352–360. [CrossRef]
20. Diéguez, A.; Romano-Rodríguez, A.; Vilà, A.; Morante, J.R. The Complete Raman Spectrum of Nanometric SnO_2 Particles. *J. Appl. Phys.* **2001**, *90*, 1550–1557. [CrossRef]
21. Scott, J.F. Raman Spectrum of SnO_2. *J. Chem. Phys.* **2003**, *53*, 852–853. [CrossRef]
22. Stanoiu, A.; Somacescu, S.; Calderon-Moreno, J.M.; Teodorescu, V.S.; Florea, O.G.; Sackmann, A.; Simion, C.E. Low Level NO_2 Detection under Humid Background and Associated Sensing Mechanism for Mesoporous SnO_2. *Sens. Actuators B Chem.* **2016**, *231*, 166–174. [CrossRef]
23. Maftei, A.E.; Buzatu, A.; Damian, G.; Buzgar, N.; Dill, H.G.; Apopei, A.I. Micro-Raman—A Tool for the Heavy Mineral Analysis of Gold Placer-Type Deposits (Pianu Valley, Romania). *Minerals* **2020**, *10*, 988. [CrossRef]
24. Yu, K.N.; Xiong, Y.; Liu, Y.; Xiong, C. Microstructural Change of Nano-SnO_2 Grain Assemblages with the Annealing Temperature. *Phys. Rev. B* **1997**, *55*, 2666–2671. [CrossRef]
25. Shek, C.H.; Lin, G.M.; Lai, J.K.L. Effect of Oxygen Deficiency on the Raman Spectra and Hyperfine Interactions of Nanometer SnO_2. *Nanostructured Mater.* **1999**, *11*, 831–835. [CrossRef]
26. Li, T.H.; Liu, L.Z.; Li, X.X.; Wu, X.L.; Chen, H.T.; Chu, P.K. Oxygen Vacancy Density-Dependent Transformation from Infrared to Raman Active Vibration Mode in SnO_2 Nanostructures. *Opt. Lett.* **2011**, *36*, 4296–4298. [CrossRef] [PubMed]
27. Liu, L.Z.; Wu, X.L.; Gao, F.; Shen, J.C.; Li, T.H.; Chu, P.K. Determination of Surface Oxygen Vacancy Position in SnO_2 Nanocrystals by Raman Spectroscopy. *Solid State Commun.* **2011**, *151*, 811–814. [CrossRef]
28. Abdelkrim, A.; Rahmane, S.; Nabila, K.; Hafida, A.; Abdelouahab, O. Polycrystalline SnO_2 Thin Films Grown at Different Substrate Temperature by Pneumatic Spray. *J. Mater. Sci. Mater. Electron.* **2017**, *28*, 4772–4779. [CrossRef]
29. Naumkin, A.V.; Kraut-Vass, A.; Gaarenstroom, S.W.; Powell, C.J. *NIST X-ray Photoelectron Spectroscopy Database*; NIST Standard Reference Database 20, version 4.1; National Institute of Standards and Technology: Gaithersburg, MD, USA, 2000. [CrossRef]
30. Stanoiu, A.; Simion, C.E.; Sackmann, A.; Baibarac, M.; Florea, O.G.; Osiceanu, P.; Teodorescu, V.S.; Somacescu, S. Networked mesoporous SnO_2 nanostructures templated by Brij®35 with enhanced H2S selective performance. *Microporous Mesoporous Mater.* **2018**, *270*, 93–101. [CrossRef]
31. Stanoiu, A.; Ghica, C.; Somacescu, S.; Kuncser, A.C.; Vlaicu, A.M.; Mercioniu, I.F.; Florea, O.G.; Simion, C.E. Low temperature CO sensing under infield conditions with in doped Pd/SnO_2. *Sens. Actuators B Chem.* **2020**, *308*, 127717. [CrossRef]
32. Dascalu, I.; Somacescu, S.; Hornoiu, C.; Calderon-Moreno, J.M.; Stanica, N.; Stroescu, H.; Anastasescu, M.; Gartner, M. Sol-Gel Zn, Fe Modified SnO_2powders for CO Sensors and Magnetic Applications. *Process Saf. Environ. Prot.* **2018**, *117*, 722–729. [CrossRef]
33. Gong, H.; Hu, J.Q.; Wang, J.H.; Ong, C.H.; Zhu, F.R. Nano-Crystalline Cu-Doped ZnO Thin Film Gas Sensor for CO. *Sensors Actuators B Chem.* **2006**, *115*, 247–251. [CrossRef]

gels

Article

A SEM-EDX Study on the Structure of Phenyl Phosphinic Hybrids Containing Boron and Zirconium

Petru Merghes [1], Narcis Varan [1], Gheorghe Ilia [2], Iosif Hulka [3] and Vasile Simulescu [2,*]

[1] University of Life Sciences "King Michael I" from Timisoara, 119 Calea Aradului, 300645 Timisoara, Romania; petrumerghes@usvt.ro (P.M.); narcisvaran@usvt.ro (N.V.)
[2] Faculty of Chemistry, Biology, Geography, West University of Timisoara, 16 Pestalozzi Street, 300115 Timisoara, Romania; gheorghe.ilia@e-uvt.ro
[3] Research Institute for Renewable Energies, Politehnica University Timisoara, 38 Gavriil Musicescu, 300501 Timisoara, Romania; iosif.hulka@upt.ro
* Correspondence: vasile.simulescu@e-uvt.ro

Abstract: The SEM-EDX method was used to investigate the structure and morphology of organic–inorganic hybrids containing zirconium, boron and phosphorus compounds, synthesized by the sol–gel method. We started by using, for the first time together, zirconyl chloride hexa-hydrate ($ZrOCl_2 \cdot 6H_2O$), phenyl phosphinic acid and triethyl borate as precursors and reagents, at different molar ratios. The obtained hybrids showed a very high thermal stability and are not soluble in water or in organic solvents. As a consequence, such hybrid solid materials are suitable for applications at high temperatures. The obtained hybrids have complex 3D structures and form organic–inorganic networks containing Zr-O-Zr, Zr-O-P and Zr-O-B bridges. Such organic–inorganic networks are also expected to form supramolecular structures and to have many potential applications in different fields of great interest such as catalysis, medicine, agriculture, energy storage, fuel cells, sensors, electrochemical devices and supramolecular chemistry.

Keywords: sol–gel synthesis; organic–inorganic hybrids; phenyl phosphinic acid; zirconium; boron; SEM-EDX

1. Introduction

The sol–gel process is part of the green chemistry concept, due to the fact that the syntheses take place at room temperature, in mild conditions and by using green solvents (in general, water is used, as in the present work, but sometimes, alcohol or mixtures of alcohol:water at different volume ratios could also be involved). The sol–gel technique is a modern synthesis, which was developed and used in the last decades for obtaining hybrid compounds containing different functional groups (alcoholic, carboxylic, sulphonic groups and so on) [1–3]. The sol–gel process is called „Chimie douce" due to the mild conditions in which it takes place and also due to the use of green solvents. Organic–inorganic hybrid materials, sometimes called inorganic polymers or organic–inorganic networks, contain an inorganic part and organic moieties. This makes these hybrid compounds very interesting materials because such networks possess the properties of both, that is, the rigid inorganic backbone and the greater flexibility of organic groups. Due to the surface protons of these organic groups and due to their complex structures, the obtained organic–inorganic hybrids show solid state proton conduction [4–7]. Furthermore, the organic–inorganic hybrids are not soluble in most organic solvents or in water and have high thermal stability [8]. All of these properties make them very useful for a lot of applications of great interest, as fuel cells, energy storage, different sensors, catalysts, water electrolysis units, other electrochemical devices and so on. By using the sol–gel method, several compounds could be synthesized, starting from phosphonic acids, with a lot of potential applications.

Due to their potential application, such organic–inorganic hybrids containing phosphorus are currently of huge interest. The organic–inorganic hybrid networks are novel materials with promising applications in electrochemical devices and in the fields of catalysis, medicine, or agriculture [1,9–22], as previously mentioned. Hybrids containing phosphorus and zirconium have already been synthesized by using the sol–gel method in different studies, and the results have been published and are available in the literature [1–4,8,13–21]. Alberti et al. [2,4], for example, synthesized different alkyl and phenyl phosphonates containing zirconium, as porous hybrid organic–inorganic materials, by using metal phosphonates as precursors [1–4]. It was proved that the P/M ratio has a significant influence on the hybrid material porosity (pore shape, pore number, pore size, pore surface distribution and so on) [23–32].

Other examples of using the sol–gel synthesis are the studies performed and published by Vioux et al. [8,14,16,18,19,31]. They used a two-step sol–gel method for the synthesis of ZrO_2- and TiO_2-phenylphosphonate hybrids, as follows:

- The condensation reaction of a phosphonic acid with a metal alkoxide under anhydrous conditions.
- The hydrolysis–condensation reactions of the remaining metal alkoxide groups [27,31,32].

As a consequence, it was observed and proved that M-O-P (metal–oxygen–phosphorus) bridges are obtained during the condensation step and M-O-M (metal–oxygen–metal) bridges are obtained during the hydrolysis–condensation reaction, in the structure of the organic–inorganic hybrids [21,25–40]. On the other hand, it should be mentioned that the condensation of two or more P-O-H groups, in order to obtain P-O-P bridges, is not possible under the mild conditions used for the sol–gel synthesis [27,29,31,33].

The sol–gel synthesis of hybrid materials containing phosphorus involves the reaction of metal precursors with different phosphonate molecules (phosphonic acids, phosphonic acid derivatives, biphosphonic acids, polyphosphonic acids and derivatives, or phosphonate sodium or ammonium salts in aqueous medium or anhydrous conditions) [27]. These phosphonate molecules are called "coupling molecules" due to their role in the sol–gel process. A templating agent could also be used in order to control and/or change the pore size and, therefore, the materials' porosity [34–39]. Small amounts of some organic additives (for example, different co-polymers containing ethyleneoxide groups such as PEG, with different molar mass and polydispersity, and also non-ionic surfactants or cationic surfactants, cyclodextrins) could be very useful for the improvement of the hybrid's texture [40].

The structure of the hybrid materials containing phosphorus compounds was studied, as reported in the literature, by several spectroscopic methods, such as FTIR [27,34], X-ray photoelectron spectroscopies [35], SEM, TEM, TG, EDX and XRD.

2. Results and Discussion

The organic–inorganic hybrid materials synthesized in the present work were obtained by using the sol–gel method, at room temperature, by using water as solvent. It was possible to use a green solvent as water is used because all the precursors and reagents involved in the sol–gel syntheses from the present work are water soluble. Five organic–inorganic hybrids (S1–S5) were synthesized by using zirconyl chloride hexa-hydrate, triethyl borate and phenyl phosphinic acid (Figure 1) for the first time together in a sol–gel process. All the hybrids were synthesized by using the sol–gel method in mild conditions, at room temperature. The products were subsequently analyzed by SEM-EDX technique.

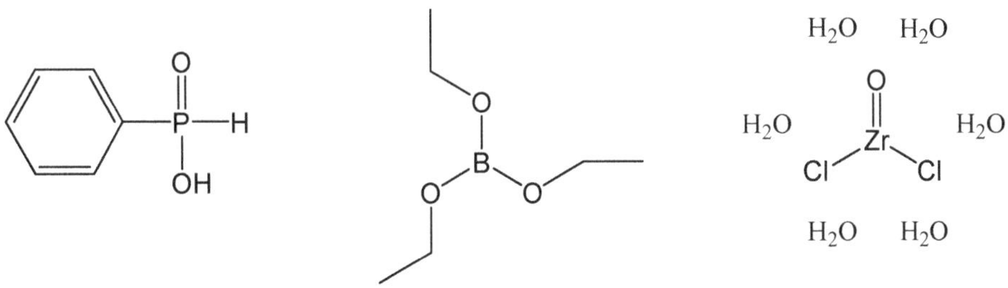

(a) (b) (c)

Figure 1. The structure of the used reagents: (**a**) phenyl phosphinic acid, (**b**) triethyl borate $B(OEt)_3$; (**c**) zirconyl chloride hexa-hydrate $ZrOCl_2 \cdot 6H_2O$.

The sol–gel method was chosen to be used in the present work because it is the most efficient, yet also the greenest method (it is part of the green chemistry concept) for obtaining such organic–inorganic hybrids in mild conditions. The efficiency of the sol–gel method as a green synthesis was already proved (including its mechanism) by Vioux et al., Mutin et al., Guerero et al. Clearfield et al., Maillet et al. and Mehring et al. [6,11,12,14,16,18,19,21,31,38,39], as well as many more researchers involved in the fields of organic–inorganic hybrid synthesis and characterization.

All the syntheses (S1–S5) were carried out for 6 h at room temperature, under continuous stirring, by using water as solvent. The used reagents are soluble in water, but the obtained organic–inorganic hybrids are not. This is the first indication that the synthesis takes place, and this is also very helpful for the final separation step (sedimentation, filtration, washing and drying). At the beginning, only the soluble reagents are present, and therefore, in the flask, a homogeneous solution is obtained. When the hybrid appears during the synthesis, due to its insolubility, it leads first to the obtaining of a suspension (with a visible turbidity) and finally a sediment on the bottom of the used flask.

Therefore, the sol–gel syntheses were performed at room temperature, in water, as follows:

- Hybrids S1–S3 were obtained from phenyl phosphinic acid and $ZrOCl_2 \cdot 6H_2O$, in water.
- Hybrids S4 and S5 were synthesized from phenyl phosphinic acid, $ZrOCl_2 \cdot 6H_2O$ and triethyl borate at different molar ratios, also in aqueous solution.

When the sol–gel synthesis was complete, the obtained products were separated by filtration and subsequently washed several times with water. The used volume of water for washing was minimal. The water, for the sol–gel syntheses and washing, was double distilled just before use. The washing process removed all the possible unreacted (if excessive) precursors, which are highly soluble in water, as mentioned already. The reagents in excess were thereby recovered and the synthesized hybrids were purified. Then, the organic–inorganic hybrids were dried for 6 h at 80 °C in an oven, in order to completely remove the water retained on the products (due to their porosity).

The washing, filtration and drying steps took place consecutively. Considering that the drying took place over 6 h at a temperature of 80 °C, no burning occurred and there was no thermal decomposition. The hybrid compounds were very stable from a thermal point of view. The main aim of the above-mentioned procedures (washing, filtration, drying) was the separation and purification of the insoluble hybrid product, from (if it is the case) unreacted precursors and reagents (usually if in excess). The reagents and precursors are soluble, while the hybrids are not. Finally, the pure dried organic–inorganic hybrid was obtained. This product (for each of the syntheses, S1–S5) was further analyzed by SEM-EDX. On the other hand, the recovered reagents, when necessary (excessive reagents, unreacted precursors and so on), were further separated and purified in order to be used for other syntheses.

The phenyl groups from the structure of phenyl phosphinic acid are very stable in the mild conditions used for the sol–gel process, and they do not react, as we observed in our previous work [8,9,15], except when using a different phosphorus source and therefore a different organic moiety (phenyl phosphonic acid or vinyl phosphonic acid, in these other cases). In other words, the mild conditions used are not sufficient to make the phenyl groups react. Therefore, the organic moiety from the hybrid's structures was derived from the phenyl phosphinic acid, containing the unreacted and very stable phenyl radicals.

On the other hand, phosphorus, zirconium and boron atoms are always connected via an oxygen atom to the hybrid's structure and finally lead to the obtaining of a 3D organic–inorganic network. Bridges like P-O-Zr, Zr-O-Zr or B-O-Zr finally lead to the obtaining of the organic–inorganic hybrid, forming the bond between the organic and inorganic moieties. Hybrid S1 was obtained with a phenyl phosphinic acid:$ZrOCl_2$ 1:1 molar ratio, while for hybrid S2, phosphinic acid was in excess, and similarly for hybrid S3, zirconium chloride was in excess. The zirconium excess (S3) leads to the formation of more Zr-O-Zr bridges. It is expected that hybrid S2 obtained at a P/Zr 2:1 ratio contains both P-O-Zr and Zr-O-Zr bridges in approximately equal ratio. Hybrid S3 obtained with a zirconium excess at a P/Zr 1:2 ratio has more Zr-O-Zr bridges than hybrids S1 and S2.

In general, it is good to have an excess of phosphorus, and as previously mentioned, the ideal P/M ratio is 2 [11]. When the ratio P/Zr is 2:1, as it is for hybrid S2, it is more likely to obtain both P-O-Zr and Zr-O-Zr bridges in more or less equal ratios. Thus, it is good to have a P/Zr ratio of 2:1 if we want to obtain a hybrid with both P-O-Zr and Zr-O-Zr bridges in an equivalent proportion. But if a hybrid with more P-O-Zr bridges is needed, or one with more Zr-O-Zr bridges is the goal of the sol–gel process, we can change the ratio of the precursors used (for example, an excess of Zr will lead to more Zr-O-Zr bridges). As we can observe, bonds like Zr-O-Zr are also present, but no P-P, Zr-Zr, P-Zr or P-O-P bridges occurred (such bonds could not be obtained in the mild conditions used in the present work for the sol–gel process, as it is also not possible to involve them in the sol–gel reactions or to decompose the phenyl radicals from phenyl phosphinic acid).

Therefore, the excess of one component or the other significantly influenced the structure of the synthesized hybrid and also its properties, first, by changes to the ratio and the number of P-O-Zr and Zr-O-Zr bridges, and then as a consequence, the pore number, the pore shape and, therefore, the entire material porosity. Thus, the organic–inorganic hybrids S1–S3 contain phosphorus and zirconium. In addition, hybrids S4 and S5 synthesized in the presence of triethyl borate, which also contains boron. As already mentioned, phosphorus, zirconium and boron atoms are always connected via an oxygen atom on the hybrid's structure and finally lead to the obtaining of 3D organic–inorganic networks (because in the mild conditions used for the sol–gel process, no direct bonds like P-P, P-Zr, P-B or Zr-B could be obtained). When boron is added at different ratios (S4 and S5), Zr-O-B bridges are also obtained. As can be observed, only zirconium could be involved in bridges with phosphorus (P-O-Zr), because at room temperature and in the mild conditions used for this sol–gel process, it is not possible to obtain a P-O-B bond. Therefore, while it is expected that there will be competition between boron and zirconium for phosphorus atoms on the hybrids' structure (for hybrids S4 and S5, both containing boron), in fact, only zirconium could be connected to phosphorus via an oxygen atom (P-O-Zr bridges). Then, no P-O-B and also no B-O-B are obtained. Practically boron is involved only in Zr-O-B bridges. Moreover, phosphorus is contained only in Zr-O-P bridges. In this case of obtaining hybrid materials containing P, Zr and B, an excess of zirconium is needed, as in the case of hybrid S5 synthesized at a phenyl phosphinic acid:$ZrOCl_2$:$B(OEt)_3$ ratio of 1:2:1. In this way, zirconium could be involved in P-O-Zr, Zr-O-Zr and also in B-O-Zr bridges. If there is an excess of phosphorus and/or boron (or even at equal ratio, as for hybrid material S4), there is expected to be strong competition between phosphorus and boron atoms for zirconium atoms, in order to obtain P-O-Zr or B-O-Zr bridges, or more likely both. The P-O-Zr bridges are in this case the only way to incorporate the organic moieties in the structure of the hybrid materials. When zirconium is excessive (S3 and S5), it is expected to obtain more

Zr-O-Zr bridges than P-O-Zr bridges, and as a consequence, these hybrids contain in this case a lower ratio of organic moieties in comparison with the inorganic part. If they are regarded as inorganic co-polymers, there will be more inorganic parts at one organic moiety in their structures. For hybrid S5, the ratio of the organic moieties is even lower, because it also contains boron, which will lead to obtaining B-O-Zr bridges and will be in direct competition with phosphorus and zirconium as well. What is very interesting in all the synthesized hybrids in the present work, is the fact that even the presence of zirconium will lead to obtaining bridges with zirconium. Therefore, the ratios of organic–inorganic moieties are in general higher for the inorganic part.

The phenyl groups from the structure of phenyl phosphinic acid are very stable, as we observed in our previous work when using different phosphonic acids [8,15]. It should be pointed out and given due consideration that the sol–gel syntheses were performed in mild conditions, in water as solvent and at room temperature. In these mild conditions, the phenyl groups do not react, and as a consequence, they will be found on the structure of the synthesized hybrids. Changing the phosphorus source makes a significant difference when compared with our previous work, where we used, for example, phenyl phosphonic acid (it contains two OH groups). The phenyl phosphinic acid used in the present work contains only one OH group directly connected to a phosphorus atom (Figure 1).

The porosity of the obtained materials and the pore shape and number on their surface is strongly influenced by the phosphorus source and by the ratio of the used reagents. Most of the SEM images generally showed a similar morphology for the obtained hybrids, with several compact structures (Figures 2–6).

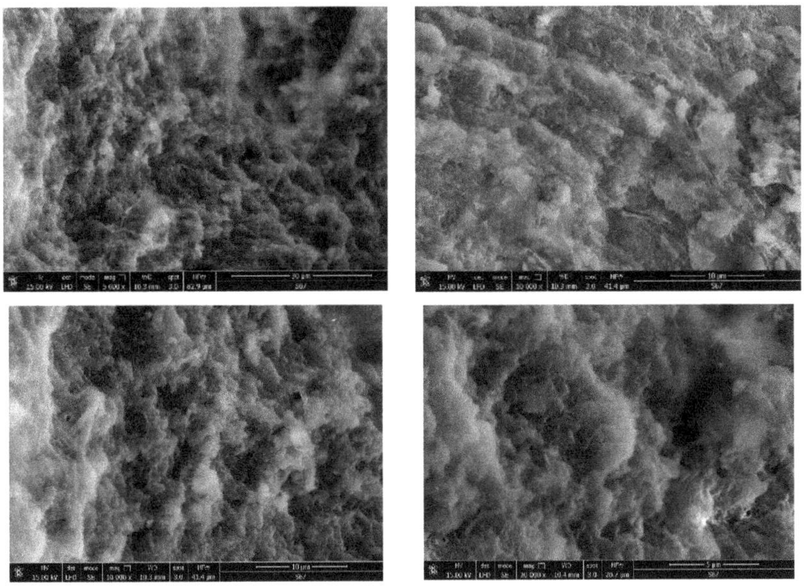

Figure 2. Hybrid material S1 analysed by SEM.

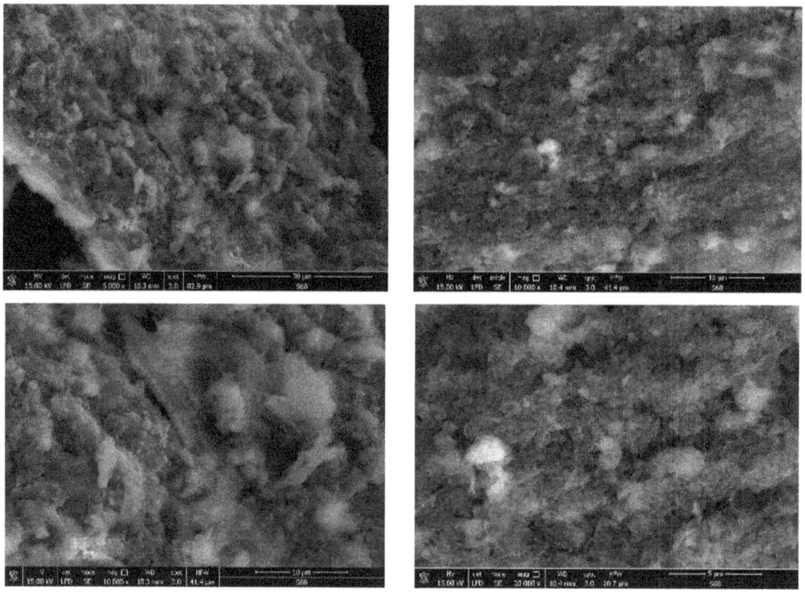

Figure 3. SEM images of the hybrid compound S2.

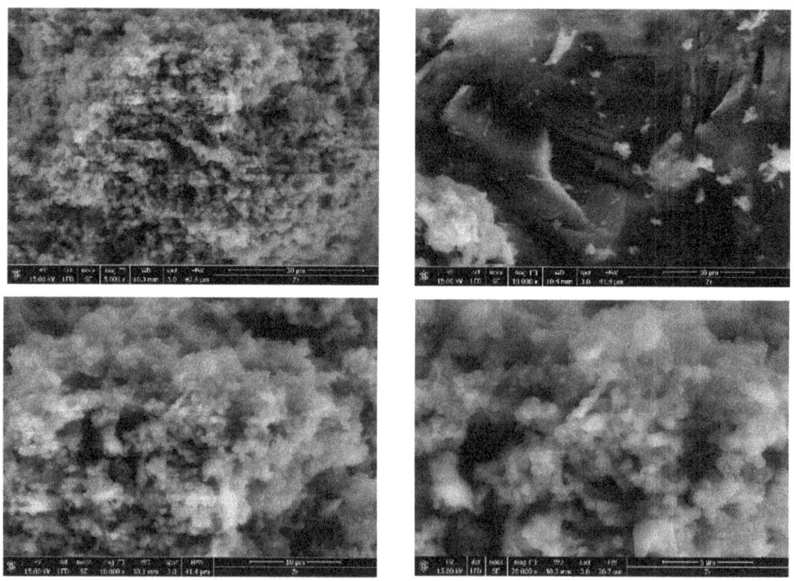

Figure 4. Results obtained by using SEM for hybrid S3.

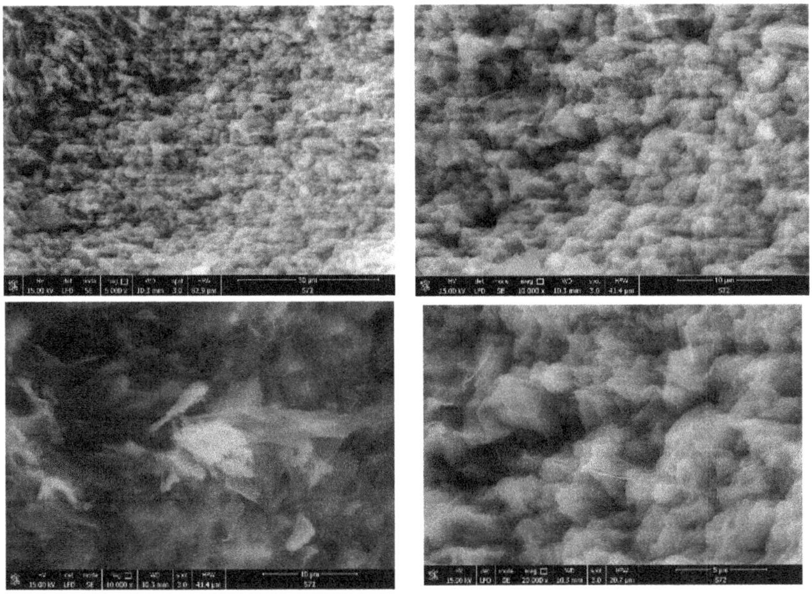

Figure 5. Hybrid S4 morphology observed with SEM.

Figure 6. SEM results of hybrid S5 surface conformation.

Starting from the same reagents and precursors and using the same conditions, except the molar ratio and the presence (S4 and S5) or absence (S1–S3) of boron, we expected to obtain a relatively similar morphology from the SEM images at this resolution. But a difference was still observed for hybrids S4 and S5 (Figures 5 and 6) containing boron, in addition to phosphorus and zirconium, in terms of surface morphology, in comparison with hybrids S1–S3 (Figures 2–4). The SEM images of hybrids S4 and S5, especially at the maximum resolution possible in these determinations, showed a higher surface porosity.

This is the effect of boron. It is practically due to the B-O-Zr bridges that occurred. Table 1 shows the bridges formed and expected, and the bonds which could not be obtained in the mild conditions used are shown for each of the synthesized hybrids.

Table 1. The bridges from the structures of the synthesized materials.

Bridge	S1	S2	S3	S4	S5
P-O-Zr	Expected	Expected	Expected	Expected	Expected
Zr-O-Zr	Expected	Expected	Expected	Expected	Expected
P-O-P	Not possible	Not possible	Not possible	Not possible	Not possible
Zr-O-B	-	-	-	Expected	Expected
P-O-B	-	-	-	Not possible	Not possible
P-P	Not possible	Not possible	Not possible	Not possible	Not possible
Zr-Zr	Not possible	Not possible	Not possible	Not possible	Not possible
B-B	Not possible	Not possible	Not possible	Not possible	Not possible
P-Zr	Not possible	Not possible	Not possible	Not possible	Not possible
P-B	Not possible	Not possible	Not possible	Not possible	Not possible
Zr-B	Not possible	Not possible	Not possible	Not possible	Not possible

As can be easily observed from Table 2, Zr and Zr, P and Zr, or B and Zr, are always connected via an oxygen atom in the structure of the obtained hybrids, and moreover, it is not possible to obtain direct bonds between these elements in the mild conditions used. It is also obvious from Table 2 that it was not possible to obtain bridges such as P-O-P and P-O-B. As a consequence, only zirconium could be involved in bridges with phosphorus and boron, via an oxygen atom, as already explained. For that reason, when boron is also present (S4 and S5), an excess of zirconium will be necessary, due to its possible presence in the bridges mentioned above.

Table 2. Carbon and zirconium content (Wt %) from EDX data for the synthesized materials.

Organic–Inorganic Hybrid	C%	Zr%
S1	52.07	17.26
S2	60.55	15.25
S3	53.74	12.9
S4	62.25	11
S5	58.59	11.72

On the other hand, if we compare the SEM images of hybrids S1–S3 which did not contain boron, it can be observed that the materials of S1 and S2 are rather similar, but hybrid S3 showed a somewhat different surface morphology, due to the zirconium excess and therefore due to the presence of more Zr-O-Zr bridges.

The structure of such hybrids is very complex and could be assimilated to a randomly coiled polymer, with a different number and different order of the X-O-P bridges (where X could be P, Zr and B). For that reason, such hybrids are also called inorganic polymers (or inorganic co-polymers). Such organic–inorganic networks are highly suited to forming supramolecular structures and could be used for several applications in different fields, such as catalysis, energy storage, medicine or supramolecular chemistry. The metal used has a significant influence on the properties of the synthesized materials, in addition to the P/M ratio. When using tetravalent metal phosphonates (as is the case for zirconium), a phosphorus excess is needed [11]. Then, M-O-P and M-O-M bridges are obtained and the presence of residual M-O-H, P-OH and P=O bonds can be observed. On the other hand, boron is not a metal. It is at the border between metals and non-metals (practically, it is the only non-metal from its group, which contains aluminium, for instance). Boron is very often called metalloid, but it still forms B-O-Zr bridges in the structure of hybrids S4 and S5. Therefore, this B-O-Zr bridge is obviously not a M-O-M bond, but it is obtained according to the same mechanism. On the other hand, the presence of boron could not lead to the obtaining of P-O-B bridges (as was observed in the case of zirconium, i.e., Zr-O-P). Boron is just before carbon in the periodic table of the elements. As expected, boron is similar to carbon in its ability to form stable covalent bonds in molecular structures and networks (but, of course, their number is different). It forms regular boron icosahedra. Boron is also a very hard material, and therefore, it is expected to bring even more mechanical resistance to the obtained hybrids S4 and S5, in addition to the presence of zirconium.

The organic–inorganic hybrid materials were already involved in many applications of great interest, such as ion-exchange and catalysis [11,29,40–46], photocatalysis [15,30,34–38,41], medicine, or for electrochemical devices and membranes with specific transport properties in separation and sensor technologies, or as a stationary phase in chromatography [40–49]. The applications of organic–inorganic hybrid materials containing phosphorus are in general determined by their high specific surface area, from the presence of residual acidic P-OH groups and/or of the flexible organic functional groups, and from the properties of the rigid inorganic part (strongly influenced by the used metal) [41–45].

The synthesized materials are also very stable from a thermal point of view and insoluble in organic solvents and in water. In addition to these properties, which make the material very stable from a chemical and physical point of view, the synthesized organic–inorganic hybrids (S1–S5) are also mechanically resistant due to the presence of zirconium. Zirconium has started to replace titanium in the last decades in hard-tissue medical prostheses due to its mechanical properties. It is found in natural sources that contain a small amount (1–2%) of hafnium, and together with titanium, these three elements appear in the same group of the periodic table (the fourth group). Therefore, it is expected to have similar properties (including mechanical resistance). Although titanium was already applied and used, including in such organic–inorganic hybrids, the interest in using zirconium increased in the last decade and is currently still growing.

The EDX data (Figures 7–11) confirmed the presence of phosphorus and zirconium on the structure of the obtained hybrids. EDX analysis showed the presence of C, O, P and Zr for the performed sol–gel syntheses S1–S5.

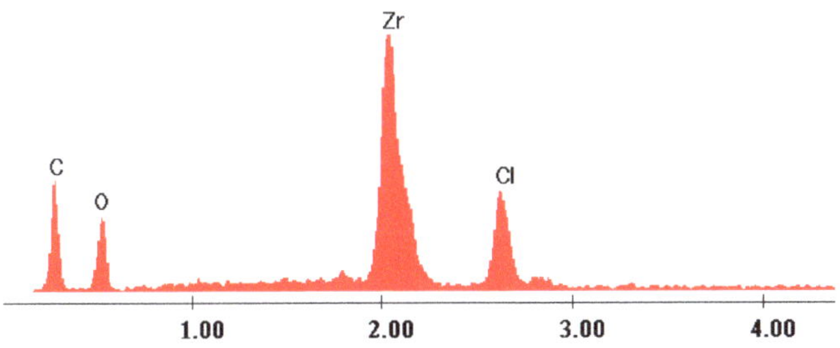

Figure 7. EDX data of hybrid S1.

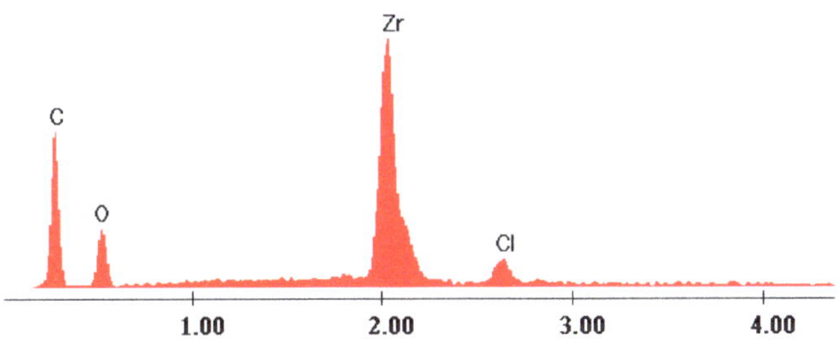

Figure 8. EDX data of hybrid S2.

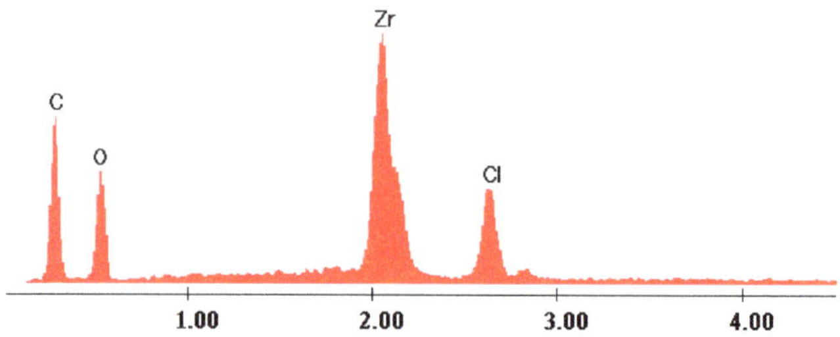

Figure 9. EDX data of hybrid S3.

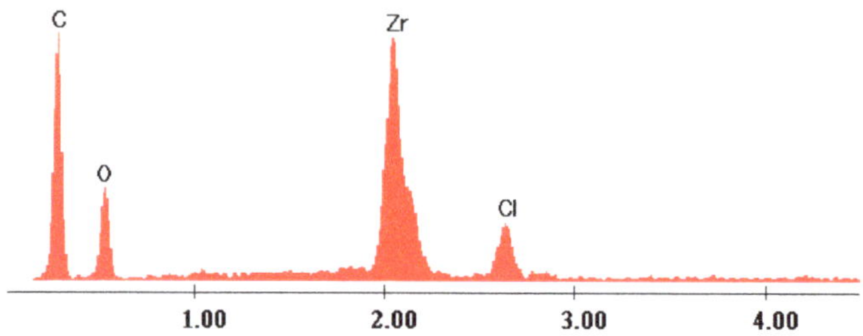

Figure 10. EDX data of hybrid S4.

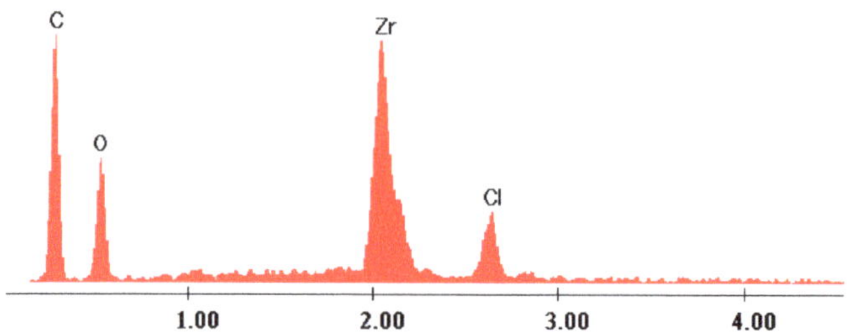

Figure 11. EDX data of hybrid S5.

The EDX data proved and confirmed that the chemical reactions performed by sol–gel method took place and the hybrid products were obtained as organic–inorganic networks. Table 2 presents the percentages of the content of the elements on the synthesized hybrids, from EDX data, according to Figures 7–11.

The elemental composition was changed function of the molar ratio, as it is obvious from Table 3, and this is also confirmed from both SEM and EDX analyses. Moreover, a clear indication that the hybrid structures were formed is the fact that the product is not soluble and, during the synthesis, it started to form a sediment (and subsequently form a suspension). Moreover, all the used precursors are water soluble. Therefore, the initial mixture is a homogeneous solution, and the resulting product is a heterogeneous suspension containing the insoluble product, water and the dissolved (but not reacted) precursors and reagents (if it is the case, usually when an excess is used).

Table 3. The used reagents for the sol–gel syntheses S1–S5.

Organic–Inorganic Hybrid	Phenyl Phosphinic Acid:$ZrOCl_2$:$B(OEt)_3$
S1	1:1:0
S2	2:1:0
S3	1:2:0
S4	1:1:1
S5	1:2:1

3. Conclusions

In the present work, novel hybrids containing zirconium and phosphorus compounds were synthesized by using the sol–gel method, starting from zirconyl chloride hexa-hydrate ($ZrOCl_2{\cdot}6H_2O$) and phenyl phosphinic acid. All the syntheses were performed at room temperature by using water as solvent. The sol–gel syntheses were completed in 6 h. Afterwards, the synthesized hybrids were washed, subsequently filtered and finally dried in an oven at 80 °C for 8 h. The used reagents and precursors are water soluble, but the hybrid compounds are not. The obtained products are very stable from a thermal point of view and are not soluble. The sol–gel method is a green alternative for obtaining organic–inorganic hybrids containing phosphorus, zirconium and boron.

The chemical reaction takes place between phosphorous and zirconium or two zirconium atoms, always connected by an oxygen atom (P-O-Zr or Zr-O-Zr). The phenyl groups from the structure of phenyl phosphinic acid are very stable, very hydrophobic and do not react in the mild conditions used. These phenyl groups are found on the structures of the synthesized hybrids as well, connected in the phosphorus atom, as in the structure of phenyl phosphinic acid.

For hybrids S1–S3, EDX data confirmed the presence of phosphorus and zirconium on their structures. In addition, for hybrids S4 and S5, the presence of boron was proved. Moreover, SEM images showed similar morphology for all the hybrids synthesized in the present work, but some differences were observed when boron was used. When Zr was excessive, it was expected to have more Zr-O-Zr bridges than P-O-Zr bonds. The obtained hybrids have complex 3D structures and form organic–inorganic networks containing Zr-O-Zr, Zr-O-P and Zr-O-B bridges. It is very interesting that boron could be involved only in Zr-O-B bridges, while zirconium could be involved in all Zr-O-Zr, P-O-Zr and B-O-Zr bridges. Such organic–inorganic networks are expected to form supramolecular structures and to have many potential applications in different fields of great current interest such as catalysis, medicine (especially for hard-tissue reconstruction), agriculture, energy storage, fuel cells, sensors, electrochemical devices and supramolecular chemistry. In addition, all the hybrids synthesized here showed high thermal stability, which makes them very attractive materials for different applications and processes that take place at high temperatures.

4. Materials and Methods

Zirconyl chloride hexa-hydrate ($ZrOCl_2{\cdot}6H_2O$), phenyl phosphinic acid and triethyl borate $B(OEt)_3$ were purchased from Fluka.

The reagents zirconyl chloride hexa-hydrate ($ZrOCl_2{\cdot}6H_2O$), triethyl borate and phenyl phosphinic acid were used at different molar ratios (Table 3).

The sol–gel method was used for obtaining the hybrid materials containing zirconium, boron and phosphorus compounds. All the syntheses were performed at room temperature by using water as solvent. After the syntheses were carried out, the obtained hybrids were filtered, washed and dried in an oven for 6 h at 80 °C. All the hybrid solid materials were further analyzed and characterized by using SEM and EDX methods. SEM and EDX measurements and determinations (including the images of the material surfaces and the element content) were performed by using a Jeol JSM 6400 Scanning Microscope coupled with an X-ray microanalyzer EXL II System Link Analytical, with a detector of 133 eV.

Author Contributions: Conceptualization: G.I. and V.S.; data curation: P.M. and N.V.; formal analysis: I.H. and V.S.; investigation: V.S. and N.V.; methodology: G.I. and P.M.; software: I.H.; validation: P.M. and N.V.; writing—original draft: G.I. and V.S.; writing—review and editing: G.I. All authors have read and agreed to the published version of the manuscript.

Funding: This research was funded by grant number 6 PFE from University of Life Sciences "King Michael I" from Timisoara.

Institutional Review Board Statement: Not applicable.

Informed Consent Statement: Not applicable.

Data Availability Statement: The data presented in this study are available from the corresponding author upon request.

Conflicts of Interest: The authors declare no conflict of interest. The funders had no role in the design of the study; in the collection, analyses, or interpretation of data; in the writing of the manuscript; or in the decision to publish the results.

References

1. Gómez-Alcántara, M.M.; Cabeza, A.; Martínez-Lara, M.; Aranda, M.A.G.; Suau, R.; Bhuvanesh, N.; Clearfield, A. Layered microporous tin(iv) bisphosphonates. *Inorg. Chem.* **2004**, *43*, 5283–5293. [CrossRef] [PubMed]
2. Alberti, G.; Costantino, U. *Inclusion Compounds*; Atwood, J.L., Ed.; Oxford University Press: Oxford, UK, 1992.
3. Jang, M.Y.; Yamazaki, Y. Preparation, characterization and proton conductivity of membrane based on zirconium tricarboxybutylphosphonate and polybenzimidazole for fuel cells. *Solid State Ion.* **2004**, *167*, 107–112. [CrossRef]
4. Alberti, G.; Casciola, M.; Donnadio, A.; Piaggio, P.; Pica, M.; Sisani, N. Preparation and characterisation of α-layered zirconium phosphate sulfophenylenphosphonates with variable concentration of sulfonic groups. *Solid State Ion.* **2005**, *176*, 2893–2898. [CrossRef]
5. Hogarth, W.H.G.; da Costa, D.J.C.; Lu, M.G.Q. Solid acid membranes for high temperature (140 °C) proton exchange membrane fuel cells. *J. Power Source* **2005**, *142*, 223–237. [CrossRef]
6. Clearfield, A. Inorganic ion exchangers, past, present, and future. *Solvent Extr. Ion Exch.* **2000**, *18*, 655–678. [CrossRef]
7. Xu, Z.P.; Jin, Y.; da Costa, D.J.C.; Lu, M.G.Q. $Zr(HPO_4)_2$ based organic/inorganic nanohybrids as new proton conductors. *Solid State Ion.* **2008**, *178*, 1654–1659. [CrossRef]
8. Gheonea, R.; Crasmareanu, E.; Plesu, N.; Sauca, S.; Simulescu, V.; Ilia, G. New hybrid materials synthesized with different dyes by sol-gel method. *Adv. Mater. Sci. Eng.* **2017**, *2017*, 4537039. [CrossRef]
9. Macarie, L.; Pekař, M.; Simulescu, V.; Plesu, N.; Iliescu, S.; Ilia, G.; Tara-Lunga-Mihali, M. Properties in aqueous solution of homo- and copolymers of vinylphosphonic acid derivatives obtained by UV-curing. *Macromol. Res.* **2017**, *25*, 214–221. [CrossRef]
10. Gómez-Alcántara, M.M.; Cabeza, A.; Olivera-Pastor, P.; Fernandez-Moreno, F.; Sobrados, I.; Sanz, J.; Morris, R.E.; Clearfield, A.; Aranda, M.A.G. Layered microporous tin(iv) bisphosphonates. *Dalton Trans.* **2007**, *23*, 2394–2404. [CrossRef]
11. Mutin, P.H.; Guerrero, G.; Alauzun, J.G. Sol-gel processing of phosphonate-based organic-inorganic hybrid materials. *J. Ceram. Soc. Jpn.* **2015**, *123*, 709–713. [CrossRef]
12. Mutin, P.H.; Guerrero, G.; Vioux, A. Organic-inorganic hybrid materials based on organophosphorus coupling molecules: From metal phosphonates to surface modification of oxides. *Comptes Rendus Chim.* **2003**, *6*, 1153–1164. [CrossRef]
13. Sanchez, C.; Julian, B.; Belleville, P.; Popall, M. Applications of hybrid organic–inorganic nanocomposites. *J. Mater. Chem.* **2005**, *15*, 3559–3592. [CrossRef]
14. Mutin, P.H.; Guerrero, G.; Vioux, A. Hybrid materials from organophosphorus coupling molecules. *J. Mater. Chem.* **2005**, *15*, 3761–3768. [CrossRef]
15. Gheonea, R.; Mak, C.; Crasmareanu, E.; Simulescu, V.; Plesu, N.; Ilia, G. Surface modification of SnO_2 with phosphonic acids. *J. Chem.* **2017**, *2017*, 2105938. [CrossRef]
16. Guerrero, G.; Mutin, P.H.; Vioux, A. Mixed Nonhydrolytic/Hydrolytic Sol-Gel Routes to Novel Metal Oxide/Phosphonate Hybrids. *Chem. Mater.* **2000**, *12*, 1268–1272. [CrossRef]
17. Caprita, R.; Caprita, A.; Ilia, G.; Cretescu, I.; Simulescu, V. Laboratory procedures for assessing quality of soybean meal. In Proceedings of the World Congress on Engineering and Computer Science, San Francisco, CA, USA, 20–22 October 2010; Volume 2, pp. 791–794.
18. Mutin, H.; Delenne, C.; Medoukali, D.; Corriu, R.; Vioux, A. Introduction of Organic Moieties into Transition-Metal Oxide Matrices via Phosphonate Groups. *Mat. Res. Soc. Symp. Proc.* **1998**, *519*, 345–350. [CrossRef]
19. Guerrero, G.; Mutin, P.H.; Vioux, A. Organically modified aluminas by grafting and sol-gel processes involving phosphonate derivatives. *J. Mater. Chem.* **2001**, *11*, 3161–3165. [CrossRef]
20. Kato, M.; Sakamoto, W.; Yogo, T. Proton-conductive sol-gel membranes from phenylvinylphosphonic acid and organoalkoxysilanes with different functionalities. *J. Membr. Sci.* **2008**, *311*, 182–191. [CrossRef]
21. Clearfield, A. *Metal Phosphonate Chemistry: From Synthesis to Applications*; Clearfield, A., Demadis, K., Eds.; Royal Society of Chemistry: London, UK, 2012.
22. Gagnon, K.J.; Perry, H.P.; Clearfield, A. Conventional and Unconventional Metal-Organic Frameworks Based on Phosphonate Ligands: MOFs and UMOFs. *Chem. Rev.* **2012**, *112*, 1034–1054. [CrossRef] [PubMed]
23. Vasylyev, M.V.; Wachtel, E.J.; Popovitz-Biro, R.; Neumann, R. Titanium Phosphonate Porous Materials Constructed from Dendritic Tetraphosphonates. *Chem. Eur. J.* **2006**, *12*, 3507–3514. [CrossRef]
24. Vasylyev, M.; Neumann, R. Preparation, Characterizaton, and Catalytic Aerobic Oxidation by a Vanadium Phosphonate Mesoporous Material Constructed from a Dendritic Tetraphosphonate. *Chem. Mater.* **2006**, *18*, 2781–2783. [CrossRef]
25. Zhu, Y.-P.; Ma, T.-Y.; Liu, Y.-L.; Ren, T.-Z.; Yuan, Z.-Y. Metal phosphonate hybrid materials: From densely layered to hierarchically nanoporous structures. *Inorg. Chem. Front.* **2014**, *1*, 360–383. [CrossRef]

26. Ma, T.-Y.; Liu, L.; Deng, Q.-F.; Lin, X.-Z.; Yuan, Z.-Y. Increasing the H^+ exchange capacity of porous titanium phosphonate materials by protecting defective P-OH groups. *Chem. Commun.* **2011**, *47*, 6015–6017. [CrossRef]
27. Ma, T.-Y.; Zhang, X.-J.; Shao, G.-S.; Cao, J.-L.; Yuan, Z.-Y. Ordered macroporous titanium phosphonate materials: Synthesis, photocatalytic activity, and heavy metal ion adsorption. *J. Phys. Chem. C* **2008**, *112*, 3090–3096. [CrossRef]
28. Lin, X.-Z.; Ren, T.-Z.; Yuan, Z.-Y. Mesoporous zirconium phosphonate materials as efficient water-tolerable solid acid catalysts. *Catal. Sci. Technol.* **2015**, *5*, 1485–1494. [CrossRef]
29. Zhu, Y.-P.; Liu, Y.-L.; Ren, T.-Z.; Yuan, Z.-Y. Hollow manganese phosphonate microspheres with hierarchical porosity for efficient adsorption and separation. *Nanoscale* **2014**, *6*, 6627–6636. [CrossRef] [PubMed]
30. Ma, T.-Y.; Lin, X.-Z.; Zhang, X.-J.; Yuan, Z.-Y. Hierarchical mesostructured titanium phosphonates with unusual uniform lines of macropores. *Nanoscale* **2011**, *3*, 1690–1696. [CrossRef]
31. Mehring, M.; Lafond, V.; Mutin, P.H.; Vioux, A. New Sol-Gel Routes to Organic-Inorganic Hybrid Materials: Modification of Metal Alkoxide by Phosphonic or Phosphinic Acids. *J. Sol-Gel Sci. Technol.* **2003**, *26*, 99–102. [CrossRef]
32. Ma, T.-Y.; Yuan, Z.-Y. Organic-Additive-Assisted Synthesis of Hierarchically Meso-/Macroporous Titanium Phosphonates. *Eur. J. Inorg. Chem.* **2010**, *19*, 2941–2948. [CrossRef]
33. Ma, T.-Y.; Yuan, Z.-Y. Metal Phosphonate Hybrid Mesostructures: Environmentally Friendly Multifunctional Materials for Clean Energy and Other Applications. *Chem. Sus. Chem.* **2011**, *4*, 1407–1419. [CrossRef]
34. Zhu, Y.-P.; Ren, T.-Z.; Yuan, Z.-Y. Co^{2+}-loaded periodic mesoporous aluminum phosphonates for efficient modified Fenton catalysis. *RSC Adv.* **2015**, *5*, 7628–7636. [CrossRef]
35. Ma, T.-Y.; Yuan, Z.-Y. Functionalized periodic mesoporous titanium phosphonate monoliths with large ion exchange capacity. *Chem. Commun.* **2010**, *46*, 2325–2327. [CrossRef]
36. Zhu, Y.-P.; Ma, T.-Y.; Ren, T.-Z.; Li, J.; Du, G.-H.; Yuan, Z.-Y. Highly dispersed photoactive zinc oxide nanoparticles on mesoporous phosphonated titania hybrid. *Appl. Catal. B* **2014**, *156*, 44–52. [CrossRef]
37. Guizard, C.; Bac, A.; Barboiu, M.; Hovnanian, N. Hybrid organic-inorganic membranes with specific transport properties, Applications in separation and sensors technologies. *Sep. Purif. Tech.* **2001**, *25*, 167–180. [CrossRef]
38. Maillet, C.; Janvier, P.; Bertrand, M.-J.; Praveen, T.; Bujoli, B. Phosphonate-Based Hybrid Materials for Catalysis? Supported Rhodium/2,2′-Bipyridine Complexes as Reduction Catalysts Under Hydrogen Pressure. *Eur. J. Org. Chem.* **2002**, *2002*, 1685–1689. [CrossRef]
39. Maillet, C.; Janvier, P.; Pipelier, M.; Praveen, T.; Andres, Y.; Bujoli, B. Hybrid Materials for Catalysis? Design of New Phosphonate-Based Supported Catalysts for the Hydrogenation of Ketones under Hydrogen Pressure. *Chem. Mater.* **2001**, *13*, 2879–2884. [CrossRef]
40. Lin, X.-Z.; Yuan, Z.-Y. Synthesis of mesoporous zirconium organophosphonate solid-acid catalysts. *Eur. J. Inorg. Chem.* **2012**, *2012*, 2661–2664. [CrossRef]
41. Ma, T.Y.; Qiao, S.Z. Acid-Base Bifunctional Periodic Mesoporous Metal Phosphonate Materials for Synergistically and Heterogeneously Catalyzing CO_2 Conversion. *Acs Catal.* **2014**, *4*, 3847–3855. [CrossRef]
42. Ma, T.-Y.; Lin, X.-Z.; Yuan, Z.-Y. Cubic Mesoporous Titanium Phosphonates with Multifunctionality. *Chem. Eur. J.* **2010**, *16*, 8487–8494. [CrossRef]
43. Pramanik, M.; Bhaumik, A. Self-Assembled Hybrid Molybdenum Phosphonate Porous Nanomaterials and Their Catalytic Activity for the Synthesis of Benzimidazoles. *Chem. Cat. Chem.* **2014**, *6*, 2577–2586. [CrossRef]
44. Zhu, Y.-P.; Liu, Y.-L.; Ren, T.-Z.; Yuan, Z.-Y. Mesoporous nickel phosphate/phosphonate hybrid microspheres with excellent performance for adsorption and catalysis. *RSC Adv.* **2014**, *4*, 16018–16021. [CrossRef]
45. Ma, T.-Y.; Lin, X.-Z.; Yuan, Z.-Y. Periodic mesoporous titanium phosphonate hybrid materials. *J. Mater. Chem.* **2010**, *20*, 7406–7415. [CrossRef]
46. Kato, M.; Katayama, S.; Sakamoto, W.; Yogo, T. Synthesis of organosiloxane-based inorganic/organic hybrid membranes with chemically bound phosphonic acid for proton-conductors. *Electrochim. Acta* **2007**, *52*, 5924–5931. [CrossRef]
47. Mhrohs, T.B.; Weichold, O. Multivalent Allylammonium-Based Cross-Linkers for the Synthesis of Homogeneous, Highly Swelling Diallyldimethylammonium Chloride Hydrogels. *Gels* **2022**, *8*, 100. [CrossRef]
48. Lungu, R.; Paun, M.A.; Peptanariu, D.; Ailincai, D.; Marin, L.; Nichita, M.V.; Paun, V.A.; Paun, V.P. Biocompatible Chitosan-Based Hydrogels for Bioabsorbable Wound Dressings. *Gels* **2022**, *8*, 107. [CrossRef] [PubMed]
49. Jabbari, E.; Sepahwandi, A. Decellularized Articular Cartilage Microgels as Microcarriers for Expansion of Mesenchymal Stem Cells. *Gels* **2022**, *8*, 148. [CrossRef]

gels

MDPI

Article

Synthesis of Highly Luminescent Silica-Coated Upconversion Nanoparticles from Lanthanide Oxides or Nitrates Using Co-Precipitation and Sol–Gel Methods

Ana Iglesias-Mejuto [1,*], Alyne Lamy-Mendes [2], João Pina [3], Benilde F. O. Costa [4], Carlos A. García-González [1,*] and Luisa Durães [2]

[1] AerogelsLab, I + D Farma Group (GI-1645), Department of Pharmacology, Pharmacy and Pharmaceutical Technology, Faculty of Pharmacy, iMATUS and Health Research Institute of Santiago de Compostela (IDIS), Universidade de Santiago de Compostela, E-15782 Santiago de Compostela, Spain

[2] University of Coimbra, CIEPQPF—Centro de Investigação em Engenharia dos Processos Químicos e Produtos da Floresta, Department of Chemical Engineering, 3030-790 Coimbra, Portugal; luisa@eq.uc.pt (L.D.)

[3] Coimbra Chemistry Centre—Institute of Molecular Sciences, Department of Chemistry, University of Coimbra, 3004-535 Coimbra, Portugal; jpina@qui.uc.pt

[4] University of Coimbra, CFisUC, Physics Department, 3004-516 Coimbra, Portugal; benilde@ci.uc.pt

* Correspondence: ana.iglesias.mejuto@rai.usc.es (A.I.-M.); carlos.garcia@usc.es (C.A.G.-G.); Tel.: +34-881815063 (A.I.-M.); +34-881814882 (C.A.G.-G.)

Abstract: Upconversion nanoparticles (UCNPs) are under consideration for their use as bioimaging probes with enhanced optical performance for real time follow-up under non-invasive conditions. Photostable and core-shell $NaYF_4$:Yb^{3+}, Er^{3+}-SiO_2 UCNPs obtained by a novel and simple co-precipitation method from lanthanide nitrates or oxides were herein synthesized for the first time. The sol–gel Stöber method followed by oven or supercritical gel drying was used to confer biocompatible surface properties to UCNPs by the formation of an ultrathin silica coating. Upconversion (UC) spectra were studied to evaluate the fluorescence of UCNPs upon red/near infrared (NIR) irradiation. ζ-potential measurements, TEM analyses, XRD patterns and long-term physicochemical stability were also assessed and confirmed that the UCNPs co-precipitation synthesis is a shape- and phase-controlling approach. The bio- and hemocompatibility of the UCNPs formulation with the highest fluorescence intensity was evaluated with murine fibroblasts and human blood, respectively, and provided excellent results that endorse the efficacy of the silica gel coating. The herein synthesized UCNPs can be regarded as efficient fluorescent probes for bioimaging purposes with the high luminescence, physicochemical stability and biocompatibility required for biomedical applications.

Keywords: upconversion luminescence; co-precipitation method; sol–gel Stöber method; lanthanide nitrates; lanthanide oxides

1. Introduction

Bioimaging enables the tracking of biological functions in real time by using technologies such as X-ray, magnetic resonance or fluorescence for detecting biological probes employed for the monitoring and examination of anatomy and physiology [1]. Namely, optical sensing is one of the most commonly used technologies for biodetection due to its sensitivity and simplicity, but efficient labeling molecules are needed for these purposes [2]. Among them, fluorescent dyes and semiconductor quantum dots (QDs) are being used as bioimaging tools, but the photobleaching of organic dyes, toxicity, poor photostability and broad emission of QDs limit their performance [3]. Moreover, fluorescent probes are generally excited by ultraviolet light, which induces DNA damage and cell death [4]. Thus, the development of materials with improved and highly efficient photochemical performance using biologically tolerable exciting radiation sources is required for in vivo bioimaging applications.

Upconversion nanoparticles (UCNPs) are considered as the new generation of fluorophores due to their unique optical properties such as sharp emission bands, large anti-Stokes shifts and low photobleaching and blinking [5]. These nanocomplexes also present good signal-to-noise ratio and sensitivity due to the low scattering, autofluorescence and absorption from living tissues (background) in the red/near infrared (NIR) region (therapeutic window or NIR window) while presenting high tissue penetration depth (up to 2 mm) [2,5–7]. Upconversion (UC) luminescence is an "anti-Stokes emission", where the emitted fluorescence (visible light from 400 to 700 nm) has shorter wavelengths and higher energy than the incident light, NIR (from 808 to 980 nm), which is a radiation usually regarded as not biologically harmful [8,9] at short exposure levels and with good transmittance through tissues [10–12]. The absorption of low-energy photons leads to the emission of high-energy photons by a non-linear UC process that enables the photomediated biosensing and bioimaging in a non-invasive process [13]. Based on this principle, UCNPs present long photoluminescence lifetimes, as well as low toxicity and photodamage for living organisms [3,14]. Indeed, UCNPs allowed the transformation of fluorescence imaging from microscopic to macroscopic levels with potential applications in bioimaging, biodetection, and cell and tissue labeling [4,10]. Moreover, UCNPs are one of the most promising luminescent probes for monitoring whole-body small animals, but in vivo UC luminescence still has some limitations [15]. Among them, sensitivity and resolution are affected by the absorption and scattering effects of organs with a rapid decrease in the signals detected from in vivo experiments involving UCNPs [16]. New luminescence imaging tools based on UCNPs must be developed to address all these concerns.

UCNPs are composed of a host matrix doped with a sensitizer (light absorber) and an activator (light-emitter) as guest molecules [17]. $NaYF_4$ nanocrystals are the most efficient UC host materials to date, being detected by confocal microscopy with a size of 27 nm [18] due to the low phonon energy and good absorption [19,20]. The $NaYF_4$ lattice forms two crystalline phases, the cubic (α phase) and the hexagonal (β phase) ones [8]. Thermodynamically stable hexagonal phase yields bright luminescence, but the polymorphic change from the cubic to the hexagonal phase requires high energy, e.g., hydrothermal or solvothermal treatment [21]. A decrease in particle size may also cause a decrease in emitter numbers and a high surface quenching [18]. Therefore, the process parameter control in the UCNPs synthetic methods should allow the tuning of the particle size and shape, nanocrystal phase, and doping concentration to ensure effective properties for bioimaging [19]. Regarding the ideal sensitizers, they should have the same NIR excited energy state as activators. Specifically, rare-earth doped UCNPs are photostable and have low photoblinking and photobleaching [12,22]. Namely, ytterbium (Yb^{3+}) is used as a sensitizer due to its high absorption and broad excitation spectrum. Its single excited state matches with activators like erbium (Er^{3+}), so Yb^{3+} is commonly employed as a sensitizer of Er^{3+}. These types of UCNPs contain thousands of emission center ions and can be of interest for bioimaging as they can be excited with NIR, thus enabling the deep penetration in tissues with non-invasive treatments and a low interference with the weak autofluorescence from the biological background [12,23]. It is important to highlight that luminescence intensity and efficiency are strongly dependent on the crystal phase and morphology of the UCNPs, so an environmentally friendly synthetic approach able to yield non-toxic nanosystems with a controlled shape and phase is required [24,25].

Lanthanide-doped UCNPs are usually obtained by a thermal decomposition, hydrothermal or solvothermal method, sol-gel processing or a co-precipitation approach [17]. Thermal decomposition consists of pyrolyzing the organometallic precursors with an organic solvent in an oxygen-free environment [26]. Hydrothermal synthesis is carried out with lanthanide nitrates or chlorides as precursors and under a high pressure and temperature environment, often above the critical point of the solvent [19,21,26]. The sol–gel method starts with a solution of precursors that forms a gel by hydrolysis and polycondensation [4]. After calcination at high temperatures, highly crystalline UCNPs are synthesized, but this strategy is not used for biomedical applications because of the lack of control of particle size [19]. In contrast, the co-precipitation method consists of the simultaneous and rapid precipitation of multiple species under supersaturation conditions in a short time [27]. This UCNPs processing method is regarded as advantageous since, in contrast to the previously mentioned techniques, it does not require complex procedures or harsh conditions (such as high temperatures or the use of solvents), while the obtained UCNPs are chemically homogeneous, and easily obtained with high yields after short reaction times [16,21].

UCNPs with biocompatible surface properties and uniform morphologies are prone to be employed as bio-labels, so ligands like polyethylene glycol, oleic acid or silica are used to functionalize UCNPs, to increase water solubilization and stability and to modulate cellular interactions [4,14,17,28]. Silica modification is a well-established and commonly employed technique, easy to obtain, effective and able to improve the biocompatibility and stability of the resulting structures [7]. By the sol–gel Stöber method [29], proposed in 1968 by Werner Stöber and coworkers, TEOS (alkoxide precursor) is hydrolyzed to produce silicic acid, which undergoes a condensation reaction in mild basic conditions to form an amorphous silica coating around the previously synthesized lanthanide core, in the presence of ammonia as a catalyst [30,31]. This eco-friendly alternative sol–gel method avoids the use of potentially cytotoxic surfactants and organic solvents that would require several washing steps prior to be employed as biological labels. Moreover, the ultrathin silica shell synthesized is non-toxic, chemically inert, and optically transparent, which are important advantages to render photoluminescent UCNPs for bioimaging applications [32].

In this work, core-shell $NaYF_4:Yb^{3+}$, Er^{3+}-SiO_2 UCNPs obtained from lanthanide nitrates and, alternatively, from lanthanide oxides, were synthesized. A novel and simple co-precipitation method was applied, followed by a silica coating step based on TEOS hydrolysis by the sol–gel Stöber process. The obtained UCNPs formulations were characterized regarding their crystalline structure and photophysical properties by XRD, SEM-EDS and UC fluorescence spectroscopy. TEM and confocal microscopy were performed to unveil the morphological and luminescent properties of UCNPs just after synthesis and, comparatively, after a 2-year mid-term storage. Cell viability tests with murine fibroblasts and hemolytic activity tests with human blood were also performed for biocompatibility evaluation. Lanthanide-derived core-shell $NaYF_4:Yb^{3+}$, Er^{3+}-SiO_2 UCNPs yielded biocompatible nanocomplexes with high crystallinity, a controllable transition from cubic α-$NaYF_4$ to hexagonal β-$NaYF_4$ and long-term UC fluorescence emission and physicochemical stability. To the best of our knowledge, it is the first time that silica gel-coated UCNPs were synthesized using a procedure involving lanthanide nitrates or oxides as precursors, in a room-temperature (RT) facile, versatile, and non-hazardous co-precipitation method. Also, precursors were dissolved in aqueous solutions, thus avoiding the use of solvents or surfactants that could render toxic by-products. Overall, the novel technological combination of co-precipitation and sol–gel methods herein studied is regarded as environmentally benign and very convenient for obtaining highly photoluminescent UCNPs in a shape- and phase-controlling approach.

2. Results and Discussion

2.1. Physicochemical Properties of UCNPs

Different formulations of nanoparticles (NPs)were obtained (Table 1) and denoted differently depending on the lanthanide precursors (lanthanide nitrates—N or lanthanides oxides—O), drying procedure (supercritical drying—SCD, or oven drying—OD) and when the drying was performed (before or after the silica coating—BC or AC, respectively). A formulation without coating (O, OD) was also obtained for the sake of particle size and morphology comparison.

Table 1. Different UCNPs formulations studied.

Formulation	Lanthanide Precursors	Drying Method	Drying Moment
O, OD-AC	Lanthanide oxides	Oven drying	After silica coating
O, OD-BC	Lanthanide oxides	Oven drying	Before silica coating
O, OD	Lanthanide oxides	Oven drying	-
N, OD-AC	Lanthanide nitrates	Oven drying	After silica coating
N, OD-BC	Lanthanide nitrates	Oven drying	Before silica coating
N, SCD-AC	Lanthanide nitrates	SCD	After silica coating
N, SCD-BC	Lanthanide nitrates	SCD	Before silica coating

All the tested synthetic procedures led to the production of UCNPs formulations in the form of a whitish and fine powder. Similar yields were observed for all the studied methods, with a slightly lower amount of UCNPs obtained after procedures involving supercritical drying (SCD) due to some mass losses during the drying step. Synthesis was simpler in the case of the use of nitrates of the lanthanides than in the case of oxides of the lanthanides due to the pH adjustment needed in the latter case. Near-zero and negative zeta potential (ζ-potential) values were obtained for all the formulations studied (Table 2). These values are good since highly negative charged NPs could be passively delivered to certain tissues [33] while highly positive charged NPs induce cell electrostatic interactions that increase endocytic uptake and are thus more toxic [4,34].

Table 2. ζ-potential measurements of different UCNPs formulations. For notation of the formulations, please refer to Table 1.

UCNPs Formulation	ζ-Potential (mV)
O, OD-AC	-7.28 ± 1.99
O, OD-BC	-17.72 ± 0.12
O, OD	-6.39 ± 2.03
N, OD-AC	-9.64 ± 1.03
N, OD-BC	-7.07 ± 1.32
N, SCD-AC	-9.04 ± 1.09
N, SCD-BC	-8.78 ± 0.52

The formation of UCNPs with uniformity in shape was observed by TEM imaging (Figure 1) in all cases, regardless of the gel drying method used (SCD or oven drying) or the silica coating (presence or absence). This is a very good result since all the formulations are highly spherical (Figure 1), thus demonstrating the shape-controlling ability of the co-precipitation approach herein employed. The two formulations dried by SCD (Figure 1 N, SCD-AC; N, SCD-BC) had an aspect and size close to that of the oven-dried UCNPs (Figure 1 N, OD-AC; O, OD-AC; O, OD; O, OD-BC; N, OD-BC), suggesting the absence of an impact of the drying procedure employed during synthesis in the morphology of UCNPs. It should also be noted that the formulation without a silica coating (Figure 1 O, OD) had a similar aspect and size to the same formulation with the coating (Figure 1 O, OD-AC; O, OD-BC), suggesting the very low thickness of the silica coating (2.75 ± 0.49 nm). This ultrathin silica coating is beneficial to the energy transfer between UCNPs [35]. The mid-term storage (2 years) of UCNPs under controlled conditions had no effect on particle morphology (Figure 1 O, OD-BC just after synthesis vs. 2 years after synthesis, and N, OD-BC just after synthesis vs. 2 years after synthesis). Regarding UCNPs sizes, the major population of the O, OD-BC formulation obtained just after synthesis was lost after 2-years of mid-term storage, in contrast to the N, OD-BC formulation, where the UCNP size was preserved after mid-term storage. Therefore, results suggest that the UCNPs obtained from the lanthanide nitrates are more stable in terms of size than the ones obtained from the lanthanide oxides. This morphological stability is adequate and essential to satisfy the requirements of biological applications [35].

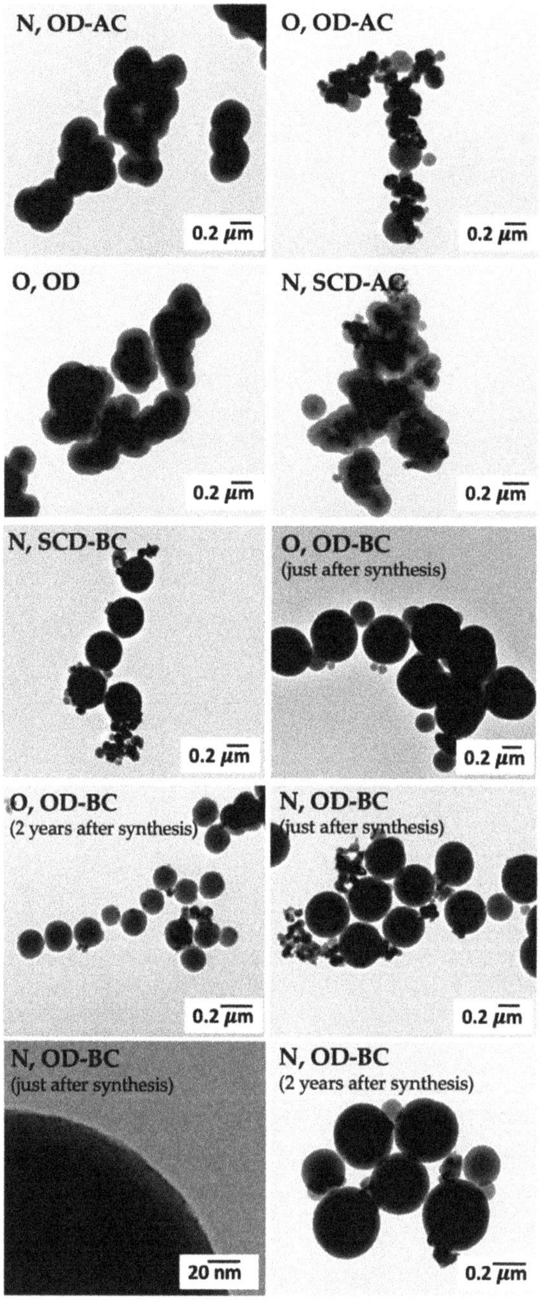

Figure 1. TEM images of different UCNPs formulations. For notation of the formulations, please refer to Table 1.

Aggregates were detected for some UCNPs formulations (Figure 1 N, OD-AC; O, OD-AC; O, OD and N, SCD-AC) with a decrease in their formation after mid-term storage (O, OD-BC just after synthesis vs. O, OD-BC 2 years after synthesis and N, OD-BC just after synthesis vs. N, OD-BC 2 years after synthesis). UCNPs tended to aggregate after annealing at higher temperatures and also due to the carbonization of the capping reagent EDTA after annealing, thus reducing UCNPs hydrophilicity, in agreement with observations by Wang and coworkers [21]. The aggregation may also be a consequence of the interparticle chemical bonding between the silica shells of the UCNPs [31]. It was previously described that there is an optimal size for efficient uptake of each type of nanomaterials into cells [36]. UCNPs of a very low size (<45 nm) were reported with the same sensitizers (Yb^{3+} and Er^{3+}) [20,37,38], which could be easily internalized by cells and induce high toxicity afterwards. In contrast, this particle size range just represents a very minor population in all UCNPs formulations (Figure 1) and adequate sizes for biomedical applications were obtained with this novel co-precipitation approach herein developed. Furthermore, an increase in the size of the nanocrystals obtained with the same trivalent lanthanide ions was reported to increase the luminescence emission [39].

All the synthesized UCNPs emitted visible radiation when excited at 980 nm, showing three different bands at 520, 540 and 657 nm in the UC luminescence spectra (Figure 2a,b), except for the oxide-derivative O, OD-AC formulation where no meaningful UC emission was detected (Figure 2b). The lower efficiency of the UC process for the O, OD-AC formulation can be attributed to the type of host matrix phase (cubic α-$NaYF_4$ or hexagonal β-$NaYF_4$) [18,19,21], which will be discussed later. The peaks at 520, 540 and 657 nm correspond to $^2H_{11/2} \rightarrow {}^4I_{15/2}$, $^4S_{3/2} \rightarrow {}^4I_{15/2}$, $^4F_{9/2} \rightarrow {}^4I_{15/2}$ optical radiative transitions of Er^{3+} ions, respectively, as previously reported for UCNPs composed of the same trivalent lanthanide ions [39,40] and in a study of Er^{3+} and Yb^{3+} UC photoluminescence on films [41]. It must be noted that an enhancement of the photoluminescence was observed when structures where co-doped with both ions, with a photoluminescence intensity more than one order-of-magnitude higher with respect to materials doped only with Er^{3+} [41].

Some formulations (N, OD-AC; N, SCD-AC and N, SCD-BC) exhibited structured green and red bands maybe due to the saturation of the detector and not to the emission from different excited states [42]. Similar structured bands were also reported for different $NaYF_4$:Yb^{3+}, Er^3 formulations [38,39,43]. In the comparative UC spectra performed under constant experimental conditions (laser intensity), the N, OD-BC and O, OD-BC formulations exhibited a higher fluorescence intensity in each peak, achieving the highest values for N, OD-BC. It has been recently published that with a very low Er^{3+} concentration (close to 1%) UC photoluminescence was not detected, so the conversion of NIR into visible light comes from the interaction between neighboring Er^{3+} ions [44]. This reaction was associated with the energy transfer from one excited Er^{3+} to another Er^{3+} and with the upgrade of the latter Er^{3+} into a higher-energy level. In this cooperative UC, Er^{3+} luminescence comes from two centers, one related to the Er^{3+} UC luminescence (520 and 540 nm peaks) and the other one related to the Er^{3+}-Yb^{3+} ion-ion interaction and its UC luminescence (657 nm peak). It is also important to note that the green band/red band UC emission intensity ratio could be related to different "excitation environments" in terms of ion size, ionic dependence of site distortions and ion concentration due to a concentration-quenching effect between Er^{3+} and Yb^{3+} ions [40,44].

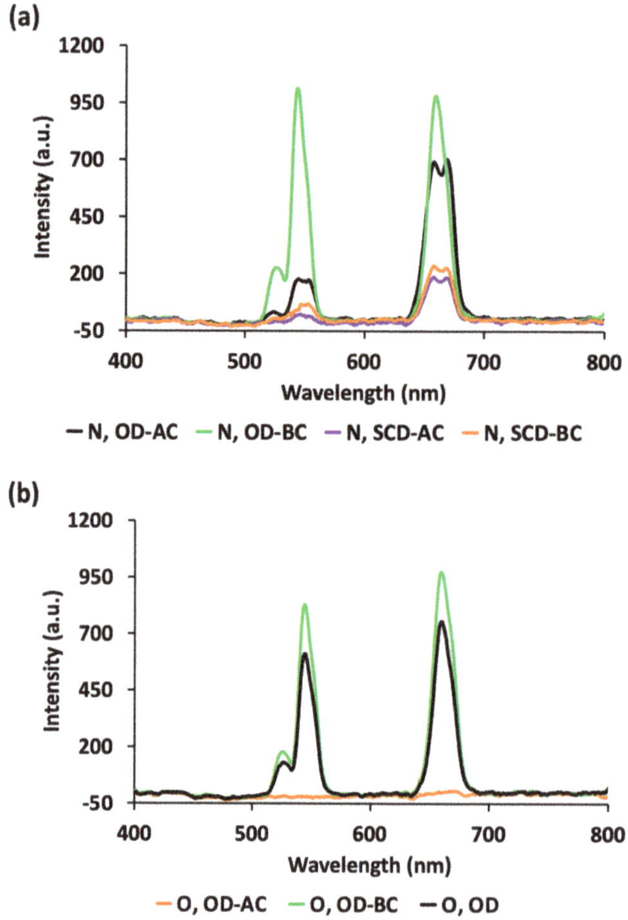

Figure 2. UC spectra for the (**a**) lanthanide-nitrates-based UCNPs and (**b**) lanthanide-oxides-based UCNPs in the solid state, collected with λ_{exc} = 980 nm. For notation of the formulations, please refer to Table 1.

The crystalline nature of UCNPs was confirmed by the corresponding XRD pattern (Figure 3a,b), with similar diffractograms to those previously reported for UCNPs with the same sensitizer–activator system (Yb^{3+}-Er^{3+}) [45]. The UC process is improved with a highly crystalline host lattice holding a low-energy environment that enables the energy transfer from the sensitizer to the activator. The essential requirements for an efficient energy transfer are the distance between the sensitizer and the activator, the spectral overlap and the achievement of a high absorbance by the activator at the emission wavelength of the sensitizer [46]. Both types of crystalloid phases, cubic and hexagonal, exist at the ambient pressure being the cubic phase the dominant one, i.e., the thermodynamically metastable. However, the green emission of the crystalline hexagonal phase is 10 times stronger than that of the crystalline cubic phase and the overall emission is ca. 5-fold higher in the thermodynamically stable β-$NaYF_4$:Yb^{3+}, Er^{3+}. Nevertheless, there is a high-energy barrier between α- and β-phases, so the phase transition needs temperatures higher than 280 °C [37]. It has also been reported that both phases coexist from 290 to 300 °C and that very long reaction times are needed for the complete α → β transition at 300 °C, so 400 °C was herein selected as the reaction temperature. Consistently, the formation of β-phased UCNPs did not involve a direct phase transformation process but a dissolution/recrystallization process. As expected, the fluorescence intensity increased according to the UCNP conversion from the host matrix cubic to the hexagonal phase after annealing at 400 °C. Nevertheless, for N, SCD-AC, N, OD-AC and O, OD-AC formulations, no peaks related to the hexagonal phase were observed (Figure 3a,b). The hexagonal phase peaks were detected for N, OD-BC, O, OD-BC and O, OD formulations, thus suggesting that the oven drying is a more appropriate method for the synthesis of UCNPs with respect to the SCD. Finally, the highest fluorescence intensity appeared when drying and annealing steps were performed before the silica coating step, so the best formulations in terms of fluorescence intensity (O, OD; O, OD-BC; N, OD-BC) were selected for further EDS characterization. The full accomplishment of the crystalline hexagonal phase under certain experimental conditions unveils the phase-controlling ability of the co-precipitation method herein employed.

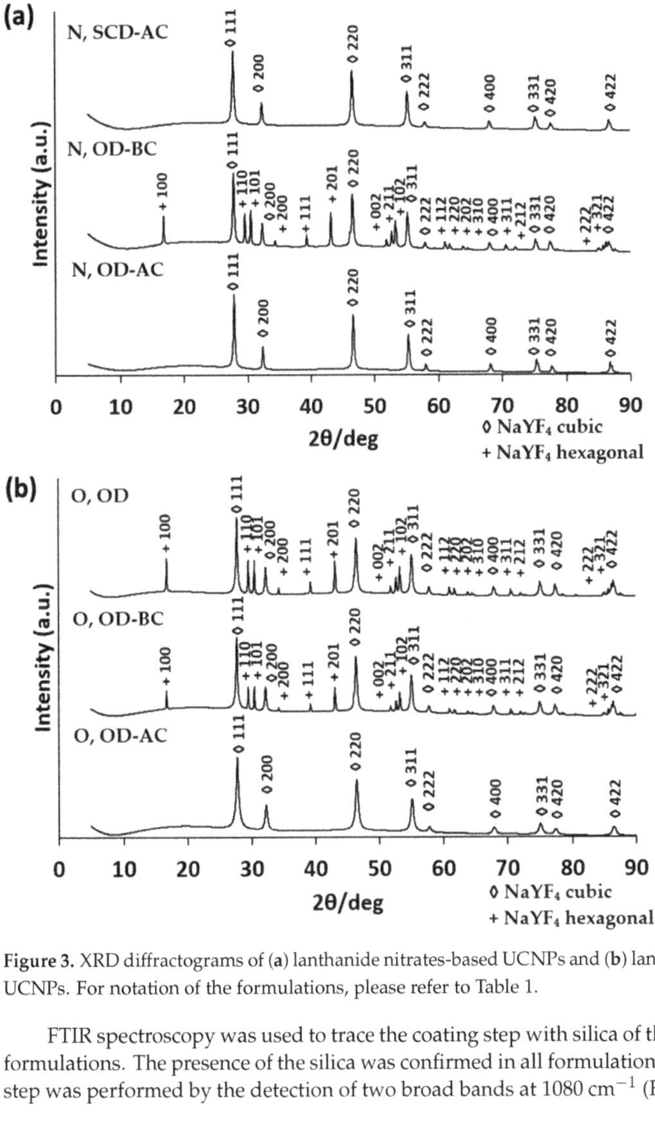

Figure 3. XRD diffractograms of (**a**) lanthanide nitrates-based UCNPs and (**b**) lanthanide oxides-based UCNPs. For notation of the formulations, please refer to Table 1.

FTIR spectroscopy was used to trace the coating step with silica of the different UCNPs formulations. The presence of the silica was confirmed in all formulations where the coating step was performed by the detection of two broad bands at 1080 cm^{-1} (Figure 4, red circles) and 1175 cm^{-1} (green circles) corresponding to vSi-O-Si vibrations. Moreover, bands at 970 cm^{-1} (blue circles) and 800 cm^{-1} (purple circles), corresponding to a vSi-O vibration of the ≡Si-O-Si≡ bond, were detected and associated with the silica functionalization [47,48]. FTIR spectra of all samples are almost completely smooth in the 1400–4000 cm^{-1} region, without exhibiting characteristic peaks that could indicate the presence of functional groups not related to silica. Results confirmed that the functionalization of UCNPs with silica was successfully achieved regardless of the drying procedure used (SCD or oven drying) or the moment when the coating was performed (before or after the drying step).

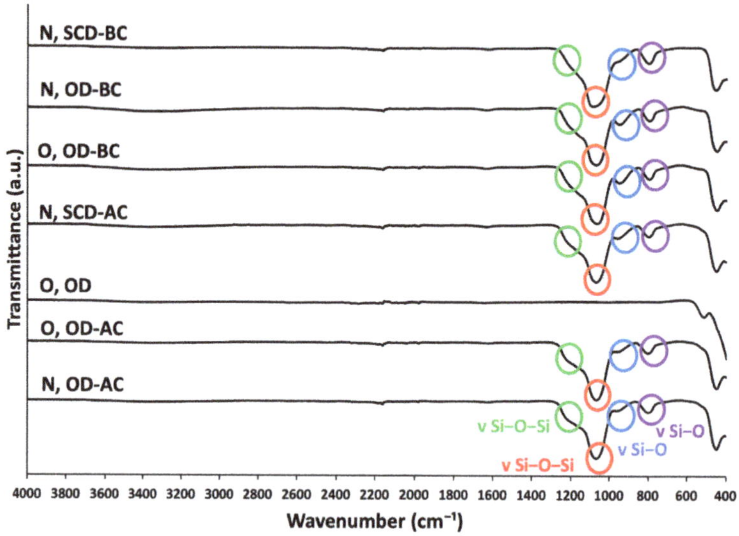

Figure 4. FTIR spectra of different formulations of UCNPs. For notation of the formulations, please refer to Table 1.

The elemental mapping obtained by the EDS technique (Figure 5a) confirmed the presence of all the employed chemical elements, including the peaks related to ions present at low concentrations (Na and F). UCNPs showed the peaks related to the host material constituents and lanthanides homogeneously distributed throughout the analyzed sample, thus confirming their expected composition and that they were congruent with the obtained UC fluorescence results. EDS results also unveiled the decoration of UCNPs with similar intensity for the elements from the host matrix (Na, F) and for the rare-earth elements (Y, Yb, Er). This elemental mapping suggests their simultaneous and homogeneous distribution throughout the sample, which is attributed to the successful connection between the rare-earth elements and the host matrix. Similar elemental mapping was previously found for UCNPs doped with the same rare-earth elements [35]. The background material with lower fluorescence is mainly silica. The more homogeneous formulations in terms of fluorescence are the lanthanide nitrate-based UCNPs. This is a very beneficial result since the nitrate-derived UCNPs are much easier to synthesize than their oxide-derived counterparts and only involve non-hazardous steps. Since high crystallinity and the presence of chemical elements were confirmed for all formulations studied, the best formulation in terms of fluorescence intensity (i.e., the formulation with the highest intensity in each peak, N, OD-BC) was selected for biological characterization (cf. Section 2.2). This formulation also exhibited physicochemical stability in terms of particle shape and fluorescence properties after a 2-year mid-term storage.

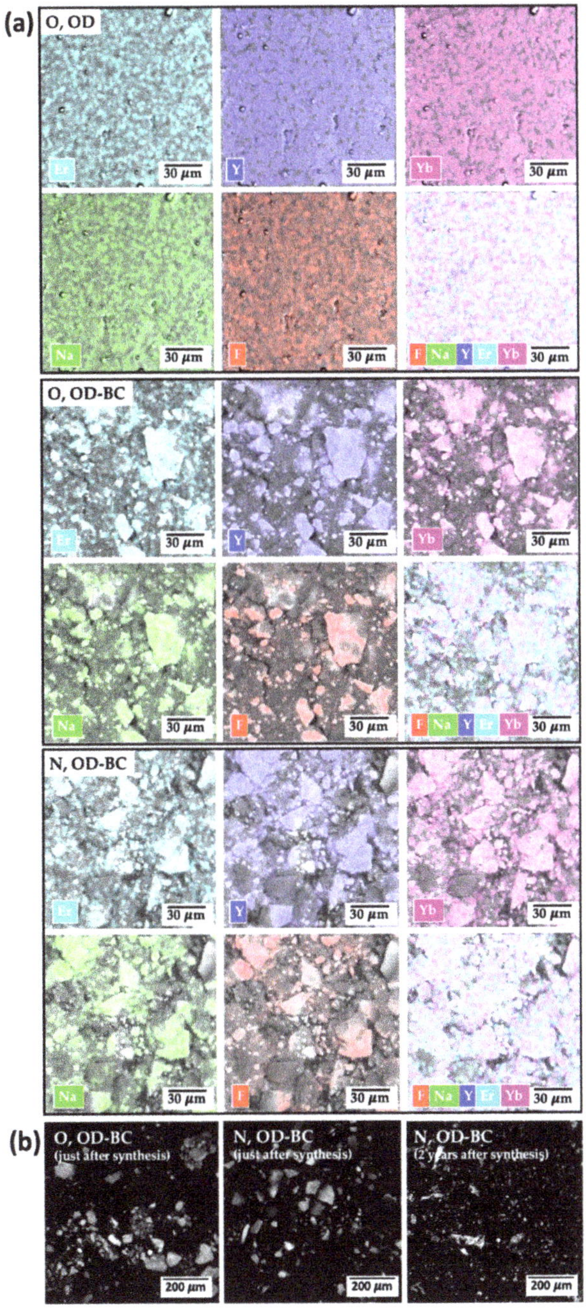

Figure 5. (**a**) EDS mapping and (**b**) confocal microscopy images of different UCNPs formulations. For notation of the formulations, please refer to Table 1.

Confocal microscopy is the main approach used for localizing UCNPs by in vitro cell-based tests and by in vivo bioimaging experiments. For this reason, the best formulations of UCNPs were also analyzed by this technique (Figure 5b), with a good fluorescence signal obtained just after synthesis (for both formulations studied) and after 2-year storage, thus endorsing the possibility of their mid-to-long-term storage without detrimental in fluorescence properties. Long-term photostability was previously studied for UCNPs [32] with good results in terms of UC spectra but after shorter time periods (5 months) than those herein evaluated.

2.2. Biocompatibility and Hemocompatibility Evaluation

The biocompatibility of UCNPs was confirmed by the high viability of mouse embryo fibroblasts in the presence of the N, OD-BC UCNP formulation for 24 and 48 h, without significant differences with respect to the positive controls (Figure 6a). These results are better than those previously reported for silica-coated $NaYF_4$:Yb^{3+}, Er^{3+} UCNPs with a low size (21 ± 5 nm) after contact with bone marrow-derived stem cells (79.5% after 24 h and 66.8% after 48 h) or skeletal myoblasts (87.8% after 24 h and 68.2% after 48 h) at 100 µg/mL [49]. Concentrations (0–5 mg/mL) close to those herein reported (5 mg/mL) were previously tested for $KGdF_4$:Yb^{3+}, Er^{3+} UCNPs with cell viabilities higher than 60% after 20 h [50]. Moreover, lower concentrations of UCNPs (100 µg/mL) showed lower levels of biocompatibility than those herein reported (close to 75% of cell viability) and were reported as safe to be used in subcellular imaging applications [51]. A reduction in the viability after 72 h of rat glioma cells and of rat mesenchymal stem cells was observed for lower concentrations than those herein studied (1000 µg/mL) of polymer-coated UCNPs composed of the same trivalent lanthanide ions [39]. Furthermore, the relevance of the protective coating to the safety of NPs was assessed by evaluating the cell viability in $NaYF_4$:Yb^{3+}, Er^{3}UCNPs with different polymeric coatings [38]. Very low cell viabilities were obtained with neat UCNPs while the different polymer coatings studied increased rat mesenchymal stem cell viabilities, evaluated in the concentration range 0–1000 µg/ mL. It is important to note that much lower concentrations than those herein tested (100 µg/mL) showed a bright fluorescence signal on HeLa cells, with a high cell viability (90%) after 24 h [32]. These findings emphasized the safety of use of the herein synthesized N, OD-BC UCNPs.

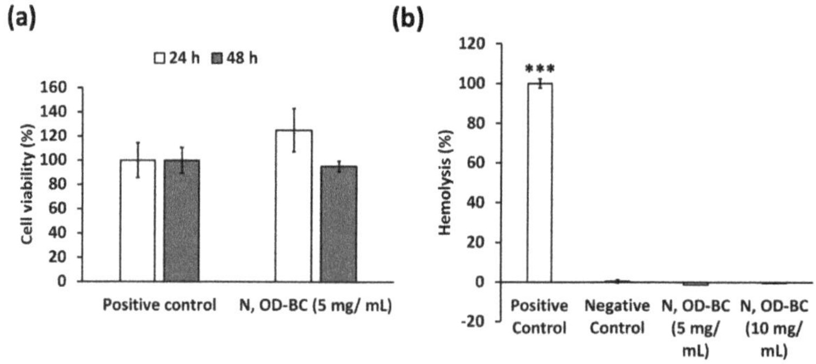

Figure 6. (**a**) Cell viability of BALB/c3T3 cells in contact with N, OD-BC UCNPs at 5 mg/mL determined by WST-1 test. Positive control: BALB/c3T3 cells. No statistically significant differences among groups were detected (post hoc Tukey HSD multiple comparison-test; $p < 0.05$). (**b**) Hemolytic activity (expressed in %) for N, OD-BC UCNPs (5 and 10 mg/mL). Positive and negative controls were 4% (v/v) Triton X-100 solution and PBS pH 7.4 solution, respectively. Statistically significant differences among groups were represented as *** (post hoc Tukey HSD multiple comparison test; $p < 0.001$). For notation of the formulation, please refer to Table 1.

No hemolysis was observed on human red blood cells after contact with N, OD-BC UCNPs (Figure 6b), with no significant differences with respect to the negative controls for the two studied concentrations (5 and 10 mg/mL). No influence of UCNPs on rat red blood cells was previously reported for $NaYF_4$:Yb^{3+}, Er^{3+},Gd^{3+} UCNPs at lower UCNPs concentrations (25–500 μg/mL) [35]. Excellent human blood compatibility was also found for $NaGdF_4:Yb^{3+}$, Er^{3+} UCNPs at a Gd content of 500 ppm [45]. These results suggest the absence of a negative impact of the N, OD-BC UCNPs, even at very high concentrations (10 mg/mL), on human red blood cells. Therefore, silica coating by sol-gel Stöber process seems to create a structurally stable nanocomplex in biological fluids that effectively protect the lanthanide core from inducing any cytotoxic event.

3. Conclusions

$NaYF_4:Yb^{3+}$, Er^{3+} core-shell nanocomplexes with a highly photostable luminescent core and a non-porous silica gel shell were successfully fabricated by a facile and environmentally friendly combination of co-precipitation (core) and sol-gel (shell) approaches. The synthetic strategy was able to control the morphology and crystalline phase of the UCNPs as well as their biocompatibility. The size and ζ-potential values achieved suggest the absence of induction of negative cell interactions. Similar morphology and luminescence of silica gel-coated UCNPs were observed after 2-year mid-term storage, thus indicating their preliminary physicochemical stability. An optimum formulation with the highest values of fluorescence intensity and coherent presence of all elements in the UCNPs was selected. Excellent bio- and hemocompatibility were obtained for these UCNPs, after assays with murine fibroblast and human blood, respectively. The protective silica coating synthesized by the sol-gel Stöber process seems to preserve over time the lanthanide core UC emission while promoting the bio- and hemocompatibility of the nanosystem. Consistently, these UCNPs represent a promising alternative to be tested as in vivo biolabels, as the concentration range could be increased until 5 mg/mL to get a detectable signal without losing bio- or hemocompatibility. These results encourage further research into potential applications of the herein synthesized silica gel-coated UCNPs as biomarkers for biological detections, in cellular and animal imaging systems by their incorporation into biomedical devices for further evaluation of these UCNPs as in vivo bioimaging probes.

4. Materials and Methods

4.1. Materials

Ytterbium (III) oxide (Yb_2O_3, 99.9% purity), erbium (III) oxide (Er_2O_3, 99.9%), ytterbium (III) nitrate pentahydrate ($Yb(NO_3)_3{\cdot}5H_2O$, 99.9%), yttrium(III) nitrate hexahydrate ($Y(NO_3)_3{\cdot}6H_2O$, 99.8%), sodium fluoride (NaF, >99%) and ethylenediaminetetraacetic acid (EDTA) were provided by Sigma Aldrich (Merck Group, Darmstadt, Germany). Yttrium (III) oxide (Y_2O_3, 99.99%), erbium (III) nitrate pentahydrate ($Er(NO_3)_3{\cdot}5H_2O$, 99.9%), and tetraethyl orthosilicate (TEOS, $Si(OC_2H_5)_4$, 98%), were obtained from Acros Organics (Thermo-Fisher Scientific, Geel, Belgium). Hydrochloric acid (HCl, 37%) was supplied by Fisher (Madrid, Spain) and ammonium hydroxide (25% NH_3 in H_2O) by Fluka Analytical (Honeywell International, Charlotte, NC, USA). CO_2 (purity > 99.9%) was obtained from Nippon Gases (Madrid, Spain) and absolute ethanol (EtOH) from VWR (Radnor, PA, USA). All reagents were used as purchased and deionized water was employed in all the procedures.

4.2. Synthesis of UCNPs by Co-Precipitation and Sol-Gel Methods

0.2 M stock solutions of Y_2O_3, Yb_2O_3, and Er_2O_3 solutions were prepared and adjusted at pH 2 by using hydrochloric acid to dissolve the oxide precursors and liberate the metal ions to the solution [52]. A total of 16 mL of 0.2 M Y_2O_3, 3.4 mL of 0.2 M Yb_2O_3 and 0.6 mL of 0.2 M Er_2O_3 was mixed with 20 mL of 0.2 M EDTA. Alternatively, another solution was prepared by mixing 16 mL of 0.2 M $Y(NO_3)_3$, 3.4 mL of 0.2 M $Yb(NO_3)_3$, 0.6 mL of 0.2 M $Er(NO_3)_3$ and 20 mL of 0.2 M EDTA. EDTA was firstly employed in the synthesis of UCNPs by Yi and coworkers [52]. It is employed as a stabilizing and chelating agent able to minimize UCNPs aggregation [53]. EDTA has been described as an efficient complexing agent for trivalent lanthanide ions because it improves the dispersibility of crystals. Complexes formed are adsorbed on the surface of NPs, limiting the growth in size of the resulting UCNPs. In fact, smaller NPs size is generally achieved by increasing the concentration of different chelators [54]. Moreover, EDTA is used to prevent precipitation before proceeding with the synthesis procedure [55].

Rare earth solution was injected into 0.05 mol of NaF dissolved in 60 mL of water and the mixture was stirred for 1 h at RT. Obtained dispersions were centrifuged and washed three times with water and once with ethanol. NPs were dried in an oven at 60 °C or, alternatively, under SCD with EtOH. For the SCD, NPs dispersed in EtOH were placed inside an autoclave filled with EtOH, sealed, and flushed with gaseous N_2. Pressure was maintained at 80 bar until reaching 260 °C with a heating rate of 80 °C/h. The reactor valve outlet was opened for EtOH release under supercritical conditions, thus dropping the pressure gradually. The system was cooled down to RT before opening the autoclave. NPs annealing was carried out under a N_2 atmosphere, for 5 h and 400 °C with a heating rate of 20 °C/min. NPs were cooled down until RT under the same atmosphere.

To obtain the silica shell, 30 mL of NPs dispersed in ethanol (3 mg/mL) were placed in a bath at 0 °C under stirring [47]. TEOS (0.2 mL) was firstly added and, 5 min later, ammonium hydroxide (3 mL) was then incorporated. The reaction took place for 2 h, under stirring and at 0 °C. The thus obtained NPs were washed three times with EtOH.

Synthesis of the lanthanide core-shell UCNPs was thus performed according to the scheme depicted in Figure 7a,b.

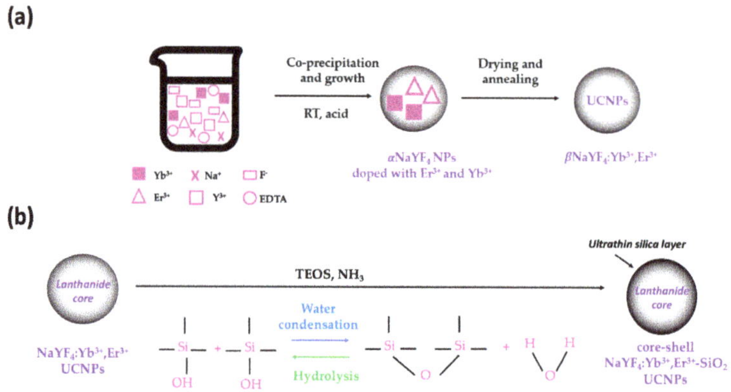

Figure 7. Experimental procedure used to obtain silica gel-coated UCNPs: (**a**) Synthesis of UCNPs by co-precipitation, and (**b**) production of core–shell $NaYF_4:Yb^{3+}$, Er^{3+}-SiO_2 UCNPs by sol–gel Stöber method. The simplified chemical reactions of the sol–gel method, hydrolysis, and condensation of TEOS, are also included. Silica gel coating method was performed either before or after the drying step.

4.3. Physicochemical Evaluation of UCNPs

ζ-potential measurements of UCNPs were performed with a ζ-potential analyzer (Malvern Panalytical, Malvern, UK). The morphology of UCNPs was evaluated by transmission electron microscopy (TEM, JEOL JEM-2010, Tokyo, Japan) operating at 200 kV. Confocal microscopy (Leica TCS-SP2 spectral confocal microscope, Leica Microsystems Heidelberg GmbH, Mannheim, Germany) was also used to qualitatively evaluate the fluorescence emission of the UCNPs powder. A 2-year physicochemical stability of UCNPs was evaluated for the most promising formulations in terms of luminescent properties by comparative TEM and confocal microscopy analysis, performed just after synthesis and after a 2-year mid-term storage of the UCNPs at RT, protected from light and stored in closed glass vials during this time.

To obtain the luminescence measurements and verify the UC process in the nanocomplexes, the UC emission spectra of the UCNPs in a 2 mm cuvette were collected in a sample holder with a ~45° angle with respect to the excitation. The excitation source consisted of a Spectra Physics Solstice-Ace laser coupled to a TOPAS Prime Amplifier, with an excitation wavelength of 980 nm (λ_{exc} = 980 nm) while the emission was recorded using an AvaSpec-ULS-TEC Avantes Senseline Fiber Optic Spectrometer System. A Newport 10SWF-800-B (BG059) short-bandpass filter was used in the emission optical path to cut the laser-scattered light. The UC spectra were collected in the 450–725 nm emission range.

Crystalline phases of the nanocomplexes were identified by X-ray diffraction (XRD) using a Bruker 8D advance diffractometer (Billerica, MA, USA). Diffractograms were obtained using Cu Kα radiation (λ = 0.154184 nm) in the 5–120° (2θ) range. The step was set at 0.03°, the recording time at 7 s per step, and the applied voltage and current at 40 kV and 40 mA, respectively. Energy-dispersive X-ray spectroscopy (EDS) was performed with a TESCAN VEGA3 SBH (Brno, Czech Republic) microscope with an EDS detector Burker XFlagh 410 M (Billerica, MA, USA) operating at 20 kV. Attenuated total reflectance/Fourier-transform infrared spectroscopy (ATR/FT-IR) was obtained with a Gladi-ATR accessory. A diamond crystal (Pike, Madison, WI, USA) was employed to analyze the chemical structure of the nanocomplexes in the mid-IR spectrum (400–4000 cm^{-1}) with 32 scans and a resolution of 2 cm^{-1}.

4.4. Viability of Mouse Embryo Fibroblasts

Cytocompatibility of the most promising UCNPs formulation (N, OD-BC) was evaluated by assessing the degradation of WST-1 into formazan, which directly correlates with the number of cells that are metabolically active. BALB/c3T3 cells (6500 cells/cm^2) were seeded in 24-well plates with 1000 µL of DMEM supplemented with 15% fetal bovine serum, penicillin 100 U/mL and streptomycin 100 g/mL. After incubation at 37 °C in a humidified atmosphere enriched with 5% CO_2 for 24 h, UCNPs (N, OD-BC) in powder form (5 mg) were UV-sterilized for 30 min and placed in a culture insert in contact with cells by triplicate. The previous literature does not evaluate concentrations above 5 mg/mL for bioimaging applications, including in vivo experiments, since lower amounts must provide a detectable fluorescence signal without compromising biocompatibility [21,28,32]. Cells alone were also tested under the same conditions as the positive controls. After 24 and 48 h of culture, inserts were removed, 25 µL of WST-1 was pipetted and plates were incubated for 2 h. Some 110 µL were then pipetted to a 96-well plate and absorbance was measured at λ = 450 nm (Infinite® M200, Tecan Group Ltd., Männedorf, Switzerland).

4.5. Hemolytic Activity of Human Red Blood Cells

The hemolytic activity of 5 and 10 mg of the most promising formulation of UCNPs (N, OD-BC) was tested by contact with human blood. Fresh human whole blood was obtained from the Galician Transfusion Center (Spain) in accordance with the Declaration of Helsinki. Firstly, blood was diluted to 3% (*v*/*v*) in 0.9% (*w*/*v*) NaCl. Then, 1000 µL of the diluted blood was poured into conical tubes containing different amounts of UCNPs (5 or 10 mg). Furthermore, 900 µL of the diluted blood was mixed with 100 µL of 4% (*v*/*v*) Triton X-100 (positive control) or 100 µL of 0.9% (*w*/*v*) PBS pH 7.4 (negative control). Samples were tested by triplicate and incubated at 37 °C and 100 rpm in an orbital shaker for 60 min. After centrifugation at 10.000g for 10 min (Sigma 2-16P, Sigma Laboratory Centrifuges, Osterode am Harz, Germany), 150 µL of the supernatant were pipetted into a 96-well plate to measure the hemoglobin absorbance at λ = 540 nm (FLUOStar Optima, BMG Labtech, Ortenberg, Germany). Hemolysis was determined by Equation (1):

$$\text{Hemolysis (\%)} = (\text{Abs}_s - \text{Abs}_n)/(\text{Abs}_p - \text{Abs}_n) \times 100 \quad (1)$$

where Abs_s is the absorbance of the samples, Abs_n is the absorbance of the negative control (0% of hemolysis), and Abs_p is the absorbance of the positive control (100% of hemolysis).

4.6. Statistical Analysis

Results of in vitro assays ($n = 3$) were reported as the mean value ± standard deviation and post hoc Tukey HSD multiple comparison tests were performed to assess the statistical significance of the differences between groups and regarding controls. Values of $p < 0.05$ were considered statistically significant.

Author Contributions: All authors contributed to the study conception, design, and investigation. Methodology, data curation, and formal analysis were performed by A.I.-M., J.P. (UC spectra) and B.F.O.C. (XRD patterns). Resources, supervision, project administration, and funding acquisition were managed by C.A.G.-G. and L.D. The original draft of the manuscript was written by A.I.-M. and all authors reviewed and edited the previous versions of the manuscript. All authors have read and agreed to the published version of the manuscript.

Funding: This work was supported by MICINN [PID2020-120010RB-I00/AEI/10.13039/501100011033; PDC2022-133526-I00/AEI/10.13039/501100011033], Xunta de Galicia [ED431C 2020/17], Agencia Estatal de Investigación [AEI] and FEDER funds. NPs synthesis and characterization at CIEPQPF, CQC-IMS and CFisUC were supported by Fundação para a Ciência e Tecnologia (FCT) [projects UIDB/EQU/00102/2020; UIDB/QUI/00313/2020; UIDB/00313/2020; UIDP/00313/2020] and TAIL-UC facility [ICT-2009-02-012-1890]. Work was carried out in the framework of the "Advanced Engineering and Research of aeroGels for Environment and Life Sciences" (AERoGELS), funded by the European Commission. A.I.-M. acknowledges the COST Action CA18125 for the granted Short Term Scientific Mission to perform the UCNPs synthesis in the Universidade de Coimbra and the Xunta de Galicia [ED481A-2020/104] for her predoctoral research fellowship.

Institutional Review Board Statement: Not applicable.

Informed Consent Statement: Not applicable.

Data Availability Statement: The data presented in this study are openly available in article.

Conflicts of Interest: The authors declare no conflict of interest.

References

1. Lahoti, H.S.; Jogdand, S.D. Bioimaging: Evolution, Significance, and Deficit. *Cureus* **2022**, *14*, e28923. [CrossRef] [PubMed]
2. Wang, C.; Li, X.; Zhang, F. Bioapplications and Biotechnologies of Upconversion Nanoparticle-Based Nanosensors. *Analyst* **2016**, *141*, 3601–3620. [CrossRef] [PubMed]
3. Hu, Y.; Wu, B.; Jin, Q.; Wang, X.; Li, Y.; Sun, Y.; Huo, J.; Zhao, X. Facile Synthesis of 5 Nm NaYF4:Yb/Er Nanoparticles for Targeted Upconversion Imaging of Cancer Cells. *Talanta* **2016**, *152*, 504–512. [CrossRef] [PubMed]
4. Gerelkhuu, Z.; Lee, Y.-I.; Yoon, T.H. Upconversion Nanomaterials in Bioimaging and Biosensor Applications and Their Biological Response. *Nanomaterials* **2022**, *12*, 3470. [CrossRef] [PubMed]
5. Malhotra, K.; Hrovat, D.; Kumar, B.; Qu, G.; Houten, J.V.; Ahmed, R.; Piunno, P.A.E.; Gunning, P.T.; Krull, U.J. Lanthanide-Doped Upconversion Nanoparticles: Exploring A Treasure Trove of NIR-Mediated Emerging Applications. *ACS Appl. Mater. Interfaces* **2023**, *15*, 2499–2528. [CrossRef] [PubMed]
6. Mahata, M.K.; De, R.; Lee, K.T. Near-Infrared-Triggered Upconverting Nanoparticles for Biomedicine Applications. *Biomedicines* **2021**, *9*, 756. [CrossRef] [PubMed]
7. Vorotnikov, Y.A.; Vorotnikova, N.A.; Shestopalov, M.A. Silica-Based Materials Containing Inorganic Red/NIR Emitters and Their Application in Biomedicine. *Materials* **2023**, *16*, 5869. [CrossRef]
8. Le, X.T.; Youn, Y.S. Emerging NIR Light-Responsive Delivery Systems Based on Lanthanide-Doped Upconverting Nanoparticles. *Arch. Pharm. Res.* **2020**, *43*, 134–152. [CrossRef]
9. Tsai, S.-R.; Hamblin, M.R. Biological Effects and Medical Applications of Infrared Radiation. *J. Photochem. Photobiol. B Biol.* **2017**, *170*, 197–207. [CrossRef]
10. Gnach, A.; Lipinski, T.; Bednarkiewicz, A.; Rybka, J.; Capobianco, J.A. Upconverting Nanoparticles: Assessing the Toxicity. *Chem. Soc. Rev.* **2015**, *44*, 1561–1584. [CrossRef]
11. Tanaka, Y.; Tsunemi, Y.; Kawashima, M.; Nishida, H. The Impact of Near-Infrared in Plastic Surgery. *Plast. Surg. Int. J.* **2013**, *2013*, 973073. [CrossRef]
12. Zhu, X.; Zhang, J.; Liu, J.; Zhang, Y. Recent Progress of Rare-Earth Doped Upconversion Nanoparticles: Synthesis, Optimization, and Applications. *Adv. Sci.* **2019**, *6*, 1901358. [CrossRef] [PubMed]
13. Yi, Z.; Luo, Z.; Qin, X.; Chen, Q.; Liu, X. Lanthanide-Activated Nanoparticles: A Toolbox for Bioimaging, Therapeutics, and Neuromodulation. *Acc. Chem. Res.* **2020**, *53*, 2692–2704. [CrossRef] [PubMed]

14. Liu, Q.; Chen, M.; Sun, Y.; Chen, G.; Yang, T.; Gao, Y.; Zhang, X.; Li, F. Multifunctional Rare-Earth Self-Assembled Nanosystem for Tri-Modal Upconversion Luminescence /Fluorescence /Positron Emission Tomography Imaging. *Biomaterials* **2011**, *32*, 8243–8253. [CrossRef] [PubMed]
15. Cheng, L.; Yang, K.; Shao, M.; Lu, X.; Liu, Z. In Vivo Pharmacokinetics, Long-Term Biodistribution and Toxicology Study of Functionalized Upconversion Nanoparticles in Mice. *Nanomedicine* **2011**, *6*, 1327–1340. [CrossRef]
16. Generalova, A.N.; Chichkov, B.N.; Khaydukov, E.V. Multicomponent Nanocrystals with Anti-Stokes Luminescence as Contrast Agents for Modern Imaging Techniques. *Adv. Colloid Interface Sci.* **2017**, *245*, 1–19. [CrossRef]
17. Min, Y.; Li, J.; Liu, F.; Padmanabhan, P.; Yeow, E.; Xing, B. Recent Advance of Biological Molecular Imaging Based on Lanthanide-Doped Upconversion-Luminescent Nanomaterials. *Nanomaterials* **2014**, *4*, 129–154. [CrossRef]
18. Zhao, J.; Lu, Z.; Yin, Y.; McRae, C.; Piper, J.A.; Dawes, J.M.; Jin, D.; Goldys, E.M. Upconversion Luminescence with Tunable Lifetime in $NaYF_4$: Yb,Er Nanocrystals: Role of Nanocrystal Size. *Nanoscale* **2013**, *5*, 944–952. [CrossRef]
19. Lingeshwar Reddy, K.; Balaji, R.; Kumar, A.; Krishnan, V. Lanthanide Doped Near Infrared Active Upconversion Nanophosphors: Fundamental Concepts, Synthesis Strategies, and Technological Applications. *Small* **2018**, *14*, 1801304. [CrossRef]
20. Homann, C.; Krukewitt, L.; Frenzel, F.; Grauel, B.; Würth, C.; Resch-Genger, U.; Haase, M. $NaYF_4$: Yb,Er/$NaYF_4$ Core/Shell Nanocrystals with High Upconversion Luminescence Quantum Yield. *Angew. Chem. Int. Ed.* **2018**, *57*, 8765–8769. [CrossRef]
21. Wang, M.; Abbineni, G.; Clevenger, A.; Mao, C.; Xu, S. Upconversion Nanoparticles: Synthesis, Surface Modification and Biological Applications. *Nanomed. Nanotechnol. Biol. Med.* **2011**, *7*, 710–729. [CrossRef]
22. Yang, L.; Shao, B.; Zhang, X.; Cheng, Q.; Lin, T.; Liu, E. Multifunctional Upconversion Nanoparticles for Targeted Dual-Modal Imaging in Rat Glioma Xenograft. *J. Biomater. Appl.* **2016**, *31*, 400–410. [CrossRef]
23. Alkahtani, M.; Alsofyani, N.; Alfahd, A.; Almuqhim, A.A.; Almughem, F.A.; Alshehri, A.A.; Qasem, H.; Hemmer, P.R. Engineering Red-Enhanced and Biocompatible Upconversion Nanoparticles. *Nanomaterials* **2021**, *11*, 284. [CrossRef]
24. Rafique, R.; Baek, S.H.; Park, C.Y.; Chang, S.-J.; Gul, A.R.; Ha, S.; Nguyen, T.P.; Oh, H.; Ham, S.; Arshad, M.; et al. Morphological Evolution of Upconversion Nanoparticles and Their Biomedical Signal Generation. *Sci. Rep.* **2018**, *8*, 17101. [CrossRef]
25. Zhou, R.; Ma, T.; Qiu, B.; Li, X. Controlled Synthesis of β-NaYF 4:Yb, Er Microphosphors and Upconversion Luminescence Property. *Mater. Chem. Phys.* **2017**, *194*, 23–28. [CrossRef]
26. Xu, D.; Li, C.; Li, W.; Lin, B.; Lv, R. Recent Advances in Lanthanide-Doped up-Conversion Probes for Theranostics. *Front. Chem.* **2023**, *11*, 1036715. [CrossRef]
27. Yan, C.; Zhao, H.; Perepichka, D.F.; Rosei, F. Lanthanide Ion Doped Upconverting Nanoparticles: Synthesis, Structure and Properties. *Small* **2016**, *12*, 3888–3907. [CrossRef]
28. Sun, Y.; Peng, J.; Feng, W.; Li, F. Upconversion Nanophosphors $Nalu_4$: Yb,Tm for Lymphatic Imaging In Vivo by Real-Time Upconversion Luminescence Imaging under Ambient Light and High-Resolution X-Ray CT. *Theranostics* **2013**, *3*, 346–353. [CrossRef]
29. Stöber, W.; Fink, A.; Bohn, E. Controlled Growth of Monodisperse Silica Spheres in the Micron Size Range. *J. Colloid Interface Sci.* **1968**, *26*, 62–69. [CrossRef]
30. Gonçalves, M.C. Sol-Gel Silica Nanoparticles in Medicine: A Natural Choice. Design, Synthesis and Products. *Molecules* **2018**, *23*, 2021. [CrossRef]
31. Muhr, V.; Wilhelm, S.; Hirsch, T.; Wolfbeis, O.S. Upconversion Nanoparticles: From Hydrophobic to Hydrophilic Surfaces. *Acc. Chem. Res.* **2014**, *47*, 3481–3493. [CrossRef]
32. Liu, F.; Zhao, Q.; You, H.; Wang, Z. Synthesis of Stable Carboxy-Terminated $NaYF_4$: Yb^{3+}, Er^{3+}@SiO_2 Nanoparticles with Ultrathin Shell for Biolabeling Applications. *Nanoscale* **2013**, *5*, 1047–1053. [CrossRef]
33. Guryev, E.L.; Smyshlyaeva, A.S.; Shilyagina, N.Y.; Sokolova, E.A.; Shanwar, S.; Kostyuk, A.B.; Lyubeshkin, A.V.; Schulga, A.A.; Konovalova, E.V.; Lin, Q.; et al. UCNP-Based Photoluminescent Nanomedicines for Targeted Imaging and Theranostics of Cancer. *Molecules* **2020**, *25*, 4302. [CrossRef]
34. Jin, J.; Gu, Y.-J.; Man, C.W.-Y.; Cheng, J.; Xu, Z.; Zhang, Y.; Wang, H.; Lee, V.H.-Y.; Cheng, S.H.; Wong, W.-T. Polymer-Coated $NaYF_4$: Yb^{3+}, Er^{3+} Upconversion Nanoparticles for Charge-Dependent Cellular Imaging. *ACS Nano* **2011**, *5*, 7838–7847. [CrossRef]
35. Chen, J.; Zhang, D.; Zou, Y.; Wang, Z.; Hao, M.; Zheng, M.; Xue, X.; Pan, X.; Lu, Y.; Wang, J.; et al. Developing a pH-Sensitive $Al(OH)_3$ Layer-Mediated UCNP@$Al(OH)_3$/Au Nanohybrid for Photothermal Therapy and Fluorescence Imaging in vivo. *J. Mater. Chem. B* **2018**, *6*, 7862–7870. [CrossRef]
36. Verma, A.; Stellacci, F. Effect of Surface Properties on Nanoparticle—Cell Interactions. *Small* **2010**, *6*, 12–21. [CrossRef]
37. Li, D.; Shao, Q.; Dong, Y.; Jiang, J. Phase-, Shape- and Size-Controlled Synthesis of $NaYF_4$:Yb^{3+}, Er^{3+} Nanoparticles Using Rare-Earth Acetate Precursors. *J. Rare Earths* **2014**, *32*, 1032–1036. [CrossRef]
38. Nahorniak, M.; Patsula, V.; Mareková, D.; Matouš, P.; Shapoval, O.; Oleksa, V.; Vosmanská, M.; Machová Urdzíková, L.; Jendelová, P.; Herynek, V.; et al. Chemical and Colloidal Stability of Polymer-Coated $NaYF_4$:Yb, Er Nanoparticles in Aqueous Media and Viability of Cells: The Effect of a Protective Coating. *Int. J. Mol. Sci.* **2023**, *24*, 2724. [CrossRef]
39. Patsula, V.; Mareková, D.; Jendelová, P.; Nahorniak, M.; Shapoval, O.; Matouš, P.; Oleksa, V.; Konefał, R.; Vosmanská, M.; Machová-Urdziková, L.; et al. Polymer-Coated Hexagonal Upconverting Nanoparticles: Chemical Stability and Cytotoxicity. *Front. Chem.* **2023**, *11*, 1207984. [CrossRef]
40. He, S.; Xia, H.; Zhang, J.; Zhu, Y.; Chen, B. Highly Efficient Up-Conversion Luminescence in Er^{3+}/Yb^{3+} Co-Doped $Na_5Lu_9F_{32}$ Single Crystals by Vertical Bridgman Method. *Sci. Rep.* **2017**, *7*, 8751. [CrossRef]

41. Lashkovskaya, E.I.; Gaponenko, N.V.; Stepikhova, M.V.; Yablonskiy, A.N.; Andreev, B.A.; Zhivulko, V.D.; Mudryi, A.V.; Martynov, I.L.; Chistyakov, A.A.; Kargin, N.I.; et al. Optical Properties and Upconversion Luminescence of $BaTiO_3$ Xerogel Structures Doped with Erbium and Ytterbium. *Gels* **2022**, *8*, 347. [CrossRef]
42. Avram, D.; Colbea, C.; Patrascu, A.A.; Istrate, M.C.; Teodorescu, V.; Tiseanu, C. Up-Conversion Emission in Transition Metal and Lanthanide Co-Doped Systems: Dimer Sensitization Revisited. *Sci. Rep.* **2023**, *13*, 2165. [CrossRef]
43. Rafique, R.; Baek, S.H.; Phan, L.M.T.; Chang, S.-J.; Gul, A.R.; Park, T.J. A Facile Hydrothermal Synthesis of Highly Luminescent $NaYF_4$:Yb^{3+}/Er^{3+} Upconversion Nanoparticles and Their Biomonitoring Capability. *Mater. Sci. Eng. C* **2019**, *99*, 1067–1074. [CrossRef]
44. Gaponenko, N.V.; Sudnik, L.V.; Vityaz, P.A.; Luchanok, A.R.; Stepikhova, M.V.; Yablonskiy, A.N.; Lashkovskaya, E.I.; Shustsikava, K.V.; Radyush, Y.V.; Zhivulko, V.D.; et al. Upconversion Luminescence of Er^{3+} Ions from Barium Titanate Xerogel Powder and Target Fabricated by Explosive Compaction Method. *J. Appl. Spectrosc.* **2022**, *89*, 238–243. [CrossRef]
45. Liu, F.; He, X.; Lei, Z.; Liu, L.; Zhang, J.; You, H.; Zhang, H.; Wang, Z. Facile Preparation of Doxorubicin-Loaded Upconversion@Polydopamine Nanoplatforms for Simultaneous In Vivo Multimodality Imaging and Chemophotothermal Synergistic Therapy. *Adv. Healthc. Mater.* **2015**, *4*, 559–568. [CrossRef]
46. Zhou, B.; Shi, B.; Jin, D.; Liu, X. Controlling Upconversion Nanocrystals for Emerging Applications. *Nat. Nanotech* **2015**, *10*, 924–936. [CrossRef]
47. Campos-Gonçalves, I.; Costa, B.F.O.; Santos, R.F.; Durães, L. Superparamagnetic Core-Shell Nanocomplexes Doped with Yb^{3+}/Er^{3+}/Ho^{3+} Rare-Earths for Upconversion Fluorescence. *Mater. Des.* **2017**, *130*, 263–274. [CrossRef]
48. Vaz-Ramos, J.; Cordeiro, R.; Castro, M.M.C.A.; Geraldes, C.F.G.C.; Costa, B.F.O.; Faneca, H.; Durães, L. Supercritically Dried Superparamagnetic Mesoporous Silica Nanoparticles for Cancer Theranostics. *Mater. Sci. Eng. C* **2020**, *115*, 111124. [CrossRef]
49. Abdul Jalil, R.; Zhang, Y. Biocompatibility of Silica Coated NaYF4 Upconversion Fluorescent Nanocrystals. *Biomaterials* **2008**, *29*, 4122–4128. [CrossRef]
50. Wong, H.-T.; Tsang, M.-K.; Chan, C.-F.; Wong, K.-L.; Fei, B.; Hao, J. In Vitro Cell Imaging Using Multifunctional Small Sized $KGdF_4$:Yb^{3+}, Er^{3+} Upconverting Nanoparticles Synthesized by a One-Pot Solvothermal Process. *Nanoscale* **2013**, *5*, 3465. [CrossRef]
51. Rostami, I. Empowering the Emission of Upconversion Nanoparticles for Precise Subcellular Imaging. *Nanomaterials* **2021**, *11*, 1541. [CrossRef]
52. Yi, G.; Lu, H.; Zhao, S.; Ge, Y.; Yang, W.; Chen, D.; Guo, L.-H. Synthesis, Characterization, and Biological Application of Size-Controlled Nanocrystalline $NaYF_4$: Yb,Er Infrared-to-Visible Up-Conversion Phosphors. *Nano Lett.* **2004**, *4*, 2191–2196. [CrossRef]
53. Milićević, B.; Periša, J.; Ristić, Z.; Milenković, K.; Antić, Ž.; Smits, K.; Kemere, M.; Vitols, K.; Sarakovskis, A.; Dramićanin, M. Hydrothermal Synthesis and Properties of Yb^{3+}/Tm^{3+} Doped Sr_2LaF_7 Upconversion Nanoparticles. *Nanomaterials* **2022**, *13*, 30. [CrossRef]
54. Zhou, X.; Wang, Z.; Li, S.; Shan, S.; Wang, X. Formation and Luminescence of Sodium Rare Earth Fluoride Nanocrystals in the Presence of Chelators. *J. Nanosci. Nanotechnol.* **2010**, *10*, 2193–2202. [CrossRef]
55. Grzyb, T.; Przybylska, D. Formation Mechanism, Structural, and Upconversion Properties of Alkaline Rare-Earth Fluoride Nanocrystals Doped with Yb^{3+}/Er^{3+} Ions. *Inorg. Chem.* **2018**, *57*, 6410–6420. [CrossRef]

gels

MDPI

Article

Electrochemical Storage Behavior of a High-Capacity Mg-Doped P2-Type $Na_{2/3}Fe_{1-y}Mn_yO_2$ Cathode Material Synthesized by a Sol–Gel Method

Mobinul Islam [1,*], Md. Shahriar Ahmed [1], Daseul Han [1], Gazi A. K. M. Rafiqul Bari [2] and Kyung-Wan Nam [1,*]

[1] Department of Energy & Materials Engineering, Dongguk University, Seoul 04620, Republic of Korea; endend42@naver.com (D.H.)
[2] School of Mechanical Smart and Industrial Engineering, Gachon University, Seongnam 13120, Republic of Korea; grafiqulbari@gachon.ac.kr
* Correspondence: mobin85@dongguk.edu (M.I.); knam@dongguk.edu (K.-W.N.); Tel.: +82-2-2260-4978 (K.-W.N.)

Abstract: Grid-scale energy storage applications can benefit from rechargeable sodium-ion batteries. As a potential material for making non-cobalt, nickel-free, cost-effective cathodes, earth-abundant $Na_{2/3}Fe_{1/2}Mn_{1/2}O_2$ is of particular interest. However, Mn^{3+} ions are particularly susceptible to the Jahn–Teller effect, which can lead to an unstable structure and continuous capacity degradation. Modifying the crystal structure by aliovalent doping is considered an effective strategy to alleviate the Jahn–Teller effect. Using a sol–gel synthesis route followed by heat treatment, we succeeded in preparing an Mg-doped $Na_{2/3}Fe_{1-y}Mn_yO_2$ cathode. Its electrochemical properties and charge compensation mechanism were then studied using synchrotron-based X-ray absorption spectroscopy and in situ X-ray diffraction techniques. The results revealed that Mg doping reduced the number of Mn^{3+} Jahn–Teller centers and alleviated high voltage phase transition. However, Mg doping was unable to suppress the P2-P′2 phase transition at a low voltage discharge. An initial discharge capacity of about 196 mAh g^{-1} was obtained at a current density of 20 mAh g^{-1}, and 60% of rate capability was maintained at a current density of 200 mAh g^{-1} in a voltage range of 1.5–4.3 V. This study will greatly contribute to the ongoing search for advanced and efficient cathodes from earth-abundant elements for rechargeable sodium-ion batteries operable at room temperature.

Keywords: sol–gel; gel precursor; sodium-ion battery; energy storage; cathode material

1. Introduction

Population growth, technological advancement, and industrialization have led to a substantial increase in energy demand. Fossil fuels have been the primary energy source for the past few decades. However, alternative energy sources that could replace fossil fuels are needed to preserve our planet and achieve green energy for sustainable human development. There is an increasing interest in renewable energy sources such as wind, solar, etc. However, such renewable energy sources have inherent limitations related to their intermittency and geographic dispersion that need large-scale electric energy (grid storage) systems. Batteries are considered the most promising option for grid storage in the future due to their low operating costs and long life cycles [1]. Lithium-ion batteries (LIBs) have the highest energy density among modern batteries. However, lithium is a scarce element in the Earth's crust, and it is not widely distributed [2]. Additionally, the use of portable electronics and electric vehicles has increased exponentially in recent years, resulting in an increased demand for lithium supplies and price hikes. Alternatively, sodium-ion batteries (SIBs) could complement LIBs due to abundant available sources and suitable redox potential of sodium ions ($E_{(Na/Na+)} = -2.71$ V vs. SHE) [3]. SIBs are ideal grid-scale storage solutions where battery energy density is not a major factor. However, the lack of appropriate and cost-effective cathode materials that can provide high capacity and excellent cycle stability has made it difficult to utilize SIBs in practical applications.

Various types of SIB cathode materials have been reported, with layered-type oxide materials ($NaxTMO_2$ = Mn, Co, Fe, Ni, Ti, Cu, Cr, or mixtures of these elements) being the most promising due to their large capacities and tunable properties [4]. Layered oxides with Fe and Mn as redox centers in positive electrode materials have special importance because both Fe and Mn are cheap, abundant, and environmentally friendly. It is important to note that although Na is ten times cheaper than Li, the per-molar cost of transition metals Co is four times higher, and Ni is 0.7 times higher than that of Li. Therefore, the economic competitiveness of SIBs will be greatly diminished if they use a significant fraction of Co and Ni. According to the nomenclature proposed by Delmas et al. [5], layered oxides are categorized into two main groups: O3 and P2. The P2-Na_xFeO_2 phase does not exist. Although O3-type $NaFeO_2$ is electrochemically active, its reversible capacity (90 mAh g^{-1}) is less than half of its theoretical capacity (242 mAh g^{-1}) [6]. P2-$Na_{0.67}MnO_2$ provides a substantial initial capacity of 170 mAh g^{-1}, which is close to its theoretical capacity [7]. Despite this, Na_xMnO_2 is subject to an irreversible P2→OP4 phase transition due to the gliding of MnO_2-slabs, which can lead to a decay of capacity with time [8]. Compared to a P2 phase, an O3-$NaMnO_2$ phase contains more sodium with a greater capacity. However, O3-type oxides are easily contaminated due to hygroscopic degradation. They are also prone to more complex phase transitions during the electrochemical process, which can cause unsatisfactory cycling performance [9]. A P2-type oxide exhibits relatively high resistance to hygroscopic degradation. Thus, it can maintain its original structure over a wide voltage range. In other words, P2-type oxides have a significant practical advantage in handling and processing.

A combination of Na_xFeO_2 and Na_xMnO_2 is also an attractive choice as a cobalt-free cathode. It is noteworthy that the P2-type $Na_{2/3}Fe_{1/2}Mn_{1/2}O_2$ has a relatively high initial capacity of 190 mAh g^{-1}, higher than all Fe or Mn-based layered oxides reported [6–10]. Several other derivatives of Fe/Mn-based combined layered oxides have also been investigated since the first report of auspicious properties of P2-$Na_{2/3}Fe_{1/2}Mn_{1/2}O_2$ [10]. $Na_{2/3}Fe_{2/3}Mn_{1/3}O_2$, a Fe-rich derivative with a P2-type structure, has an initial reversible capacity of approximately 151 $mAhg^{-1}$ [11]. In $Na_{2/3}Fe_{2/3}Mn_{1/3}O_2$, Mn is used as both a promoter and a stabilizer for P2-phase reactions because it is impossible to synthesize P2-$Na_{2/3}FeO_2$. Mn content is likely to be associated with better electrochemical performance. For example, a Mn-rich layered system P2-$Na_{2/3}Fe_{1/3}Mn_{2/3}O_2$ can deliver a reversible capacity of 193 $mAhg^{-1}$ [12]. Nevertheless, P2-type $Na_xFe_{1-y}Mn_yO_2$ cathode systems suffer from poor capacity retention, hindering their practical and commercial applications. Research has identified irreversible phase changes as a major contributing factor to its poor capacity retention [13,14]. An SIB cathode material is usually designed with the goal of either increasing energy density or seeking structural stability. By doping/substituting some foreign elements, such as Al, Ca, Ti, Cr, and Mg, the crystal structure can be stabilized, and phase transformations can be restrained in the case of P2-type layered oxide materials, resulting in a more stable cycle [15,16].

The majority of Na-Fe-Mn-O-based cathode samples have been synthesized by conventional solid-state methods [17]. Sol–gel synthesis offers more control over synthesis parameters than solid-state synthesis. It is possible to fine-tune sol–gel routes by assessing many independent synthesis parameters (e.g., gelling temperature, gel aging, chelating agent, gel formation solution composition, gel precursor nature, precursor ratio, solution environment, and annealing conditions) to prepare specific crystal morphologies and defects [18]. In the present study, we present a citric-assisted sol–gel synthesis of P2-$Na_{2/3}Mg_{2/9}Fe_{2/9}Mn_{5/9}O_2$ (NFM-Mg), an Mg-doped Mn-rich earth-abundant cathode with a P2 structure, and its ability to reversibly work in sodium-ion batteries using an aprotic electrolyte. Different studies are in progress to decouple the effect of various chelating agents and the nature of the gel precursors (chelate gel and organic polymeric gel) on particle morphology and charge storage capability of the studied material. The results will be published elsewhere. A P2-type $Na_{2/3}Fe_{1/2}Mn_{1/2}O_2$ (NFM) cathodes contain both Mn^{3+} and Mn^{4+} ions. Mn has an average oxidation state of 3.66. When Mg doping is applied to the NFM-Mg cathode, the Mn oxidation state is pushed toward 4+, which minimizes the number of Mn^{3+} Jahn–Teller centers and delays or erases the high voltage phase transition that is typically experienced on P2-$Na_{2/3}Fe_{1/2}Mn_{1/2}O_2$ and P2-$Na_{2/3}MnO_2$ cathodes [8,13]. The major goal of this manuscript is to shed light on the NFM-Mg charge compensation mechanism to understand and search for ways to enhance structural stability and minimize capacity fading in a Na-Fe-Mn-O-based cathode system upon cycling.

2. Results and Discussion

2.1. Crystal Structure

The Rietveld refinement on the synchrotron X-ray diffraction (XRD, Rigaku, Tokyo, Japan) data of the prepared sample is presented in Figure 1. The XRD pattern can be fully indexed to the hexagonal P6$_3$/mmc space group, with diffraction peaks mainly at 2θ = 15.63°, 31.67°, 35.42°, 38.86°, 43.01°, 48.30°, 61.37°, 63.48°, and 65.88°. The refinement is in good agreement with the P2-single phase [10], and all the peaks are identical to the standard values of $Na_{2/3}Fe_{1/2}Mn_{1/2}O_2$ (ICSD No. 194-731). There was no noticeable impurity or broadened peaks evident across the entire structure. Prominent and sharp diffraction lines indicated that the as-synthesized compound was highly crystalline. The refined lattice parameters were a = 2.91062 Å and c = 11.19962 Å, as shown in Table 1. The reliability R-factors were R_{wp} = 15.4 and R_{exp} = 7.62; nevertheless, this is not a highly ideal result, which may be due to the uncontrollable stacking faults in layered compounds. The crystal structure of the prepared sample consisting of ordered stacked MO_2 (M = Fe, Mn, and Mg) and Na layers is shown in Scheme 1. It could be defined as prismatic 2 (P-2) because Na^+ ions occupied two different sites (Na_f and Na_e) between prismatic AB-BA-type layers.

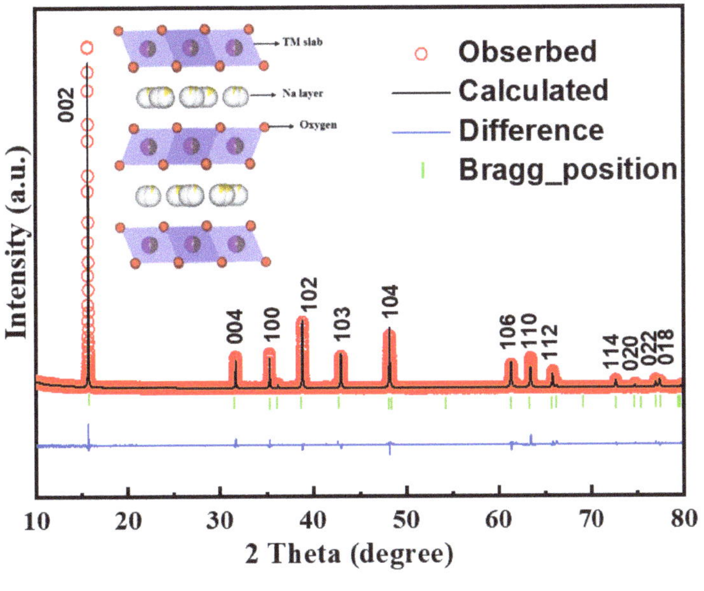

Figure 1. Rietveld refinement result for the P2-$Na_{2/3}Mg_{2/9}Fe_{2/9}Mn_{5/9}O_2$ compound on the synchrotron high-resolution powder diffraction (HRPD) data. The brag positions correspond to the hexagonal P2-$Na_{2/3}Fe_{1/2}Mn_{1/2}O_2$ phase (P6$_3$/mmc).

Table 1. Crystallographic parameters for P2-$Na_{2/3}Mg_{2/9}Fe_{2/9}Mn_{5/9}O_2$ compound refined by the Rietveld method.

Crystal System			**Hexagonal**		
Space Group			***P*6$_3$/*mmc***		
Lattice Parameter			**$a = b$ = 2.91062 Å, c = 11.19962 Å** **α = 90°, β = 90°, γ = 120°**		
Atom	*x*	*y*	*z*	Occupancy	B_{iso} (Å^3 × 10^2)
Na1	0.00000	0.00000	0.25000	0.119	6.380
Na2	0.33333	0.66667	0.75000	0.216	6.380
Fe	0.00000	0.00000	0.00000	0.110	1.215
Mn	0.00000	0.00000	0.00000	0.280	1.215
Mg	0.00000	0.00000	0.00000	0.110	1.215
O	0.33333	0.66667	0.08443	1.000	2.582
R_p: 11.0, R_{wp}: 15.4, R_{exp}: 7.62, χ^2 = 4.10					

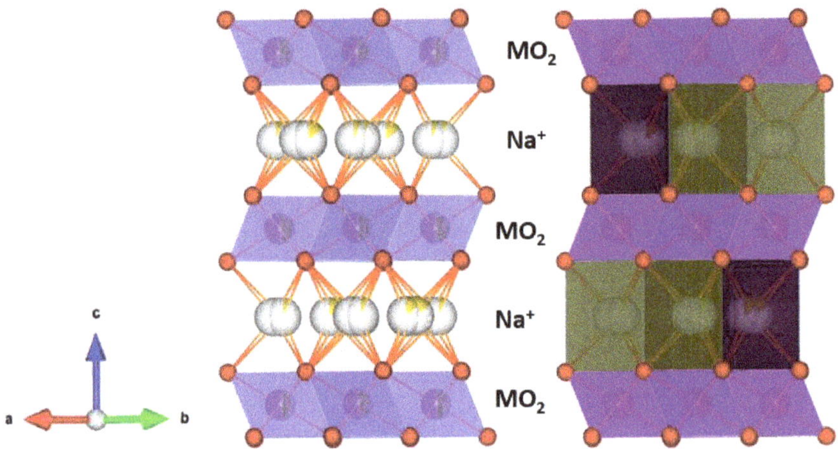

Scheme 1. Schematic illustration of P2-$Na_{2/3}Mg_{2/9}Fe_{2/9}Mn_{5/9}O_2$ structure.

Surface-sensitive X-ray photoelectron spectroscopy (XPS) was applied to clarify the valence states and composition of the as-synthesized P2-$Na_{2/3}Mg_{2/9}Fe_{2/9}Mn_{5/9}O_2$ material, and the results are shown in Figure 2. The deconvoluted XPS spectra of Mn 2p (Figure 2a) are fitted into two peaks at 643.67 eV and 655.1 eV, which can be assigned to Mn^{4+} ($2p_{3/2}$) and Mn^{4+} ($2p_{1/2}$) species, respectively [19,20]. The Fe 2p spectrum exhibits two peaks at 724.4 and 711.1 eV, corresponding to Fe $2p_{1/2}$ and Fe $2p_{3/2}$, respectively (Figure 2b). These peaks can be indexed to the Fe (+3) oxidation state [21]. The O 1s spectrum can be fitted into three components (Figure 2c), which can be attributed to the lattice oxygen (529.6 eV) and to surface-adsorbed species (531.5 and 535.6 eV) [20]. Figure 2d,e show the XPS spectra of the Na 1s and Mg 2p peaks, respectively, with binding energies of 1071.5 and 1303.75 eV, confirming the presence of Na^+ and Mg^{2+} ions in the as-synthesized P2-$Na_{2/3}Mg_{2/9}Fe_{2/9}Mn_{5/9}O_2$ material.

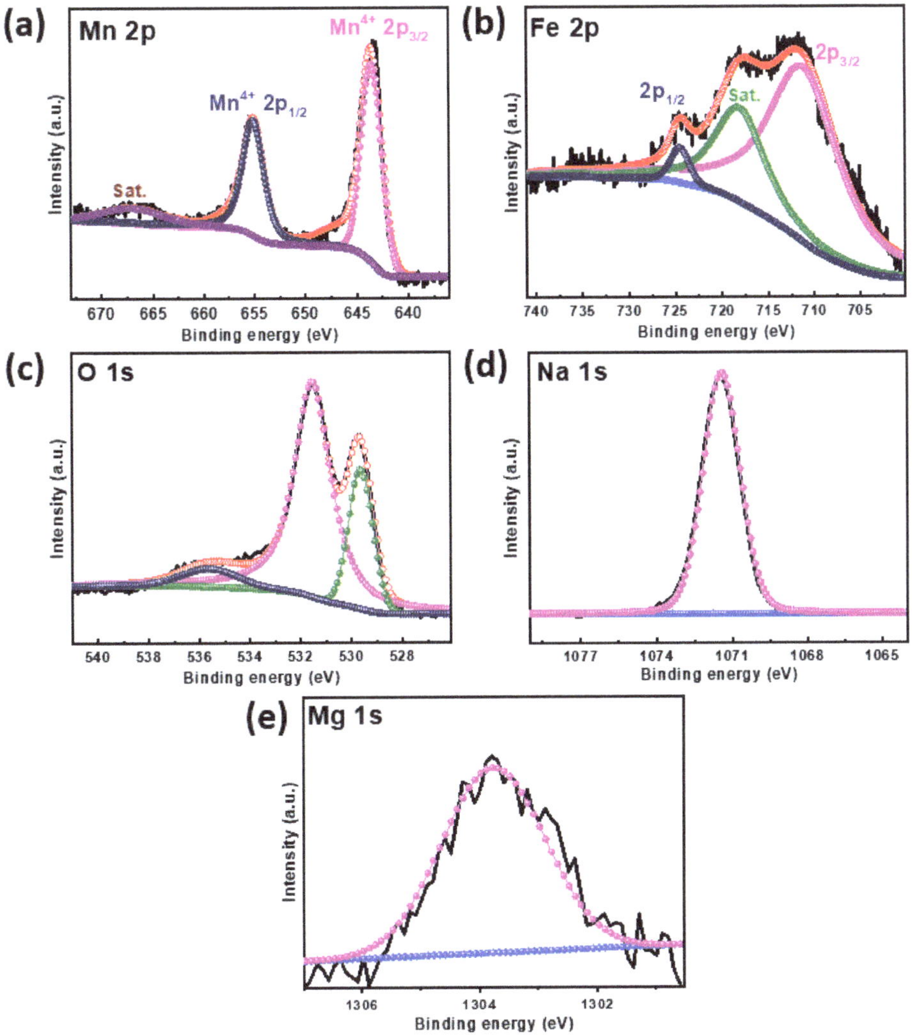

Figure 2. XPS analyses of P2-$Na_{2/3}Mg_{2/9}Fe_{2/9}Mn_{5/9}O_2$ material: the high-resolution XPS spectra of (**a**) Mn 2p, (**b**) Fe 2p, (**c**) O 1s, (**d**) Na 1s, and (**e**) Mg 1s.

Morphology of the P2-type $Na_{2/3}Mg_{2/9}Fe_{2/9}Mn_{5/9}O_2$ sample synthesized at 850 °C was examined by SEM. The SEM image is shown in Figure 3. Particles in the sample exhibited a uniform distribution. They were mainly plate-shaped sub-micron particles with sizes ranging from 1 µm to 4 µm, often seen for a P2-layered structure [16,22,23]. The atomic compositions of the sample were measured by plasma-optical emission spectroscopy (ICP-OES), and the results are listed in Table 2. The as-prepared sample has an atomic ratio that is almost identical to the designed stoichiometric ratio. In addition, we utilized energy-dispersive X-ray spectroscopy (EDS) coupled with SEM to check the elemental distribution and atomic composition of the as-prepared sample. Figure 3c shows that all elements, such as Mn, Fe, Mg, Na, and O, are homogeneously distributed within the sample. Table S1 shows that the atomic ratio of Mg, Fe, and Mn is 0.22:0.21:0.55. These values are very close to the projected values of these elements in the P2-$Na_{2/3}Mg_{2/9}Fe_{2/9}Mn_{5/9}O_2$ material.

Figure 3. SEM micrographs of as-synthesized P2-$Na_{2/3}Mg_{2/9}Fe_{2/9}Mn_{5/9}O_2$ at (**a**) low and (**b**) high magnification and (**c**) corresponding SEM-EDS elemental mapping images.

Table 2. ICP-OES results of the sample.

Theoretical Chemical Formula	Measured Atomic Ratio			
	Na	Mg	Fe	Mn
$Na_{0.67}Mg_{0.22}Fe_{0.22}Mn_{0.56}O_2$	0.684	0.228	0.220	0.570

2.2. Electrochemical Performance

Charge valance of the benchmark compound, P2-$Na_{2/3}Fe^{3+}{}_{1/2}Mn^{3+}{}_{x}Mn^{4+}{}_{1/2-x}O_2$ (NFM), indicates the presence of Jahn–Teller active Mn^{3+} ions. Our designed composition formula of P2-$Na_{2/3}Mg^{2+}{}_{2/9}Fe^{3+}{}_{2/9}Mn^{4+}{}_{5/9}O_2$ (NFM-Mg) was supposed to consist of non-Jahn–Teller Fe^{3+} and Mn^{4+} ions. Electrochemical performance of the designed NFM-Mg as a positive electrode material for Na-ion batteries was evaluated against Na metal anodes with a liquid electrolyte of 1.2 M $NaPF_6$ in 9:9:2 v/v ethylene carbonate/dimethyl carbonate (DMC)/propylene carbonate (PC) of coin cells. Figure 4a shows typical charge and discharge profiles of the NFM-Mg cathode at 20 mA g^{-1}. The NFM-Mg electrode voltage curve exhibited a slope accompanied by a well-defined voltage plateau at 4.2 V, which could be attributed to Fe^{3+}/Fe^{4+} and/or oxygen redox reactions during the first charge. Considering that NFM-Mg has tetravalent Mn to maintain charge balance, its oxidation from Mn^{4+} to Mn^{5+} would be impossible because Mn^{5+} is not stable in octahedron coordination. The NFM-Mg voltage curve during the first charge was quite different from that of NFM (Figure S2a). The NFM had a two-step slope corresponding to Mn^{3+}/Mn^{4+} and Fe^{3+}/Fe^{4+} [10,13]. The appearance of the voltage plateau (at ca.

4.2 V vs. Na) in Figure 4a reminds us about anionic redox (O^{2-}/O^{n-}) based Na cathode material's electrochemical behavior in Na-cell [24,25]. It suggests that oxide ions might contribute to the redox process. The NFM-Mg delivered ≈152.5 mAh g^{-1} of initial charge capacity and ≈196 mAh g^{-1} of discharge capacity between 1.5 V and 4.3 V. Note that the theoretical capacity based on a single-electron redox process of a Fe^{3+}/Fe^{4+} redox couple is only ≈45 mAh g^{-1}. This means that the capacity during the initial charge process might be attributed to oxygen redox (O^{2-}/O^{n-}). The second charge and discharge capacities of the NFM-Mg cathode were 198.5 and 190 mAh g^{-1}, respectively. On the other hand, the NFM delivered ≈112.5 mAh g^{-1} of initial charge capacity and ≈183 mAh g^{-1} of discharge capacity, followed by the second charge (≈180.7 mAh g^{-1}) and discharge (≈178.3 mAh g^{-1}) capacities at the same condition (Figure S2a). In comparison with the NFM performance, the NFM-Mg clearly showed a better charge storage capacity. Anodic and cathodic peaks and their evolution upon cycling were more visible in the dQ/dV plots, as shown in Figures 4b and S2b. The high voltage region showed one anodic peak at around 4.2 V during the first charge, which shifted to 4.05 V in the following charge cycle, consistent with the voltage profile shown in Figure 3a. A broad cathodic peak at around 2.8 V in the differential capacity curve corresponded to oxygen reduction [26]. It was about 1.5 V away from the oxidation peak. The reasons for this strong voltage hysteresis are unknown at this point. In contrast, the Fe^{3+}/Fe^{4+} redox peaks at around 4.1 and 3.25 V dominated the high-voltage region of the NFM cathode material (Figure S2b). The redox peaks in the low-voltage region of the NFM electrode are associated with the Mn^{3+}/Mn^{4+} redox reaction. Although there were no redox peaks associated with the Fe^{3+}/Fe^{4+} redox reaction [23], a pair of anodic/cathodic peaks at around 2.0 and 1.85 V associated with the Mn^{3+}/Mn^{4+} redox reaction dominated the low-voltage region of the NFM-Mg cathode. A detailed charge compensation mechanism is discussed in the next section. The reversible capacity gradually dropped from the initial 196 mAh g^{-1} to 162 mAh g^{-1} at the 30th cycle and 134 mAh g^{-1} at the 80th cycle, as shown in Figure 4c. After 40 cycles, the Mg-doped NFM electrode exhibited better capacity retention (≈77%) than the undoped NFM electrode (≈66%), and the 82% capacity retention of the NFM-Mg electrode after 30 cycles at 20 mA g^{-1} was marginally better than that described in the study of Yabuuchi et al. [10]. They reported that approximately 80% capacity retention was obtained at a rate of 12 mA g^{-1} (after 30 cycles) under the same voltage range. The rate capability of the NFM and NFM-Mg cathodes is shown in Figure 4d. The NFM-Mg cathode material could release average reversible discharge capacities of 158, 137, 117, and 68 mAh g^{-1} at 40, 100, 200, and 600 mA g^{-1}, respectively. It is evident that the Mg-doped NFM electrode demonstrates improved rate capability compared with the NFM electrode, particularly at high current densities. The NFM-Mg electrode could retain approximately 60% of its reversible capacity when the current density was increased ten times (from 20 to 200 mAh g^{-1}). This rate performance is impressive for a layered Fe/Mn-based Na-ion battery cathode. The interlayer spacing of the Mg-doped sample usually becomes larger due to a larger radius of Mg (2+) ions than those of Mn and Fe cations, which might lead to a faster diffusion of Na^+ ions [27].

2.3. Charge Compensation and Crystal Structure Evolution

We measured NFM-Mg electrode samples with various electrochemical states during the first one-and-a-half cycles using synchrotron ex situ spectroscopic techniques to understand better the roles played by Fe, Mn, and O in electrochemical processes and comprehend the charge compensation mechanism. Hard X-ray absorption spectroscopy (hXAS) spectra were collected on Mn and Fe K-edges to examine variations in valence state and evolution of local structure. To examine how the electronic structure of the TM3d-O2p hybrid changed under different states of sodium intercalation/deintercalation, ex situ soft X-ray absorption spectroscopy (sXAS) spectra were also collected on Mn $L_{2,3}$-edge and O K-edge. Herein, TM was defined as transitional metals (Mn, Fe).

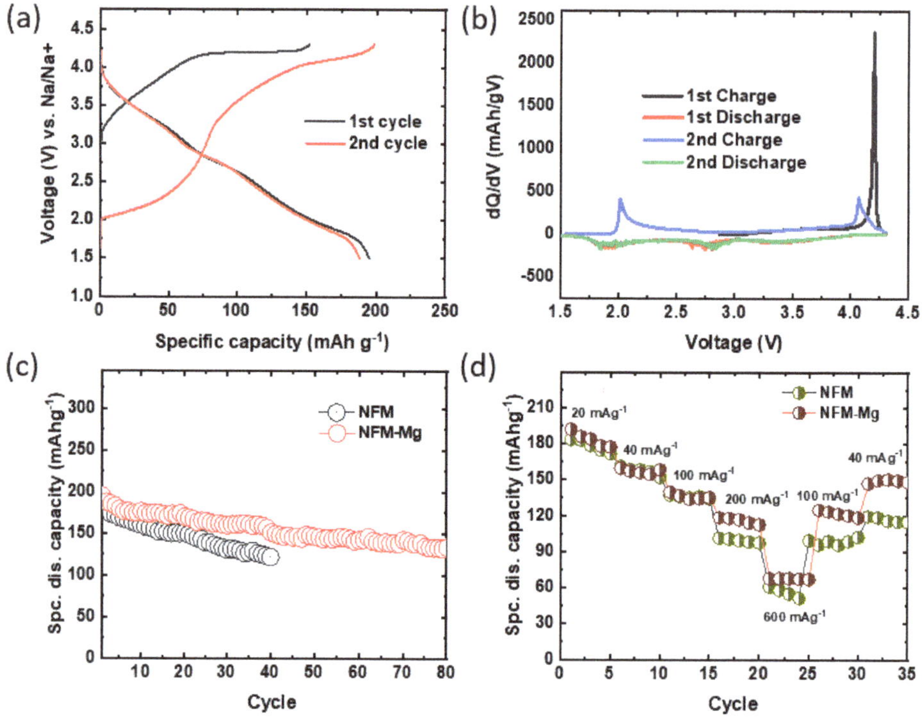

Figure 4. (**a**) Galvanostatic voltage profiles; (**b**) dQ/dV curves of the NFM-Mg cathode material; (**c**) cycling; (**d**) rate capability performance comparison between P2-$Na_{2/3}Fe_{1/2}Mn_{1/2}O_2$ (NFM) and P2-$Na_{2/3}Mg_{2/9}Fe_{2/9}Mn_{5/9}O_2$ (NFM-Mg) cathode materials in Na-ion half-cell.

Figure 5a,b present normalized X-ray absorption near edge structure (XANES) spectra of Mn and Fe K-edges, respectively, at different sodiation and desodiation states. The valence state of Fe in the pristine sample was approximately 3+ based on a comparison between Fe K-edge XAS spectra of the pristine sample and standard reference compounds. On the other hand, Mn in NFM-Mg had a valence close to 4+ in comparison with the standard reference spectrum. During initial charging from OCV (pristine) to 4.3 V (Full Ch), the Mn K-edge showed a very small change in absorption peak shape, which could not be attributed to oxidation change. It was surprising that the Fe K edge did not shift towards higher energy regions during initial charging, indicating that the Fe valence state remained unchanged ($\approx$+3). This suggested that the initial charge capacity came from only a redox reaction on oxygen anions, as discussed in the previous section. It was found that the Fe K-edge again remained static upon discharge. On the other hand, Mn K-edge substantially shifted to lower energies after discharge to 1.5 V, indicating that Mn was reduced to a lower valence state at the end of the discharge process. It appeared that the first discharge capacity mainly resulted from the reduction in Mn ions. After the second charge, the spectral position of the Mn K-edge was almost identical to that of MnO_2, indicating that the Mn^{4+} oxidation state was reestablished. In the second charge, the Fe K-edge moved to a slightly higher energy position, suggesting that Fe had a very minimal contribution to the second charge capacity. Studies of those redox behaviors during long-range cycling are in progress. Figure 5c,d show the evolution of Fourier-transformed extended X-ray absorption fine structure (FT-EXAFS) spectra with cycling. Both Fe and Mn K-edge EXAFS showed two strong peaks. The first peak between 1 and 2 Å was related to the average TM-O length of the first shell TM-O6 octahedron. The second peak between 2 and 3 Å was related to bonds between transition metals in the second shell of the TM-TM6 hexagon. From initial charge (4.3 V) to full discharge (1.5 V) to second charge up to 4.3 V, all interatomic Mn-O, Fe-O, and TM-TM bond lengths underwent a reversible decrease/increase upon desodiation/sodiation. A considerable displacement of these interatomic bond distances from the pristine sample indicated a distortion of Mn and Fe atoms' local environment during the insertion/extraction of Na^+ ions, possibly induced by irreversible oxygen loss.

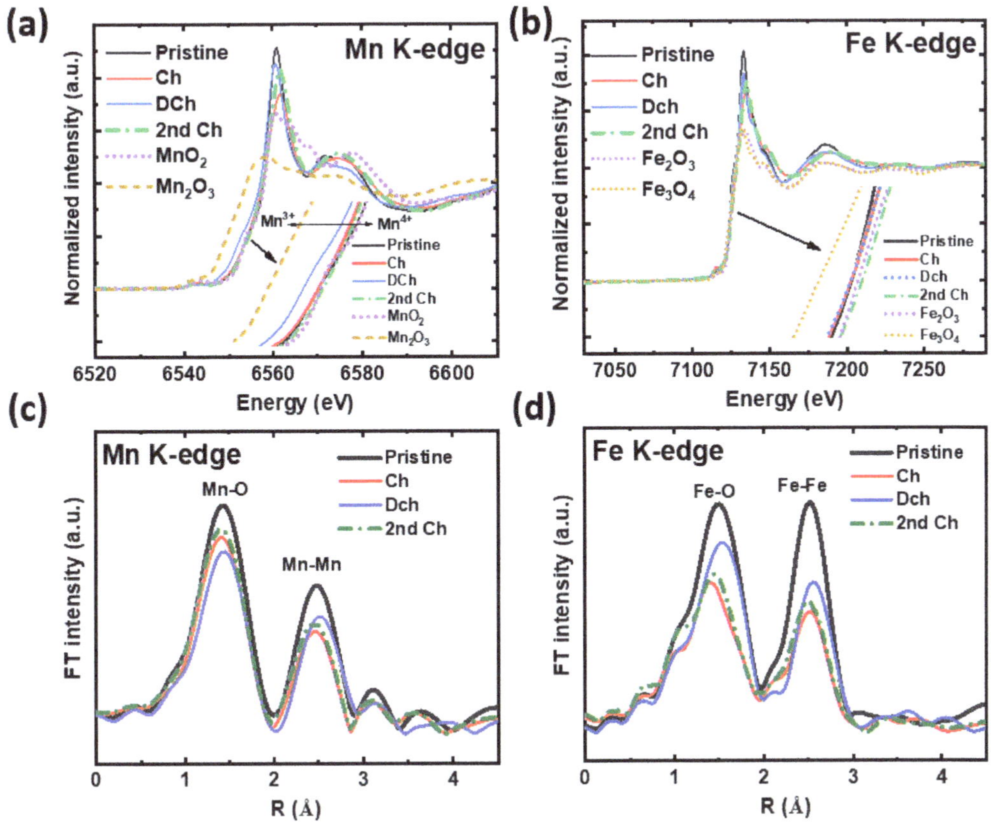

Figure 5. Ex situ XAS analysis of NFM-Mg electrodes at different state of charges during the first and half electrochemical cycle: (**a**) XANES and (**b**) corresponding EXAFS spectra at Mn K-edge; (**c**) XANES and (**d**) corresponding EXAFS spectra at Fe K-edge. Electrodes collected at the end of first charge, discharge, and second charge.

Metal $L_{2,3}$-edge absorption peaks resulting from electric dipole-allowed 2p→3d transition are sensitive to the oxidation state and spin state [28]. Total electron yield (TEY) mode was used to collect sXAS data at the Mn $L_{2,3}$-edge, as shown in Figure 6a. Splitting of the high energy range (metal L_2-edge) and low energy range (metal L_3-edge) corresponded to the interaction and respective electronic transitions from Mn $2p_{1/2}$ and $2p_{3/2}$ core levels to highly hybridized O2p-Mn3d orbitals, respectively. The Mn $L_{2,3}$ spectral feature of the pristine NFM-Mg sample resembled that of MnO_2 [29], suggesting that Mn was in its tetravalent (4+) state. The Mn L-edge sXAS did not change significantly after charging the NFM-Mg electrode to 4.3 V, indicating that Mn ions were maintained in their Mn 4+ states after desodiation. After sodiation to 1.5 V, the peak of Mn L_3-edge significantly shifted to a lower energy area, suggesting that the metal was being reduced from 4+ to a lower valence. The spectral shape of Mn $L_{2,3}$-edge after the second charge was identical to that of the first charge, which again evidenced the reversible oxidation/reduction in the Mn^{3+}/Mn^{4+} redox couple.

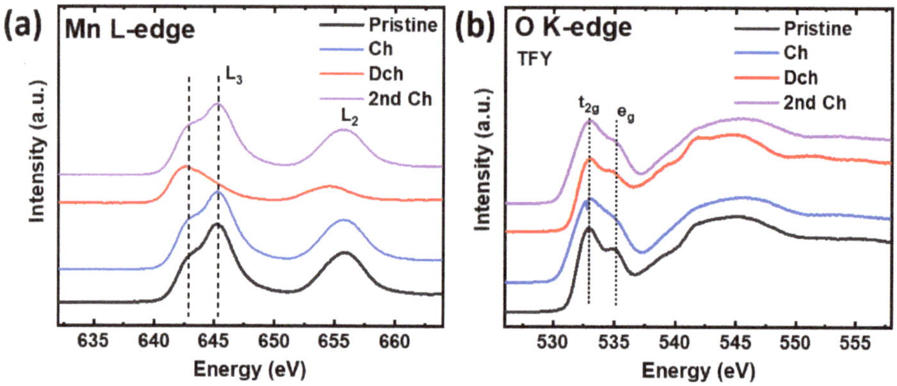

Figure 6. NFM-Mg cycled electrodes: (**a**) Mn $L_{2,3}$-edge XAS in TEY mode; (**b**) O K-edge XAS in TFY mode.

Next, O K-edge sXAS data were collected in bulk-sensitive total fluorescence yield (TFY) mode. Results are presented in Figure 6b, where two distinct regions can be distinguished from the spectra. A pre-edge feature below 535 eV corresponded to an excitation from core-level O 1s orbitals to empty and hybridized (with Metal 3d) O 2p orbitals. The broad and wide hump above 535 eV represented the transition from O 1s orbitals to O 2p-TM 4sp hybridized states. Due to this, an O K-edge sXAS feature cannot be used directly to prove oxygen redox reactions. Instead, mapping of resonant inelastic X-ray scattering (mRIXS) is a powerful tool for exploring the lattice oxygen redox because this technique disentangles the intrinsic oxygen redox from the TM-O hybridization [30–32]. The sXAS O K-edge was only a measure of the change in TM-O hybridization strength. Valence states of TM dominated spectral line shapes of two pre-edge peaks assigned to t_{2g} and e_g states [26]. In Figure 6b, the dip between t_{2g} and e_g states were filled during the first charge process. Note that the lattice oxygen redox could be confirmed by identifying the emergence of a specific feature between t_{2g} and e_g peaks in the excitation energy coordinate system in mRIXS [32]. This filled dip might indirectly indicate that lattice oxygen redox participated in charge compensation during the first charge process. Because of electron filling in the t_{2g} energy, the intensity ratio of t_{2g} and e_g states decreased after discharge. During the second charge, the intensity ratio increased again, but the dip between t_{2g} and e_g states remained unfilled. The exact explanation for this phenomenon is ambiguous at this stage. Nevertheless, in combination with the electrochemical (dQ/dV) analysis discussed earlier, it was evident that there was a substantial difference between the first cycle and the second cycle's lattice oxygen contributions to the charge compensation process.

To study structural rearrangements caused by cycling in the material, a synchrotron-based in situ XRD experiment was carried out. The experiment started with the material in the OCV state (a fresh electrode), which corresponds to the composition P2-$Na_{2/3}Mg_{2/9}Fe_{2/9}Mn_{5/9}O_2$. The results are shown in Figure 7. The P2-type hexagonal structure of the NFM-Mg cathode material was maintained during electrochemical desodiation. During the timespan of charge (desodiation), P2 (100), (102), and (112) peaks gradually changed directions towards higher angles, while P2 (002) and (110) peaks deviated towards lower angles. Moving peaks were correlated with decreasing a-axis and increasing **c**-axis. As oxygen anions are repelled between two transition-metal layers during Na^+ extraction (charge), the interlayer distance between them increases, which corresponds to **c**-axis expansion. Notably, when the cell was discharged, parameters returned in the original direction, indicating that the structure was reversible. The P2 phase (blue color) is retained throughout the whole charge process, except for a stacking fault (OP4 phase) formation. The origin of this intermediate OP4 phase can be attributed to irreversible oxygen loss. However, the detrimental "Z/OP4 phase", which is usually observed at around $2\theta \approx 17^\circ$ for the P2-$Na_{2/3}Fe_{1-y}Mn_yO_2$-type compound [13,33] when cycling above 4.0 V vs. Na^+/Na, did not appear here. A P2-to-"Z"/OP4-type phase transition associated with large volume changes is accountable for capacity fading [34]. In short, the introduction of Mg^{2+} ions into the crystal structure successfully suppressed unknown "Z" phase formation. It appears that the P2 phase (green color) continues at the initial stage of the discharge process. Nevertheless, discharging toward a lower voltage of ~1.5 V, an additional phase (orthorhombic P′2, space group: Cmcm) gradually appeared, accompanied by a reduction in Mn^{4+} ions, which was evidenced by a splitting of (002) and (100) peaks. Figure 7 clearly identifies the biphasic mechanism (red and green mixed colors) of the discharge process, which occurs after an intermediate solid solution (yellow color) region. It has been reported that P2 and P′2 phases have a significant volume difference (4.96 Å^3) in the low voltage (~1.5 V) region [35], which could cause serious structural separation. It might also lead to mechanical strain during cycles. In summary, our in situ XRD results illustrate that Mg-doping in a P2 type $Na_{2/3}Fe_{1-y}Mn_yO_2$ compound can significantly inhibit unknown "Z" phase formation during desodiation without suppressing P2-P′2 phase transition during sodiation at low voltages.

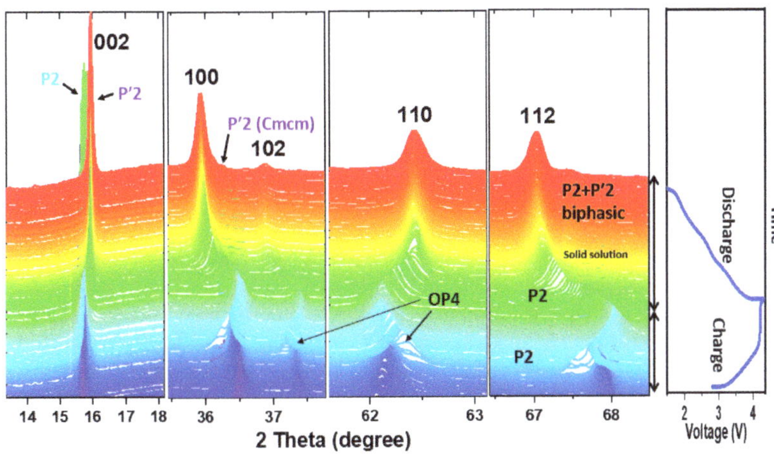

Figure 7. In situ synchrotron XRD of NFM-Mg electrode during the first charge/discharge process at a current density of 10 mA g^{-1}. The electrode charged to 4.3 V (P2 phase, blue color) and then discharged to 1.5 V. Discharge at the initial stage continues by maintaining the P2 phase (green color). Eventually, a P2 + P′2 biphasic region (red color) with an intermediate solid solution (yellow color) region appears.

Overall, the results of the above experiments indicate that oxygen is the only element involved in the initial charging process, while oxygen and manganese are both involved in the subsequent discharge process. Mg doping remediated the Jahn–Teller active Mn^{3+} ions in a P2-$Na_xFe_{1-y}Mn_yO_2$ system that caused phase transition during high voltage desodiation associated with gliding of MnO_2-slabs and decaying capacities. It was certain that the probability of Jahn–Teller active Fe^{4+} ions forming at high voltage regions was reduced because Fe^{3+} ions could not be oxidized during desodiation here. However, irreversible oxygen activity is an issue that should be resolved in future studies. The 68% capacity retention after 80 cycles during electrochemical cycling was satisfactory compared to several previous reports [11,17,23]. Nevertheless, this material's performance could be further improved through simultaneous doping and coating approaches.

3. Conclusions

In this study, we applied a sol–gel method for synthesizing a new cathode material, P2-$Na_{2/3}Mg_{2/9}Fe_{2/9}Mn_{5/9}O_2$. We then tested its electrochemical performance in Na half-cells. This material showed a high initial reversible capacity of 196 mAh g^{-1} and maintained a reversible capacity of mAh g^{-1} after 80 cycles with a capacity retention of 68% within a wide potential range of 1.5V to 4.3 V. Additionally, it had good rate capabilities, providing 117 mAh g^{-1} of reversible capacity at a current density of 200 mA g^{-1}. Synchrotron XAS results confirmed that its capacity was due to single cationic (Mn^{3+}/Mn^{4+}) redox in addition to partially reversible anionic redox (O^{2-}/O^{n-}) in short-term cycling. Desodiation at high voltage resulted in no undesired phase transition and structural distortion, as evidenced by in situ XRD analysis because of the presence of Jahn–Teller inactive Fe^{3+} ($t^3{}_{2g}e^2{}_g$) and Mn^{4+} ($t^3{}_{2g}$ $e^0{}_g$) ions. Irreversible oxygen loss can be minimized by altering dopant concentration, which is left to be resolved in future work. In terms of developing SIBs' layered cathode materials, these results are quite valuable.

4. Materials and Methods

The NFM-Mg was prepared using the sol–gel method using corresponding metal nitrate salts as source materials. All chemicals used in this study, including $NaNO_3$ (99%), $Fe(NO_3)_3 \cdot 9H_2O$ (99%), $Mn(NO_3)_2 \cdot 4H_2O$ (99%), $Mg(NO_3)_2 \cdot 6H_2O$, and citric acid, were purchased from Junsei Chemical Co. Ltd., Japan. We first solubilized nitrate salts in distilled water and then added them to a citric acid solution at room temperature at a molar ratio of 0.5:1 (citric acid:cations). The mixture was then stirred continuously for 1 h. The prepared mixture was then placed on a hot plate at a temperature of 80 °C. The solution was mixed continuously by magnetic stirring for a certain period and converted to gel, which involved a simple reaction of hydrolysis and condensation. The resultant gel was quickly transferred and placed overnight in a vacuum oven at 120 °C. For auto-combustion of citric acid, the gel was further heated at 200 °C. In a box furnace, resulting powders were pre-heated at 500 °C for 5 h in an ambient environment to remove residual organics. A hydrostatic press (Daehatech, HP-1B) was used to press the powder into pellets, which were heated in a tubular furnace at 850 °C for 12 h at a heating rate of 5 °C/min under oxygen flow. Instead of slow cooling, the sample was quenched and preserved inside an inert gas-filled glove box. We prepared the P2-$Na_{2/3}Fe_{1/2}Mn_{1/2}O_2$ (NFM) material by following a similar approach, excluding the addition of magnesium nitrate salt. For details on structural, electrochemical, in situ, and ex situ characterizations, please see Supplementary Materials.

Supplementary Materials: The following supporting information can be downloaded at: https://www.mdpi.com/article/10.3390/gels10010024/s1, Experimental details, Table S1: SEM-EDS analysis result for the P2-$Na_{2/3}Mg_{2/9}Fe_{2/9}Mn_{5/9}O_2$ cathode material. Figure S1: High-resolution powder XRD pattern for the P2-$Na_{2/3}Mg_{2/9}Fe_{2/9}Mn_{5/9}O_2$ (NFM-Mg) compound. Figure S2: (a) Galvanostatic voltage profiles, and (b) dQ/dV curves of the P2-$Na_{2/3}Fe_{1/2}Mn_{1/2}O_2$ (NFM) cathode material.

Author Contributions: Conceptualization, M.I.; Data curation, G.A.K.M.R.B.; Formal analysis, D.H.; Funding acquisition, K.-W.N.; Investigation, M.I., D.H. and M.S.A.; Methodology, M.I. and M.S.A.; Resources, G.A.K.M.R.B. and K.-W.N.; Visualization, M.I.; Writing—original draft, M.I.; Writing—review and editing, K.-W.N. All authors have read and agreed to the published version of the manuscript.

Funding: This research was supported by grants (grant No. RS-2023-00282389 and 2022M3J1A1054323) of the Nano & Material Technology Development Program through the National Research Foundation (NRF) funded by the Ministry of Science and ICT, Republic of Korea.

Institutional Review Board Statement: Not applicable.

Informed Consent Statement: Not applicable.

Data Availability Statement: All data and materials are available upon request from the corresponding author. The data are not publicly available due to ongoing research using a part of the data.

Acknowledgments: The use of beamlines of 9B (HRPD), 10C, 7D, & 8C (XAFS), 10A1, 6D (in situ XRD), and 10D (soft XAS) at Pohang Accelerator Laboratory (PAL) is greatly acknowledged.

Conflicts of Interest: The authors declare no conflicts of interest.

References

1. Chen, T.; Jin, Y.; Lv, H.; Yang, A.; Liu, A.; Chen, B.; Xie, Y.; Chen, Q. Applications of Lithium-Ion Batteries in Grid-Scale Energy Storage Systems. *Trans. Tianjin Univ.* **2020**, *26*, 208–217. [CrossRef]
2. Tarascon, J.M. Is lithium the new gold? *Nat. Chem.* **2010**, *2*, 510. [CrossRef]
3. Delmas, C. Sodium and Sodium-Ion Batteries: 50 Years of Research. *Adv. Energy Mater.* **2018**, *15*, 1703137. [CrossRef]
4. Fang, C.; Huang, Y.; Zhang, W.; Han, J.; Deng, Z.; Cao, Y.; Yang, H. Routes to High Energy Cathodes of Sodium-Ion Batteries. *Adv. Energy Mater.* **2016**, *6*, 1501727. [CrossRef]
5. Delmas, C.; Fouassier, C.; Hagenmuller, P. Structural classification and properties of the layered oxides. *Physica* **1980**, *99*, 81–85. [CrossRef]
6. Zhao, J.; Zhao, L.; Dimov, N.; Okada, S.; Nishida, T. Electrochemical and Thermal Properties of α-$NaFeO_2$ Cathode for Na-Ion Batteries. *J. Electrochem. Soc.* **2013**, *160*, A3077. [CrossRef]
7. Jung, E.; Park, Y.; Park, K.; Kwon, M.S.; Park, M.; Sinha, A.K.; Lee, B.H.; Kim, J.; Lee, H.S.; Chae, S.I.; et al. Synthesis of nanostructured P2-$Na_{2/3}MnO_2$ for high performance sodium-ion batteries. *Chem. Commun.* **2019**, *55*, 4757–4760. [CrossRef]
8. Kumakura, S.; Tahara, Y.; Sato, S.; Kubota, K.; Komaba, S. P′2-$Na_{2/3}Mn_{0.9}Me_{0.1}O_2$ (Me = Mg, Ti, Co, Ni, Cu, and Zn): Correlation between orthorhombic distortion and electrochemical property. *Chem. Mater.* **2017**, *29*, 8958–8962. [CrossRef]
9. Grépin, E.; Moiseev, I.; Abakumov, A.; Tarascon, J.M.; Mariyappan, S. Rational selection of sodium layered oxides for high performance Na-ion batteries: P2 vs O3 vs P2-O3 intergrowths. *J. Electrochem. Soc.* **2023**, *170*, 080510. [CrossRef]
10. Yabuuchi, N.; Kajiyama, M.; Iwatate, J.; Nishikawa, H.; Hitomi, S.; Okuyama, R.; Usui, R.; Yamada, Y.; Komaba, S. P2-type Nax$[Fe_{1/2}Mn_{1/2}]O_2$ made from earth-abundant elements for rechargeable Na batteries. *Nat. Mater.* **2012**, *11*, 512–517. [CrossRef]
11. Gonzalo, E.; Han, M.H.; L'opez del Amo, J.M.; Acebedo, B.; Casas-Cabanas, M.; Rojo, T. Synthesis and characterization of pure P2- and O3-$Na_{2/3}Fe_{2/3}Mn_{1/3}O_2$ as cathode materials for Na ion batteries. *J. Mater. Chem. A* **2014**, *2*, 18523–18530. [CrossRef]
12. Zhao, J.; Xu, J.; Lee, D.H.; Dimov, N.; Meng, Y.S.; Okada, S. Electrochemical and thermal properties of P2-type $Na_{2/3}Fe_{1/3}Mn_{2/3}O_2$ for Na-ion batteries. *J. Power Sources* **2014**, *264*, 235–239. [CrossRef]
13. Mortemard de Boisse, B.; Carlier, D.; Guignard, M.; Guerin, E.; Duttine, M.; Wattiaux, A.; Delmas, C. Influence of the Mn/Fe ratio on the electrochemical and structural properties of P2-$Na_xMn_{1-y}Fe_yO_2$ phases as positive electrode material for Na-ion batteries. *Chem. Mater.* **2018**, *30*, 7672–7681. [CrossRef]
14. Somerville, J.W.; Sobkowiak, A.; Tapia-Ruiz, N.; Billaud, J.; Lozano, J.G.; House, R.A.; Gallington, L.C.; Ericsson, T.; Häggström, L.; Roberts, M.R.; et al. Nature of the "Z"-phase in layered Na-ion battery cathodes. *Energy Environ. Sci.* **2019**, *12*, 2223–2232.
15. Song, T.; Kendrick, E. Recent progress on strategies to improve the high-voltage stability of layered-oxide cathode materials for sodium-ion batteries. *J. Phys. Mater.* **2021**, *4*, 032004. [CrossRef]
16. Peng, B.; Wan, G.; Ahmad, N.; Yu, L.; Ma, X.; Zhang, G. Recent progress in the emerging modification strategies for layered oxide cathodes toward practicable sodium ion batteries. *Adv. Energy Mater.* **2023**, *13*, 2300334. [CrossRef]
17. Hirsh, H.; Olguin, M.; Chung, H.; Li, Y.; Bai, S.; Feng, D.; Wang, D.; Zhang, M.; Meng, Y.S. Meso-structure controlled synthesis of sodium iron-manganese oxides cathode for low-cost Na-ion batteries. *J. Electrochem. Soc.* **2019**, *166*, A2528–A2535. [CrossRef]
18. Danks, A.E.; Hall, S.R.; Schnepp, Z. The evolution of 'sol–gel' chemistry as a technique for materials synthesis. *Mater. Horiz.* **2016**, *3*, 91–112. [CrossRef]
19. Wang, M.; Chen, K.; Liu, J.; He, Q.; Li, G.; Li, F. Efficiently enhancing electrocatalytic activity of α-MnO_2 nanorods/N-doped ketjenblack carbon for oxygen reduction reaction and oxygen evolution reaction using facile regulated hydrothermal treatment. *Catalysts* **2018**, *8*, 138. [CrossRef]
20. Liu, Q.; Zheng, W.; Liu, G.; Hu, J.; Zhang, X.; Han, N.; Wang, Z.; Luo, J.; Fransaer, J.; Lu, Z. Realizing high-performance cathodes with cationic and anionic redox reactions in high-sodium-content P2-type oxides for sodium-ion batteries. *ACS Appl. Mater. Interfaces* **2023**, *15*, 9324–9330. [CrossRef]
21. Yamashita, T.; Hayes, P. Analysis of XPS spectra of Fe^{2+} and Fe^{3+} ions in oxide materials. *Appl. Surf. Sci.* **2008**, *254*, 2441–2449. [CrossRef]
22. Qin, M.; Yin, C.; Xu, W.; Liu, Y.; Wen, J.; Shen, B.; Wang, W.; Liu, W. Facile synthesis of high capacity P2-type $Na_{2/3}Fe_{1/2}Mn_{1/2}O_2$ cathode material for sodium-ion batteries. *Trans. Nonferrous Met. Soc. China* **2021**, *31*, 2074–2080. [CrossRef]
23. Viswanatha, R.; Kishore, B.; Bharath, U.; Munichandraiah, N. Communication—Electrochemical investigation of plate-like $Na_{2/3}Fe_{1/2}Mn_{1/2}O_2$ for sodium ion cathode. *J. Electrochem. Soc.* **2018**, *165*, A263–A265. [CrossRef]
24. Maitra, U.; House, R.A.; Somerville, J.W.; Tapia-Ruiz, N.; Lozano, J.G.; Guerrini, N.; Hao, R.; Luo, K.; Jin, L.; Pérez-Osorio, M.A.; et al. Oxygen redox chemistry without excess alkali-metal ions in $Na_{2/3}[Mg_{0.28}Mn_{0.72}]O_2$. *Nat. Chem.* **2018**, *10*, 288–295. [CrossRef]
25. Islam, M.; Akbar, M.; Han, D.; Ali, B.; Choi, Y.J.; Lee, J.; Choi, G.; Park, J.H.; Kim, J.Y.; Jung, H.G.; et al. Unraveling vacancy-induced oxygen redox reaction and structural stability in Na-based layered oxides. *J. Chem. Eng.* **2022**, *431*, 133962. [CrossRef]

26. Cheng, C.; Ding, M.; Yan, T.; Dai, K.; Mao, J.; Zhang, N.; Zhang, L.; Guo, J. Exploring the charge compensation mechanism of P2-Type $Na_{0.6}Mg_{0.3}Mn_{0.7}O_2$ cathode materials for advanced sodium-ion batteries. *Energies* **2020**, *13*, 5729. [CrossRef]
27. Huang, Z.; Wang, Z.; Zheng, X.; Guo, H.; Li, X.; Jing, Q.; Yang, Z. Structural and electrochemical properties of Mg-doped nickel-based cathode materials $LiNi_{0.6}Co_{0.2}Mn_{0.2-x}Mg_xO_2$ for lithium-ion batteries. *RSC Adv.* **2015**, *5*, 88773–88779. [CrossRef]
28. Baker, M.L.; Mara, M.W.; Yan, J.J.; Hodgson, K.O.; Hedman, B.; Solomon, E.I. K- and L-edge X-ray absorption spectroscopy (XAS) and resonant inelastic X-ray scattering (RIXS) determination of differential orbital covalency (DOC) of transition metal sites. *Coord. Chem. Rev.* **2017**, *345*, 182–208. [CrossRef]
29. Cheng, S.L.; Du, C.H.; Chuang, T.H.; Lin, J.G. Atomic replacement effects on the band structure of doped perovskite thin films. *Sci. Rep.* **2019**, *9*, 7828. [CrossRef]
30. Yang, W.; Devereaux, T.P. Anionic and cationic redox and interfaces in batteries: Advances from soft X-ray absorption spectroscopy to resonant inelastic scattering. *J. Power Sources* **2018**, *389*, 188–197. [CrossRef]
31. Duda, L.-C.; Edström, K. Oxygen redox reactions in Li ion battery electrodes studied by resonant inelastic X-ray scattering. *J. Electron Spectrosc. Relat. Phenom.* **2017**, *221*, 79–87. [CrossRef]
32. Dai, K.; Jinpeng, W.; Zhuo, Z.; Li, Q.; Sallis, S.; Mao, J.; Ai, G.; Sun, C.; Li, Z.; Gent, W.E.; et al. High reversibility of lattice oxygen redox in Na-ion and Li-ion batteries quantified by direct bulk probes of both anionic and cationic redox reactions. *Joule* **2019**, *3*, 518–541. [CrossRef]
33. Mortemard de Boisse, B.; Carlier, D.; Guignard, M.; Bourgeois, L.; Delmas, C. P2-$Na_xMn_{1/2}Fe_{1/2}O_2$ phase used as positive electrode in Na batteries: Structural changes induced by the electrochemical (de)intercalation. *Inorg. Chem.* **2014**, *53*, 11197–11205. [CrossRef] [PubMed]
34. Zarrabeitia, M.; Nobili, F.; Lakuntza, O.; Carrasco, J.; Rojo, T.; Casas-Cabanas, M.; Muñoz-Márquez, M.Á. Role of the voltage window on the capacity retention of P2-$Na_{2/3}[Fe_{1/2}Mn_{1/2}]O_2$ cathode material for rechargeable sodium-ion batteries. *Commun. Chem.* **2022**, *5*, 11. [CrossRef]
35. Kulka, A.; Marino, C.; Walczak, K.; Borca, C.; Bolli, C.; Novak, P.; Villevieille, C. Influence of Na/Mn arrangements and P2/P2′ phase ratio on the electrochemical performance of Na_xMnO_2 cathodes for sodium-ion batteries. *J. Mater. Chem. A* **2020**, *8*, 6022–6033. [CrossRef]

gels

MDPI

Article

Combustion Enhancement of Gel Propellant Containing High Concentration Aluminum Particles Based on Carbon Synergistic Effect

iyuan Chen, Hui Zhao *, Weifeng Li and Haifeng Liu

National Energy Coal Gasification Technology Research and Development Center, Shanghai Engineering Research Center of Coal Gasification, East China University of Science and Technology, Shanghai 200237, China; 10182671@mail.ecust.edu.cn (J.C.); liweif@ecust.edu.cn (W.L.); hfliu@ecust.edu.cn (H.L.)
* Correspondence: zhaohui@ecust.edu.cn

Abstract: The addition of aluminum particles to gel propellants can improve combustion performance. However, the agglomeration of aluminum during the combustion process can result in a series of negative effects. In this paper, the aluminum agglomeration inhibition method of gel propellant based on carbon synergistic effect is proposed. Carbon particles exhibit excellent combustion properties, and the gaseous product CO_2 generated during combustion can mitigate the agglomeration of aluminum. The research demonstrates that incorporating carbon particles into aluminum-containing gel effectively reduces the incomplete combustion of aluminum particles and increases the volumetric calorific value of the gel. When the mass fraction of carbon is 5 wt%, the volume calorific value of the gel reaches the highest. Meanwhile, the rheological experiments show that the addition of carbon particles can improve the shear-thinning properties of the gel, which is beneficial to the atomization and combustion processes of the gel.

Keywords: metallized gel fuel; aluminum agglomeration; carbon synergistic effect; combustion enhancement; rheology

1. Introduction

In recent years, gels have been increasingly applied in energy, aerospace, chemistry, material, medicine, etc. [1–6]. Gels are colloids in which the liquid medium has become viscous enough to behave more or less as a solid. When subjected to shear stress exceeding the yield stress, gels demonstrate shear thinning characteristics.

Gel propellant is a liquid propellant that incorporates a certain amount of gelling agents, such as silica [7,8], hydroxypropyl cellulose (HPC) [9], and Thixatrol-ST [10], so that it becomes a colloid mixture at normal temperature and pressure. Gel propellant not only inherits the characteristics of high specific impulse, multiple start-ups, and adjustable thrust of liquid propellant but also has the advantages of solid propellants, such as non-leakage, long-term storage, easy use, and maintenance [11]. Gel engine has broad development and application prospects in the field of missile weapons and space thrusters and has become a new propulsion technology widely studied [12–15].

Gel propellants exhibit different flow and deformation characteristics under external pressure or shearing. The rheological properties of gel propellants suitable for rocket engines should be satisfied: in the storage and movement process, it appears as a solid with a large enough yield stress to prevent flow and deformation leakage, while in the working process, it appears as a fluid, the apparent viscosity should decrease with the increase of the shear rate. Therefore, gel propellants are the typical non-Newtonian pseudoplastic fluids whose rheological properties can usually be characterized by Bingham, Power–Law (PL), Herschel–Bulkley (HB), and Carreau, etc. [9,14,16–19].

In order to enhance the energy and density of gel propellant, energetic powdery particles, such as aluminum, magnesium, boron, and carbon, are usually mixed. The combustion characteristic of gels containing particles will also be significantly affected [20,21]. Among these particles, aluminum particles are extensively employed in propellants. However, research has revealed that gels containing aluminum particles does not burn completely, and the increase in aluminum content leads to a decrease in combustion efficiency [22]. Agglomeration of aluminum particles results in the formation of large agglomerates that are tens or even hundreds of times larger than the aluminum particles. Furthermore, the combustion product of aluminum particles can be divided into two components: one is the large-size alumina formed by the combustion of agglomerated aluminum, which is called Agglomerates. The remaining part is the small particle of alumina formed by aluminum vapor during combustion, which is called SOPs (Smoke oxide particles) [23]. In order to improve the combustion characteristics of aluminum in aluminum-containing propellants, optimizing the formulation of the propellant is an effective approach. Jiang [24] added 20 wt% graphene oxide (GO) to the micron-sized aluminum powder and coated it on the surface of aluminum particles for combustion experiments. The research found that the heat generated by GO combustion, the product rGO, and gaseous products (H_2O, CO_2) all contribute to promoting the combustion of aluminum powder. Liu [25] added graphene to the propellant, and studies have demonstrated that graphene can reduce aluminum agglomeration. As the content of graphene increases, the average size of large particles in the condensed combustion products decreases. By adding 0.5 wt% graphene, large agglomerates are almost eliminated. Certain fluorides have a remarkable effect on inhibiting the agglomeration of aluminum particles. Ao [26] investigated the impact of adding a new functionalized Fluorine-containing organic substance (FCOS) to aluminized solid propellants. With the addition of FCOS, the combustion rate of the propellant increases and produces more gas-phase combustion products, thereby reducing the size of aluminum aggregates. The energy performance of propellants has also been improved. In addition, some studies have explored using alloy powders to replace aluminum powder in propellants, such as Al-Mg alloy particles [27] and Al–Li alloys [28]. Therefore, the propellant formulation has a significant impact on the combustion properties of aluminum-containing propellants.

Currently, the focus of research lies in reducing the incomplete combustion of aluminum powder in aluminum-containing solid propellants. However, there has been little discussion about how to improve the incomplete combustion of aluminum powder in gel propellants. It remains to be studied. Therefore, after referring to the method of improving the incomplete combustion of aluminum powder in solid propellant, our research is devoted to improving the incomplete combustion of aluminum particles within gels.

Our research is focused on developing gel propellants with better energy performance and rheology, which can be used in the field of aerospace propellants. Previous studies have increased the addition of aluminum powder in ethanolamine gel to 50 wt%. However, the commonly used micron-sized aluminum powder has a serious agglomeration phenomenon; the excessive content of aluminum powder leads to a reduction of the combustion heat growth rate and significant incomplete combustion within the ethanolamine gel. Our research found that adding a small amount of carbon particles into the ethanolamine gel containing 50% Al powder can enhance the calorific value of the aluminum-containing gel and improve the combustion characteristics. Ethanolamine fuel, carbon particles, and aluminum powder were selected to prepare gel fuel. To investigate the impact of carbon particles on reducing the incomplete combustion of Al-containing gel, we conducted measurements on the heat of combustion of the gel and observed its combustion products using a scanning electron microscope. Additionally, we studied the rheological properties of the gel.

2. Results and Discussion

2.1. Combustion Heat Detection

The heat of combustion of the gel was measured using a 5E-AC8018 calorimeter. After the measurement, the quartz crucible was removed, and the large-size alumina formed by the combustion of agglomerated aluminum (agglomerates) and the small particle alumina formed by aluminum vapor during combustion (SOPs) could be seen, as depicted in Figure 1. The microscopic morphology of both agglomerates and SOPs can be seen by scanning electron microscopy. SOPs are white and fluffy, resulting from the stacking of small particles of alumina. The microscopic profile of agglomerates varies with the gel formulation and will be analyzed in Section 2.2.

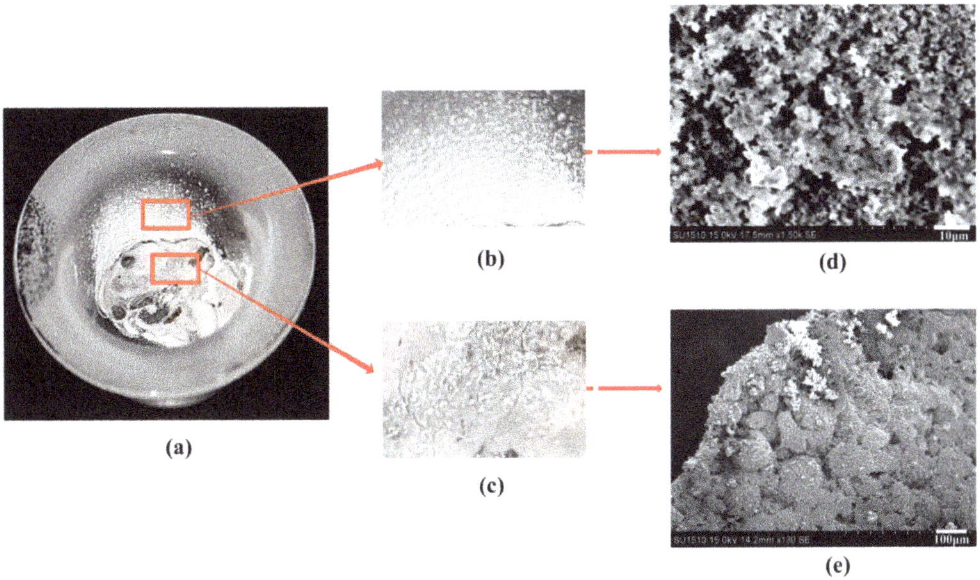

Figure 1. Combustion products of ethanolamine gel fuel and its drawing of partial enlargement. (**a**) Combustion products in a quartz crucible. (**b**) SOPs. (**c**) Agglomerates. (**d**) Microscopic topography of SOPs. (**e**) Microscopic topography of agglomerates.

According to the data from reference [29], the mass heat of the combustion of aluminum is 32 MJ/kg, and the volume calorific value is 85 MJ/kg. Ethanolamine, identified as a novel propellant [21], has a mass heat of combustion of 25.18 MJ/kg. As an organic gelling agent, the mass heat of combustion of agarose is 4.97 MJ/kg. Based on the density and combustion heat of each component within the gel, it can be calculated that the theoretical combustion heat of the gel is 27.77 MJ/kg.

The mass heat of combustion and the volume calorific value of the gel under different formulations are presented in Table 1. During the experiment, 0.3 g of gel is combusted in an oxygen bomb with a volume of 300 cm^3 and a pressure of 3.5 MPa. The oxygen content in the combustion environment surpasses the requirements for complete fuel combustion, thereby eliminating the possibility of incomplete combustion resulting from inadequate oxygen supply. Due to incomplete combustion of aluminum, the actual measured results are smaller than theoretical values.

Table 1. The mass heat of combustion and volume calorific value of the gel fuel.

Sample	Mass Heat of Combustion/MJ·kg^{-1}	Volume Calorific Value/MJ·L^{-1}	Density g/cm^3
50Al	25.578	46.193	1.805
5C45Al	26.423	46.754	1.769
10C40Al	26.787	46.421	1.732
15C35Al	26.930	45.686	1.696
20C30Al	27.128	45.031	1.659

It is evident that an increase in carbon content leads to a corresponding increase in the mass heat of combustion of the gel fuel. After calculating the volume calorific value of the gel fuel through the density of the sample, it can be seen in Figure 2 that the volume calorific value of the gel fuel first increases and then decreases. The volume calorific value of 50Al gel is measured at 46.193 MJ·L^{-1}, and after mixing with 5% carbon particles, it reaches 46.755 MJ·L^{-1}, which is 1.21% higher than 50Al gel. When the amount of carbon particles added reaches 10%, the volume calorific value remains higher than that of 50Al gel at 46.421 MJ·L^{-1}, indicating that mixing an appropriate amount of carbon particles can enhance the volume calorific value of the gel. The phenomenon can be attributed to two factors: Firstly, the carbon particles themselves possess a relatively high calorific value (33 MJ/kg), thereby enhancing the overall combustion heat of the gel fuel. Secondly, the aluminum powder utilized in this study exhibits severe agglomeration, resulting in incomplete combustion of the aluminum-containing gel during combustion, and the addition of carbon particles can mitigate this issue. The specific principles will be discussed in Section 2.3. With the addition of more carbon particles, the overall density of the gel system decreases, consequently leading to a reduction in the volume calorific value of the gel fuel. This phenomenon can be attributed to the low density exhibited by carbon particles; an excessive amount of these particles diminishes the gel's density and subsequently lowers its mass heat of combustion. Therefore, the quantity of carbon particles introduced must be carefully controlled; otherwise, it will reduce the combustion performance of the gel fuel.

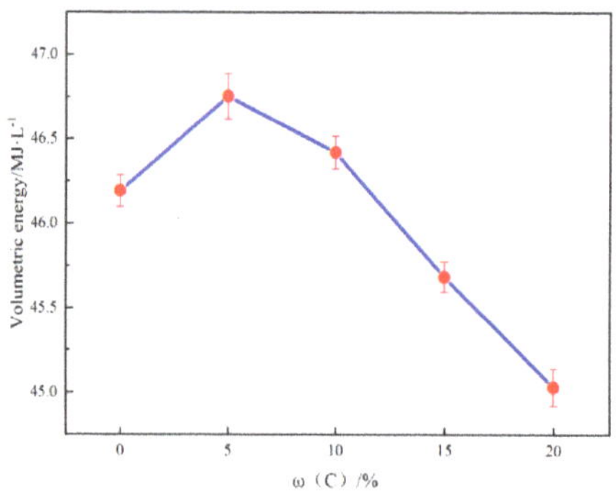

Figure 2. Relationship between carbon particle mass fraction and volume calorific value of gel fuel. Bars donate S.D.

2.2. *Scanning Electron Microscopy Analysis of Combustion Products*

The scanning electron microscope image in Figure 3 clearly reveals the solidified black-gray alumina aggregates resulting from burning, as well as the fluffy white small particle alumina formed through the combustion of aluminum vapor on the surface (SOPs). When the mass fraction of aluminum powder is 50 wt%, the alumina agglomerates formed after combustion exhibit a relatively smooth surface, and the surface is covered with a small number of white alumina particles (SOPs). Compared to the gel containing 50 wt%

aluminum powder, the surface of alumina agglomerates formed by gel combustion is rougher and has a small number of pores upon incorporation of 5% carbon particles. This phenomenon can be attributed to the CO_2 gas generated by the mixed carbon particles after combustion, which causes damage to alumina aggregates. There are also more SOPs on the surface, indicating that the production of carbon dioxide promotes the flow of aluminum vapor during combustion, resulting in more SOPs. When the mixed carbon particles content reaches 10 wt%, the surface of alumina aggregates formed by gel combustion exhibits increased roughness, and the number of pores and SOPs also increased. Upon addition of 20 wt% carbon particles, the surface structure is more complex, a large number of pores are distributed on the surface, and more SOPs with larger volumes can be observed.

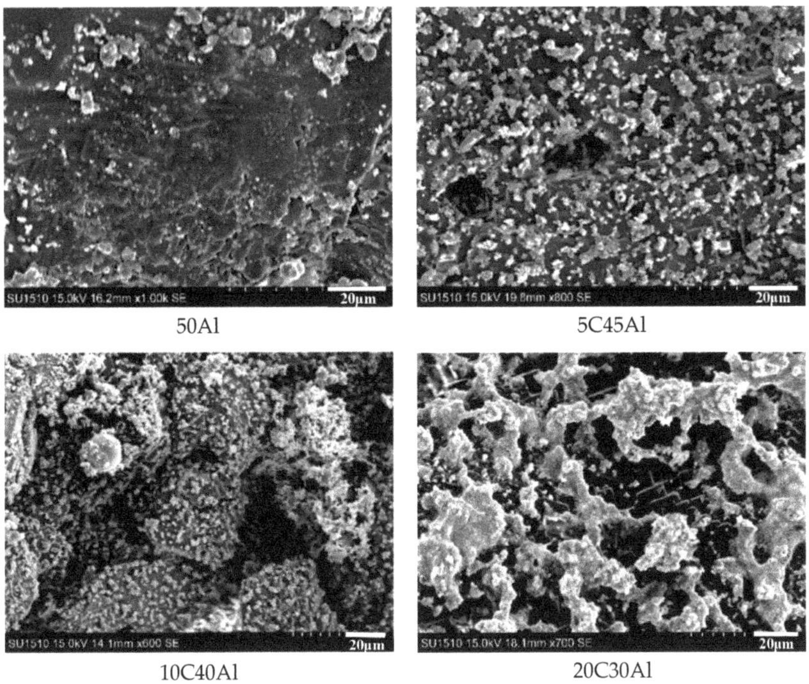

Figure 3. Scanning electron microscope images of combustion products of gel fuel.

Through electron microscopy in Figure 4, it can be observed that the combustion products of carbon-containing gel fuel exhibit the presence of multiple holes, which are uncommon in 50Al gel. The carbon particles in the gel burn and produce CO_2, which expands and breaks down the alumina agglomerates, resulting in these holes. These pores facilitate the release of unreacted aluminum, thereby enabling continuous combustion reactions. The presence of SOPs distributed within these pores indicates the complete combustion of aluminum particles.

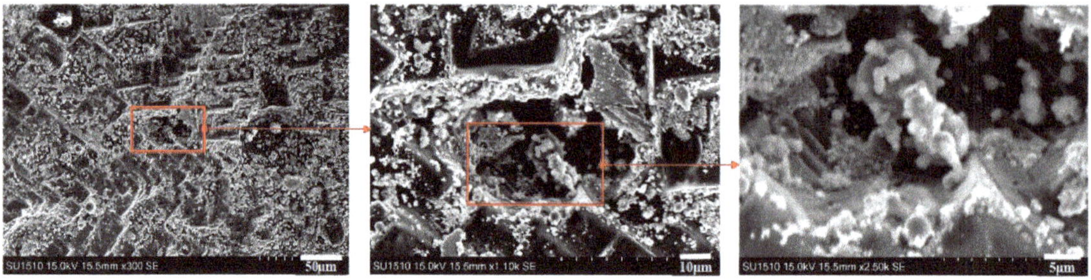

(a) (b) (c)

Figure 4. Scanning electron microscope image of holes (5C45Al). (**a**) Combustion products of gel fuel (5C45Al). (**b**) The hole image magnified by a factor of 1100. (**c**) The hole image magnified by a factor of 2500.

The aforementioned results demonstrate that as the quantity of carbon particles increases, the CO_2 gas can effectively destroy the interior of alumina aggregates, diminish the agglomeration of aluminum particles while generating a multitude of holes, and enhance the combustion efficiency of aluminum powder.

2.3. *Mechanism of Carbon Particles Improving Aluminum Agglomeration*

The schematic diagram in Figure 5 illustrates the combustion process of a gel containing carbon particles and aluminum particles in an oxygen environment. Carbon particles and aluminum particles are distributed inside the gel. After ignition, the surface of the gel undergoes combustion, gradually propagating towards its interior, resulting in a rapid increase in temperature within a short period (Figure 5a). As the combustion continues, the aluminum particles gradually melt and gather together with the temperature rising. With the rupture of the outer layer of the gel, the internal particles come into contact with oxygen and undergo combustion. Alumina is formed by the combustion of aluminum particles and covered on the surface of aluminum. The aluminum coalesces into sizable alumina aggregates following combustion, impeding the further combustion of internal aluminum. At the same time, carbon particles also burn and produce carbon dioxide (Figure 5b). As combustion progresses, these carbon particles undergo a complete reaction, generating a significant amount of carbon dioxide that rapidly expands at high temperatures and disrupts alumina aggregates. This effectively retards aluminum agglomeration during combustion. At the same time, the unreacted aluminum is released (Figure 5c), which comes into contact with the oxygen in the surrounding environment and undergoes combustion.

2.4. *Rheological Study*

2.4.1. Apparent Viscosity Test

As a shear-thinning non-Newtonian fluid, the apparent viscosity of gel decreases with the increase of shear rate. The investigation into the shear thinning characteristics of gel propellant has significant reference value for its storage, flow, and atomization. Jyoti [30] investigated the shear-thinning behavior of ethanolamine-based gel propellant across shear rates ranging from 1 to 1000 s^{-1}. The experimental results show that the viscosity of the gel increases gradually at a very low shear rate and does not begin to flow until it reaches the yield stress. In this case, the viscosity is higher than 100 Pa·s. Subsequently, the gel exhibits the characteristic of shear thinning as the shear rate increases. In addition, Jyoti [31] discovered that when aluminum particles with a mass fraction of 20 wt% were added to the ethanol gel, the viscosity of the system increased from approximately 100 Pa·s to over 1000 Pa·s. Consequently, compared to pure gel systems, the addition of metal particles can greatly improve the viscosity.

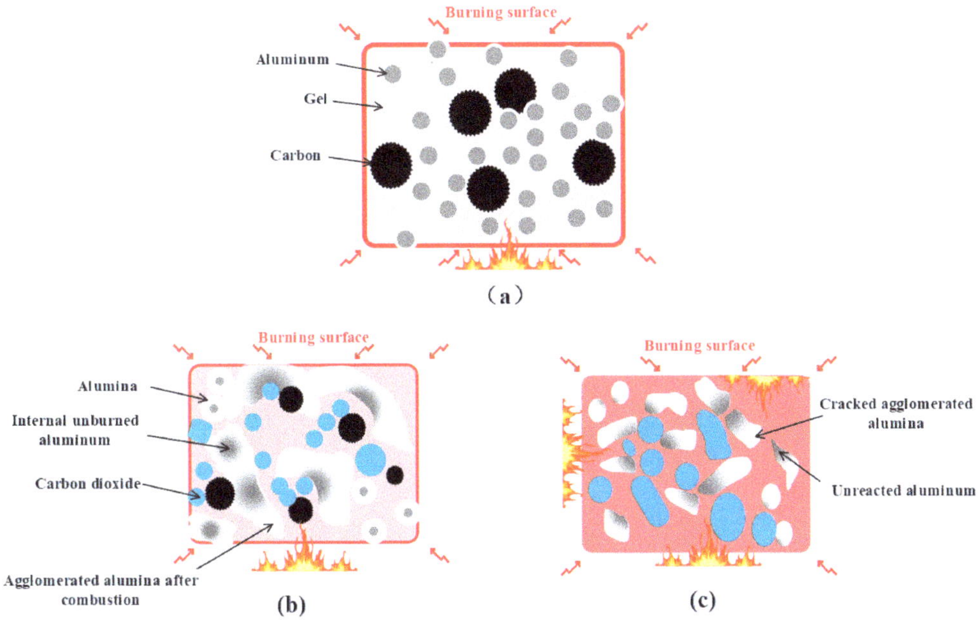

Figure 5. Mechanism diagram of carbon particles reducing incomplete combustion of aluminum particles. (**a**) The gel is ignited and forms a burning surface. (**b**) Melting, agglomeration, and oxidation of aluminum particles at high temperatures. (**c**) Large agglomerates of aluminum are destroyed by carbon dioxide.

The relationship between viscosity and shear rate of gels containing different energetic particles is illustrated in Figure 6. It can be observed from Figure that as the shear rate increases, the gels in all four formulations will first shear and thicken to the yield point, followed by a gradual decrease with a further increase in shear rate. These are consistent with references [30,31]. The difference in viscosity is primarily attributed to the type of gelling agent and the mass fraction of energetic particles within the gel formulation. When subjected to lower shear rates, substituting carbon particles for aluminum particles has minimal impact on the viscosity of the gel, and the viscosity of three gels containing carbon powders is approximately equivalent to that containing 50 wt% aluminum.

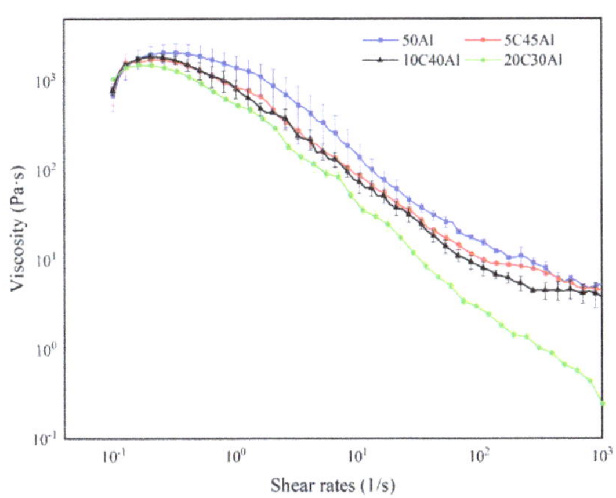

Figure 6. Relationship between gel viscosity and shear rate. Bars donate S.D.

According to the characteristics of the viscosity curve, the Power-Law fluid constitutive equation is further employed to depict the correlation between the viscosity of the gel fuel and the shear rate in the range of medium shear rate (1–10^3 s^{-1}). The Power–Law equation [32] can be expressed as follows:

$$\eta = K\gamma^{\,n-1} \tag{1}$$

where η is the viscosity of the gel, Pa·s; K is the flow consistency index, and Pa·s^n; n is the power–law exponent. The fitting results are shown in Table 2.

Table 2. Power–Law fluid constitutive equation fitting table for gel samples.

Sample	K	n	R^2
50Al	847.682	0.3602	0.99
5C45Al	738.579	0.3551	0.99
10C40Al	655.582	0.2729	0.99
20C30Al	516.438	0.1953	0.99

The size of the viscosity coefficient K reflects the level of viscosity in the gel system. As the carbon content increases, K gradually decreases, indicating that the viscosity of the system decreases, which aligns with the conclusion drawn from the viscosity curve. In Newtonian fluid, the Power-Law index n is equal to 1. Non-Newtonian fluid with $n < 1$ is called pseudoplastic fluid, which exhibits shear thinning behavior. And the more n deviates from 1, the more significant the shear thinning characteristic of pseudoplastic fluid is. It can be observed from Table 2 that the addition of carbon particles enhances the shear-thinning properties of the gel. At high shear rates, an increase in the amount of carbon particles leads to a decrease in viscosity for the gel system. When more carbon particles (20%) replace aluminum particles, the viscosity of the gel system will decrease significantly. This is attributed to the lower density and loose particle microstructure of carbon particles compared to aluminum powder, resulting in weaker bonding between particles and gels. As the content of carbon particles increases relative to pure aluminum gel, internal resistance within the gel decreases, leading to a reduced viscosity.

2.4.2. Thixotropy

When the gel system is subjected to the shear force, the internal network of the gel is also sheared, and the gel system will be destroyed under the shear force. However, upon removal of the shear force, the internal network can regenerate and reform the gel system through thixotropy. In this study, the 3ITT (three Interval Thixotropic Test) was used to evaluate the thixotropy of ethanolamine gel. The gel was exposed to a constant shear rate of 300 s^{-1} during 0–100 s. Then, the shear rate between 100–300 s was removed, and the gel was allowed to recover by itself for 200 s. Finally, a constant shear rate of 300 s^{-1} was applied again from 300–400 s. The change curve of the viscosity of the gel with time is shown in Figure 7.

It can be observed from Figure 7 that, under a constant shear rate of 300 s^{-1}, the viscosity of the gel in all four formulations first decreased rapidly and then gradually stabilized. After undergoing self-recovery for 200 s, the four gels displayed varying degrees of recovery. When the shear stress was applied again, the viscosity exhibited a similar trend as observed during the first 100 s, but none of them could return to their initial states. Based on the recovery ratio of the gel in Table 3, it can be concluded that the recovery ratio of the gel increases with the carbon content. Compared to aluminum powder, carbon particles exhibit a lower density and looser particle microstructure, and the bonding between particles and gel is weaker than aluminum, so the gel mixed with carbon particles demonstrates enhanced recovery capabilities. Since the recovery of the gel structure will increase the viscosity and energy loss of the gel fuel during practical application, the lower thixotropy levels are beneficial to the subsequent shear thinning and atomization process of the gel fuel. The excessive addition of carbon particles in practical applications is therefore deemed inappropriate.

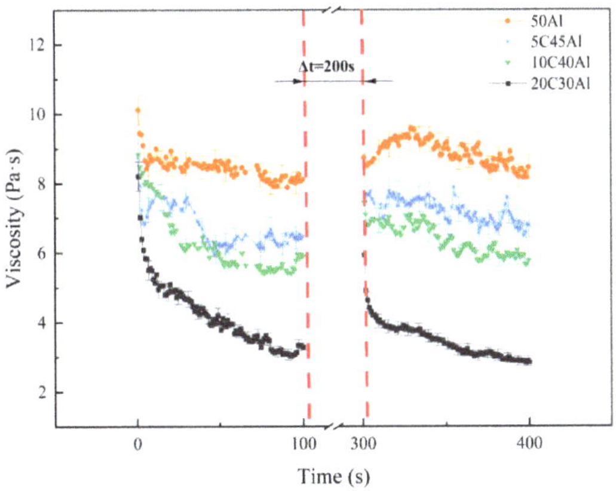

Figure 7. Thixotropy curves of gels with different formulations.

Table 3. Viscosity parameters of each sample before and after recovery.

Sample	Viscosity before Recovery/Pa·s	Viscosity after Recovery/Pa·s	Recovery Ratio/%
50Al	8.248	8.517	3.26
5C45Al	6.461	7.532	16.57
10C40Al	5.897	7.435	26.07
20C30Al	3.276	5.928	80.95

2.4.3. Amplitude Sweep

The result of the amplitude sweep is illustrated in Figure 8. With the increase of gel strain, the elastic modulus G′ initially stabilizes and subsequently exhibits a gradual decline. The range where G′ and G″ vary within 5% is called the linear viscoelastic region (LVR) of the gel fuel, and the point where the G′ of the gel decreases is called the critical strain. As the critical strain increases, so does the extent of the linear viscoelastic region (LVR).

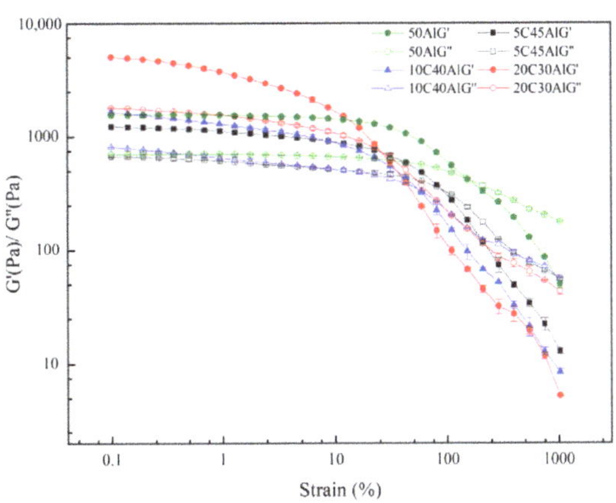

Figure 8. Amplitude scan curves of different gel fuel samples.

The critical strain of gels with different formulations is presented in Table 4. As the amount of carbon particles increases, there is a gradual decrease in the critical strain of gel fuel, accompanied by a reduction in the corresponding linear viscoelastic region. Upon reaching the critical strain, the molecular chains within the gel undergo elongation to their maximum extent along the flow direction, resulting in structural breakdown. Aluminum particles can form a strong binding force with the gel, leading to a higher critical strain. However, due to the loose microstructure of the carbon particles, the bonding between the particles and the gel is weak; with the addition of carbon particles and the reduction of aluminum particles, the bonding force between particles and molecular chains is reduced, making the gel more prone to relative slippage. The point at which the intersection of G′ and G″ occurs is commonly referred to as the yield point. It can be observed from Figure 8 that the G′ of all four gels is higher than the G″ before the yield point, indicating their solid characteristics during this stage. Beyond the yield point, the G″ of the gel is higher than the G′, and the gel transitions to fluid properties. The yield point strain of gel fuel can be found in Table 4. The yield point strains in all samples exceed the critical strain, indicating that the internal grid structure of the four gels is disrupted during flow. The incorporation of carbon particles reduces the strain at the gel's yield point, suggesting that compared to aluminum particles, carbon particles diminish the stability of gel fuel and render it more susceptible to structural failure under external forces. Moreover, when subjected to strong external forces, the shear-thinning characteristics become more pronounced.

Table 4. The critical strain and yield point strain of the gel.

Sample	Critical Strain/%	Yield Point Strain/%
50Al	22.80	163.00
5C45Al	16.60	96.00
10C40Al	4.67	51.25
20C30Al	0.50	27.05

2.4.4. Frequency Sweep

The frequency sweep can be utilized to acquire the correlation between the viscoelasticity of the gel and the time scale. Higher frequencies correspond to strains on the sample with shorter time scales, while lower frequencies represent long timescale strains on the sample. The relationship between storage modulus and loss modulus of four gels was investigated through a frequency sweep analysis. Figure 9 illustrates the relationship between G′/G″ and the frequency of four kinds of gels at 298 k and 1% strain. Within the frequency range of 1–100 Hz, the storage modulus of all four samples is higher than the loss modulus, indicating that the gel samples exhibit solid characteristics at lower strains. Furthermore, it is evident that both storage modulus and loss modulus values for 5C45Al gel are lower than those for pure 50Al gel, suggesting that incorporating a small amount of carbon particles reduces the mechanical strength of the gel. When more carbon particles are added, the storage modulus and loss modulus of 10C40Al and 20C30Al exhibit an increase compared to 50Al, and this increase is further enhanced with the addition of more carbon particles. It can be observed that the mechanical strength of the gel has been improved by mixing more carbon particles. Furthermore, as the oscillation frequency increases, all four samples demonstrate an elevation in their storage modulus. This trend suggests that at higher frequencies, the gel samples will exhibit pronounced elasticity, which will significantly impact subsequent gel atomization processes.

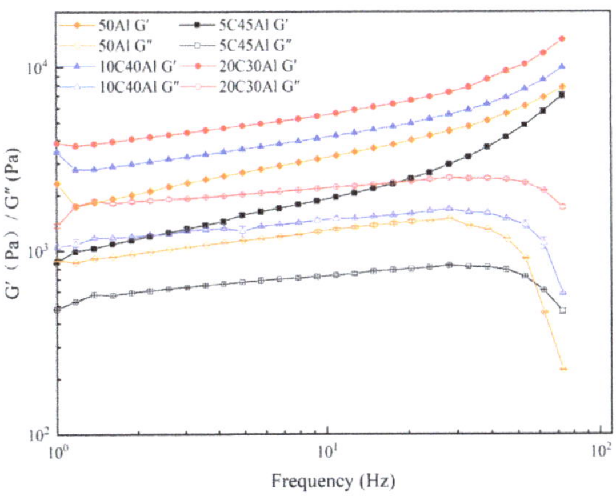

Figure 9. Relationship of storage modulus and loss modulus with frequency for different gel fuels.

Through the rheological test, it can be found that adding a small amount of carbon particles (about 5%) can slightly improve the rheological properties of the gel propellant, making it more conducive to the subsequent atomization and other processes. However, an increase in the proportion of blended carbon particles leads to adverse effects. Although the gel with a carbon content of 20% exhibits improved shear thinning properties, it displays high thixotropy, poor stability, and excessive mechanical strength, which will affect subsequent atomization.

3. Conclusions

In this study, we investigated the impact of carbon powder on the combustion and rheological properties of aluminum-containing gel fuels. Scanning electron microscope images reveal that The CO_2 gas produced by the combustion of carbon powder forms a large number of fluffy pores in the agglomerates, which makes the large-size aggregates of aluminum break so that the active aluminum wrapped by the aggregates can react. The agglomeration during aluminum combustion is mitigated. And the energy performance of gel fuel is improved, which has been confirmed in the combustion heat test.

The rheological experiments demonstrate that the addition of a small quantity of carbon particles results in a slight reduction in the viscosity of the gel system. Moreover, it diminishes the linear viscoelastic region of the gel, thereby compromising its stability and augmenting its shear-thinning properties. This makes the gel system easier to atomize. However, the excessive addition of carbon particles will result in increased thixotropy and mechanical strength of the system, thereby posing challenges to the subsequent gel atomization process.

Thus, incorporating a small amount of carbon particles (approximately 5 wt%) into the aluminized gel fuel can enhance the combustion performance of the propellant, mitigate the agglomeration effect of aluminum particles, and the rheological property plays a positive role in the subsequent atomization process of gel fuel.

4. Material and Methods

4.1. Materials

Ethanolamine fuel with a density of 1.013 g/cm^3 (C2H7NO, Analytical Pure) was purchased from Shanghai McLean Biochemical Technology Corporation, Shanghai, China. Carbon particles, with a density of 1.800 g/cm^3, were purchased from Shanghai New Platinum Chemical Technology Corporation, Shanghai, China. Agarose (Analytical Pure) was purchased from Qingdao Tenglong Microwave Technology Corporation, Qingdao, China. Aluminum particles, with a density of 2.700 g/cm^3 and a characteristic particle size D_{32} (the ratio between the total volume of all particles and the total surface area) of 1 μm (analytically pure) were purchased from Angang Industrial Fine Aluminum Particle Corporation, Anshan, China.

4.2. Equipment and Instruments

The calorimeter 5E-AC8018 was provided by Changsha Kaiyuan Instrument, Changsha, China. The precision electronic balance ME303E was provided by Mettler Toledo, Zurich, Switzerland. The scanning electron micromirrors, SU1510 and S3400-N, were provided by Hitachi-Hightech (Shanghai) International Corporation, Shanghai, China. The oscillating agitator was provided by Shanghai Yuezhong Instrument, Shanghai, China. The rotary rheometer MCR 302 was provided by the Anton Paar (Shanghai) Trading Corporation, Shanghai, China.

4.3. Preparation of Gel

Firstly, 46.00 g of ethanolamine was added to the beaker, and 4.00 g agarose was weighed. Agarose was gradually dispersed into ethanolamine using a sieve. After that, evenly mixed aluminum/carbon particles (50 g in total) were added, and the sample was placed in an oscillating agitator for high-speed stirring (2000 r/min) for 30 min, then removed and sealed for 6 h to form a gel. The composition of each sample is presented in Table 5 (The sample number represents the mass ratio of carbon particles and aluminum particles in the proportion of 50 wt% particles of gel fuel).

Table 5. Composition of each sample.

Sample	w/%			
	C_2H_7NO	Agarose	Al	C
50Al	46	4	50	0
5C-45Al	46	4	45	5
10C-40Al	46	4	40	10
15C-35Al	46	4	35	15
20C-30Al	46	4	30	20

4.4. Measurement and Characterization

4.4.1. Combustion Heat Detection

The combustion heat of gel fuel was measured by an oxygen bomb calorimeter. The experimental procedure can be found in reference [33,34]. The volume of the oxygen bomb is 300 cm^3. The 0.3 g sample was weighed in a quartz crucible, and the combustion heat of the sample was measured in an oxygen bomb tank at a pressure of 3.5 MPa. The experimental results were taken as the average value of three measurements.

4.4.2. Scanning Electron Microscope (SEM)

A proper amount of gel samples burned in a calorimeter were frozen and dried in a freeze-drying machine. The treated samples were vacuum-gilded using a vacuum pump and subsequently transferred into the scanning electron microscope cavity. The microstructure and morphology of the gel combustion products were subsequently examined using a scanning electron microscope.

4.4.3. Rheological Measurements

The rheological properties of the gel were measured by Anton Paar rheometer (MCR 302). The paddle rotor with a diameter of 22 mm is used for all measurements in this study. Before the test, gel fuel was added to the cylinder (test chamber), followed by the insertion of the paddle rotor. Then, the rheometer was operated to conduct the rheological test. The measurement result is taken as the average value of three times.

Rheological testing includes:

(1) Shear viscosity test: The rheometer operates in rotation mode. Control the shear rate of the rotor in the range of 0.1–1000 s^{-1}. By gradually increasing the shear rate of the paddle rotor according to a logarithmic law starting from 0.1 s^{-1}, we can accurately record and analyze the relationship between the shear rate and viscosity of gels;
(2) Thixotropic test: 3ITT (three Interval Thixotropic Test) [35] was used to detect the thixotropy of ethanolamine gel. A constant shear rate (300 s^{-1}) was applied to the gel during 0–100 s. Remove the shear rate within 100–300 s and let the gel recover itself for 200 s. A constant shear rate (300 s^{-1}) is still applied for 300–400 s. Finally, the curve of gel time and viscosity change is obtained;
(3) Amplitude sweeping test: Set the oscillation frequency of the rotor to 1 Hz, the strain range to 0.1–1000%, and record the storage modulus (G′) and loss modulus (G″).
(4) Frequency sweep test: constant strain is 1%, frequency is 1–100 Hz. Record the change of storage modulus (G′) and loss modulus (G″) at different frequencies.

Author Contributions: Methodology, W.L.; Investigation, J.C. and H.Z.; Funding acquisition, H.L. All authors have read and agreed to the published version of the manuscript.

Funding: This research was funded by the National Natural Science Foundation of China (22378114, U21B2088).

Institutional Review Board Statement: Not applicable.

Informed Consent Statement: Not applicable.

Data Availability Statement: All data and materials are available on request from the corresponding author. The data are not publicly available due to ongoing research using a part of the data.

Acknowledgments: This research was supported by the National Natural Science Foundation of China (22378114, U21B2088).

Conflicts of Interest: The authors declare no conflicts of interest.

References

1. Na, H.; Kang, Y.W.; Park, C.S.; Jung, S.; Kim, H.Y.; Sun, J.Y. Hydrogel-based strong and fast actuators by electroosmotic turgor pressure. *Science* **2022**, *376*, 301–307. [CrossRef]
2. Zhang, S.; Shi, W.; Wang, X. Locking volatile organic molecules by subnanometer inorganic nanowire-based organogels. *Science* **2022**, *377*, 100–104. [CrossRef]
3. Padwal, M.B.; Natan, B.; Mishra, D.P. Gel propellants. *Prog. Energy Combust. Sci.* **2021**, *83*, 100885. [CrossRef]
4. Glushkov, D.O.; Paushkina, K.K.; Pleshko, A.O.; Vysokomorny, V.S. Characteristics of micro-explosive dispersion of gel fuel particles ignited in a high-temperature air medium. *Fuel* **2022**, *313*, 123024. [CrossRef]
5. Cao, Y.; Mezzenga, R. Design principles of food gels. *Nat. Food* **2020**, *1*, 106–118. [CrossRef] [PubMed]
6. Narayanaswamy, R.; Torchilin, V.P. Hydrogels and their applications in targeted drug delivery. *Molecules* **2019**, *24*, 603. [CrossRef] [PubMed]
7. Dennis, J.D.; Willits, J.D.; Pourpoint, T.L. Performance of neat and gelled monomethylhydrazine and red fuming nitric acid in an unlike-doublet combustor. *J. Combust. Sci. Technol.* **2018**, *190*, 1141–1157. [CrossRef]
8. Arnold, R.; Santos, P.H.; Campanella, O.H.; Anderson, W.E. Rheological and Thermal Behavior of Gelled Hydrocarbon Fuels. *J. Propuls. Power* **2011**, *27*, 151–161. [CrossRef]
9. Dennis, J.D.; Kubal, T.D.; Campanella, O.; Son, S.F.; Pourpoint, T.L. Rheological Characterization of Monomethylhydrazine Gels. *J. Propuls. Power* **2013**, *29*, 313–320. [CrossRef]
10. Song, W.; Hwang, J.; Koo, J. Atomization of gelled kerosene by multi-hole pintle injector for rocket engines. *Fuel* **2021**, *285*, 119212. [CrossRef]
11. Xue, K.; Cao, J.; Pan, L.; Zhang, X.; Zou, J.-J. Review on design, preparation and performance characterization of gelled fuels for advanced propulsion. *J. Front. Chem. Sci. Eng.* **2022**, *16*, 819–837. [CrossRef]
12. Jin, Y.; Xu, X.; Yang, Q.; Dou, S.; Wang, X.; Fu, Q.; Pan, L. Combustion Behavior of Hydrocarbon/Boron Gel-Fueled Scramjet. *J. AIAA J.* **2022**, *60*, 3834–3843. [CrossRef]
13. Madlener, K.; Ciezki, H.K. Estimation of Flow Properties of Gelled Fuels with Regard to Propulsion Systems. *J. Propuls. Power* **2012**, *28*, 113–121. [CrossRef]
14. Rahimi, S.; Peretz, A.; Natan, B. Rheological matching of gel propellants. *J. Propuls. Power* **2010**, *26*, 376–379. [CrossRef]
15. Glushkov, D.; Paushkina, K.; Pleshko, A. Gel Fuels: Preparing, Rheology, Atomization, Combustion. *Energies* **2022**, *16*, 298. [CrossRef]

16. Yoon, C.; Heister, S.; Xia, G.; Merkle, C.L. Numerical Modeling of Injection of Shear-thinning Gel Propellants Through Plain-orifice Atomizer. *J. Propuls. Power* **2011**, *27*, 944–954. [CrossRef]
17. Liu, Y.; Zhang, H.; Pan, L.; Xue, K.; Zhang, X.; Zou, J.-J. High-energy-density gelled fuels with high stability and shear thinning performance. *J. Chin. J. Chem. Eng.* **2022**, *43*, 99–109. [CrossRef]
18. Jyoti, B.V.S.; Baek, S.W. Rheological characterization of hydrogen peroxide gel propellant. *Int. J. Aerosp. Eng.* **2014**, *15*, 199–204. [CrossRef]
19. Padwal, M.B.; Mishra, D.P. Interactions among synthesis, rheology, and atomization of a gelled propellant. *Rheol. Acta* **2016**, *55*, 177–186. [CrossRef]
20. Negri, M.; Ciezki, H.K. Combustion of gelled propellants containing microsized and nanosized aluminum particles. *J. Propuls. Power* **2015**, *31*, 400–407. [CrossRef]
21. Jyoti, B.V.S.; Naseem, M.S.; Baek, S.W. Hypergolicity and ignition delay study of pure and energized ethanol gel fuel with hydrogen peroxide. *Combust. Flame* **2017**, *176*, 318–325. [CrossRef]
22. Gafni, G.; Kuznetsov, A.; Natan, B. Experimental Investigation of an Aluminized Gel Fuel Ramjet Combustor. *J. Chem. Rocket Propuls.* **2017**, *31*, 297–315.
23. Ao, W.; Liu, X.; Rezaiguia, H.; Liu, H.; Wang, Z.; Liu, P. Aluminum agglomeration involving the second mergence of agglomerates on the solid propellants burning surface: Experiments and modeling. *J. Acta Astronaut.* **2017**, *136*, 219–229. [CrossRef]
24. Jiang, Y.; Deng, S.; Hong, S.; Zhao, J.; Huang, S.; Wu, C.-C.; Gottfried, J.L.; Nomura, K.-I.; Li, Y.; Tiwari, S.; et al. Energetic Performance of Optically Activated Aluminum/Graphene Oxide Composites. *J. ACS Nano* **2018**, *12*, 11366–11375. [CrossRef] [PubMed]
25. Liu, L.; Ao, W.; Wen, Z.; Zhang, Y.; Lv, X.; Qin, Z.; Liu, P. Combustion promotion and agglomeration reduction of the composite propellant using graphene. *J. Aerosp. Sci. Technol.* **2021**, *118*, 106988. [CrossRef]
26. Ao, W.; Liu, P.; Liu, H.; Wu, S.; Tao, B.; Huang, X.; Li, L.K. Tuning the agglomeration and combustion characteristics of aluminized propellants via a new functionalized fluoropolymer. *J. Chem. Eng. J.* **2019**, *382*, 122987. [CrossRef]
27. Belal, H.; Han, C.W.; Gunduz, I.E.; Ortalan, V.; Son, S.F. Ignition and Combustion Behavior of Mechanically Activated Al-Mg Particles in Composite Solid Propellants. *J. Combust. Flame* **2018**, *194*, 410–418. [CrossRef]
28. Wang, Y.; Jiang, C.; Fang, Y.; Wang, X.; Wang, Z. Mechanism of Pyrolysis Reaction of Al-Rich Al/PTFE/TiH2Active Material. *J Polym.* **2021**, *13*, 28–57.
29. Elbasuney, S.; Fahd, A. Combustion wave of metalized extruded double-base propellants. *Fuel* **2019**, *237*, 1274–1280. [CrossRef]
30. Botchu, J.; Baek, S.W. Flow and Dynamic Rheological Characterization of Ethanolamine Gel Propellant with Hybrid Gelling Agent. *Sci. Technol. Energetic Mater.* **2015**, *76*, 62–67.
31. Botchu, J.; Baek, S.W. Rheological Characterization of Metalized and Non-Metalized Ethanol Gel Propellants. *J. Propell. Explos Pyrot.* **2015**, *39*, 866–873.
32. Li, M.-G.; Wu, Y.; Cao, Q.-L.; Yuan, X.-Y.; Chen, X.; Han, J.-L.; Wu, W.-T. Rheological Properties of Organic Kerosene Gel Fuel. *Gels* **2022**, *8*, 507. [CrossRef] [PubMed]
33. Wu, J.-X.; Liu, Q.; Feng, B.; Yin, Q.; Li, Y.-C.; Wu, S.-Z.; Yu, Z.-S.; Huang, J.-Y.; Ren, X.-X. Improving the energy release characteristics of PTFE/Al by doping magnesium hydride. *J. Def. Technol.* **2022**, *18*, 219–228. [CrossRef]
34. Terry, B.C.; Gunduz, I.E.; Pfeil, M.A.; Sippel, T.; Son, S. A Mechanism for Shattering Microexplosions and Dispersive Boiling Phenomena in Aluminum-Lithium Alloy Based Solid Propellant. *J. Proc. Combust. Inst.* **2016**, *36*, 2309–2316. [CrossRef]
35. Ali, I.; Ali Shah, L.; Rehman, T.U.; Faizan, S. Investigation of the viscoelastic behavior of PVA-P (AAm/AMPS) IPN hydrogel with enhanced mechanical strength and excellent recoverability. *J. Polym. Res.* **2021**, *29*, 1–12. [CrossRef]

gels

MDPI

Article

Eco-Friendly Sol–Gel Coatings with Organic Corrosion Inhibitors for Lightweight AZ61 Alloy

Jorge Domínguez-Martínez [1], Jesús López-Sánchez [2], Federico García-Galván [3], Aída Serrano [2], Violeta Barranco [4], Juan Carlos Galván [4], Óscar Rodríguez de la Fuente [1,5] and Noemí Carmona [1,5,*]

1 Department of Materials Physics, Universidad Complutense de Madrid (UCM), 28040 Madrid, Spain; jordom02@ucm.es (J.D.-M.); osrodrig@ucm.es (Ó.R.d.l.F.)

2 Department of Electroceramics, Instituto de Cerámica y Vidrio (ICV-CSIC), 28049 Madrid, Spain; jesus.lopez@csic.es (J.L.-S.); aida.serrano@icv.csic.es (A.S.)

3 Department of Engineering, School of Architecture, Engineering and Design, Universidad Europea de Madrid, 28670 Villaviciosa de Odón, Spain; federico.garcia@universidadeuropea.es

4 National Center for Metallurgical Research (CENIM-CSIC), Avda. Gregorio del Amo 8, 28040 Madrid, Spain; violeta.barranco@csic.es (V.B.); jcgalvan@cenim.csic.es (J.C.G.)

5 Applied Magnetism Institute (IMA) "Salvador Velayos", A6 Km 22,5, Las Rozas, 28230 Madrid, Spain

* Correspondence: ncarmona@ucm.es

Abstract: The latest advances in technology and materials science have catalyzed a transformative shift towards the adoption of environmentally conscious and lightweight materials across key sectors such as aeronautics, biomedical, and automotive industries. Noteworthy among these innovations are the magnesium-aluminum (Mg-Al) alloys employed in aeronautical applications, contributing to the overall reduction in aircraft weight and subsequently diminishing fuel consumption and mitigating atmospheric emissions. The present work delves into a study of the anti-corrosive properties inherent in various sol-gel coatings, leveraging a range of environmentally friendly corrosion inhibitors, specifically tailored for samples of the AZ61 alloy. Methodologically, the work involves the synthesis and application of sol-gel coatings on AZ61 alloy containing eco-friendly inhibitors: L-cysteine, N-acetyl-cysteine, curcumin and methylene blue. Subsequently, an accelerated corrosion test in a simulated saline environment is performed. Through microstructural and compositional analyses, the best inhibitors responses are achieved with inhibitors containing S, N heteroatoms and conjugated double bonds in their structure, probably due to the creation of a continuous $MgCl_2$ layer. This research contributes to the ongoing discourse on protective eco-coatings, aligning with the broader paradigm shift towards sustainable and lightweight materials in key industries.

Keywords: AZ61 alloy; sol–gel; protective coating; corrosion eco-inhibitors

1. Introduction

In the dynamic field of the aeronautical industry, innovation is a constant driving force for progress. The industry consistently strives to redefine technological and design boundaries, as evidenced by the ongoing development of cutting-edge prototypes. Integral to this mission is the meticulous selection and understanding of optimal materials for aircraft and vehicle designs. The consequences of any oversight at the structural or material level are profound, with the potential for catastrophic outcomes [1].

Currently, there is a pressing need for aircraft materials to maintain optimal lightweight properties, a crucial factor in the ongoing efforts to minimize aircraft weight and, consequently, decrease emissions and environmental pollutants. In response to this imperative, various alternatives are under scrutiny, with magnesium alloys emerging as a prominent focus among numerous options. These alloys present the potential to be utilized either independently or in conjunction with other lightweight materials, forming composite structures that combine strength and efficiency in aerospace applications [2]. These lightweight alloys, exemplified by magnesium alloys, exhibit commendable strength, ductility, and lower densities in contrast to materials like titanium or aluminum alloys. While it is acknowledged that their malleability is relatively lower than that of aluminum, the advantage lies in the ability to attain lower temperatures and forces during machining, contributing to an extended service life of the material. Additionally, the exceptional recyclability of these alloys further underscores their environmental efficacy [3].

Nevertheless, the utilization of these alloys introduces another critical consideration—the potential occurrence of corrosion phenomena resulting from their interaction with humid environments. Furthermore, factors such as grain size, dislocation density, and texture emerge as crucial parameters that can significantly influence the corrosion resistance of these alloys [4–6]. Hence, to mitigate the potential corrosion risks, it becomes imperative to employ surface treatments or apply protective coatings. Common approaches involve the use of paints or employing anodizing processes. Notably, the sol–gel process emerges as one viable option, offering economic efficiency and versatility. Its cost-effectiveness makes it a preferred choice, while its adaptability allows for application across a spectrum of alloy types. Furthermore, it is noteworthy that the development of the sol–gel process has experienced substantial growth over the years, underscoring its evolving significance in corrosion protection methodologies [7].

The sol–gel process comprises three distinct stages: hydrolysis, polycondensation, and thermal densification. It initiates in the sol phase, characterized by a colloidal suspension of small particles in a solution, and subsequently progresses into the gel phase, exhibiting a solid–liquid biphasic structure. An advantageous aspect of the sol–gel process lies in the flexibility of the precursor materials, which can be either organic or inorganic, distinguishing it from other methods. Noteworthy attributes of the sol–gel process include its economic viability, minimal toxicity, and the high purity of the resultant products achieved through meticulous control of the precursors' composition. The coatings generated exhibit commendable chemical resistance, functioning as potent corrosion inhibitors for both metals and alloys. Moreover, the synthesis can be conducted at room temperature and atmospheric pressure [8,9], adding to its practicality. The distinctive nanoporosity of sol–gel coatings introduces an additional dimension to their functionality. This feature allows them to actively combat corrosion by harboring new molecules while preserving their initial properties, paving the way for innovative applications in diverse fields [10,11].

In the pursuit of enhancing corrosion protection in coatings, the incorporation of inhibitors offers a strategic approach. These inhibitors play a dual role by either retarding the degradation of alloys, forming a barrier against corrosion, or engaging in a self-repairing mechanism for active protection. In the wake of the ban on chromate treatments, various inorganic salts, including Ce^{3+}, Zr^{3+}, or La^{3+} ions, have historically served as inhibitors. However, a contemporary shift toward organic molecules and their derivatives is evident, driven by their superior biodegradability and reduced toxicity and environmental impact [12]. The recent transition to organic inhibitors is motivated by their enhanced inhibitory capacity, attributed to the presence of multiple heteroatoms and double bonds. This shift reflects a commitment to advancing corrosion protection methods in a more sustainable and environmentally conscious direction [13–18].

The present work is dedicated to the application and characterization of diverse sol–gel coatings utilizing environmentally friendly corrosion inhibitors. These coatings will be applied to different samples of AZ61 alloy, specifically targeting applications in sustainable transport and determining their anticorrosive capacity. The overarching objective is to contribute to the advancement of environmentally conscious coatings for the promotion of sustainability in transportation.

2. Results and Discussion

2.1. Coatings Characterization

Table 1 displays the obtained coatings with their corresponding nomenclature, preparation, and test conditions.

Table 1. Detailed description of the samples obtained.

Sample Name	Sol–Gel Coating	Corrosion Inhibitor Added	Sol's Color	Sol's pH	Corrosion Test Performed
A	NO	-	-	-	NO
REF	NO	-	-	-	YES
BLA	YES	-	Pink	2	YES
L-CYS	YES	L-cysteine	Pink	3–4	YES
N-A-CYS	YES	N-acetyl-cysteine	Red	2	YES
CUR	YES	Curcumin	Brown	2	YES
MB	YES	Methylene Blue	Blue	3	YES

The thickness of the coatings was determined through the interference of reflection spectra in the ultraviolet-visible and near-infrared ranges [19]. Specific values for each sol–gel coating are presented in Table 2. The approximate thickness of all coatings ranged between 150 and 300 nm.

Table 2. Calculated thicknesses of the coatings from the interference of reflection spectra.

Sample	BLA	L-CYS	N-A-CYS	CUR	MB
Thickness (nm)	195 ± 30	291 ± 40	281 ± 57	164 ± 30	268 ± 42

The transmission spectrum of the coated glass samples is performed in order to determine the wavelength of each coating at 50% transmission. For this purpose, a line is drawn parallel to the X-axis from the 50% transmission point until the cut-off with the corresponding curve occurs. After this, a line is drawn perpendicular to the X-axis up to this point, thus obtaining the corresponding wavelength values. As depicted in Figure 1, the sample containing the MB inhibitor exhibits the closest to the visible wavelength value, approximately at 410 nm. This suggests strong transmission within this segment of the visible spectrum, with absorption occurring in the range of 600–650 nm, resulting in a blue appearance. Following this, the sample with the L-CYS inhibitor shows an absorption edge wavelength around 310 nm. Finally, the last three samples (BLA, N-A-CYS, CUR) demonstrate a wavelength very close to 50% transmission, with values around 290 nm, highlighting the high transparency of the coatings within the visible spectrum and part of the near UV range.

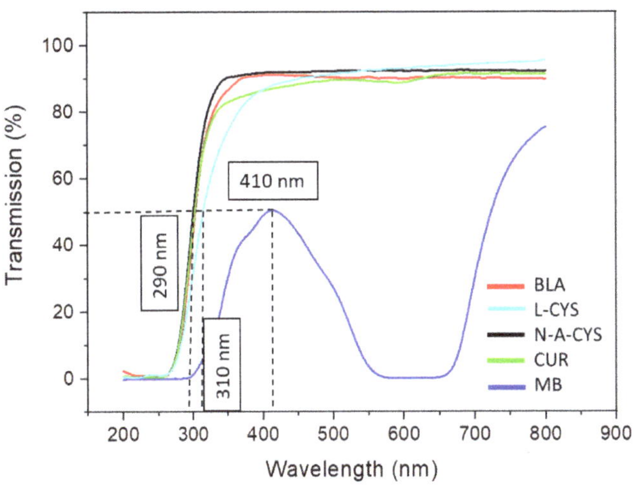

Figure 1. Transmission spectra related to N-A-CYS, BLA, CUR, MB, and L-CYS sol–gel coatings.

To determine the hydrophilic character of the coatings, the samples are characterized by contact angle measurements before the corrosion test. The procedure consists of depositing a drop of water on the coating surface and illuminating it with diffuse light in order to project the image through a camera onto a computer screen, thanks to which the tangent of the angle formed by the drop of water with the surface of each coating can be measured. In this work, different measurements were made using 1 μL drops of distilled water deposited on the surface of the coatings. Three different measurements along the surface of the sample were made to average the final result. Figure 2 lists the final contact angle values for each sample along with their respective uncertainty.

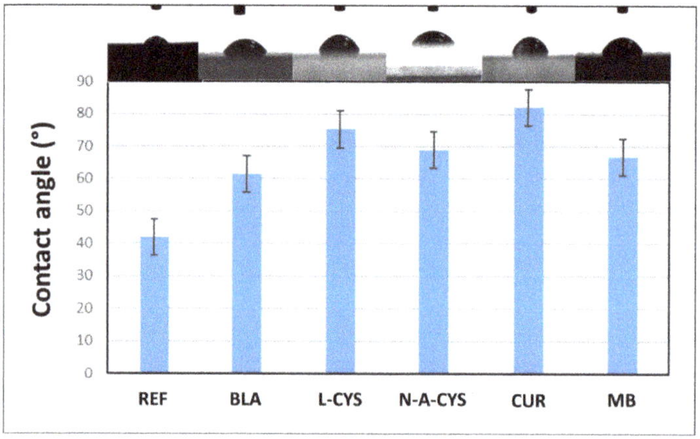

Figure 2. Contact angle values obtained for the samples REF, BLA, L-CYS, N-A-CYS, CUR, and MB with distilled water. Images of the water droplets on the surfaces on top of each sample.

As is evident from the results, all coatings exhibit a hydrophilic nature, as indicated by contact angles less than 90°. These hydrophilic properties of the coatings have been shown to be stable with reproducible contact angle values maintained for several months. The L-CYS and CUR coatings demonstrate the most favorable outcomes due to their values closely approaching 90°, suggesting their suitability for practical coating applications. By contrast, the REF sample, lacking any coating, stands out as expected, showing a relatively low contact angle compared to the others. Following closely is the BLA sample. This implies that the surfaces with lower contact angles, such as those of REF and BLA, are more prone to retaining water droplets, as the area in contact with water is greater than surfaces with higher contact angle values.

2.2. Weathering Test

After the thermal treatment of the AZ61 alloy samples for their densification, the study of their corrosion behavior in a concentrated saline medium is started. For this purpose, a variation of the ISO 11130 standard was used [20]. Each sample is attached with a fishing line and immersed in a 0.6 M NaCl solution in 125 mL Teflon bottles without stirring for 14 days. The solution temperature is 25 ± 2 °C and the solution was not changed during the test. Once the test is finished, the samples were washed with distilled water by immersion in three different beakers to ensure that they are properly cleaned of residues and/or impurities. Afterwards, the samples were dried at 25 ± 2 °C in a desiccator at less than 10% relative humidity for a further 14 days. The pH for each salt solution was measured before and after the test (including one with distilled H_2O + NaCl, with initial and final pH = 7) obtaining values of pH = 5, except in the samples with the L-CYS and MB coating where it is pH = 4. This indicates that protons are released into the saline medium causing an acidification of the latter during the corrosion process, leading, in all probability, to the following reaction:

$$\downarrow Mg + 2\,H_2O \rightarrow \downarrow Mg(OH)_2 + 2H^+ \quad (1)$$

The different samples before and after the corrosion test are shown in Figure 3, where clear differences can be appreciated.

Figure 3. Photographs of reference and samples with sol–gel coatings before corrosion test (left image) and after corrosion test (right image).

It can be observed that all the samples show a superficial deterioration after 2 weeks immersed in saline medium. White deposits of crystalline appearance can also be seen in all of them, indicating the presence of corrosion products on their surface. The sample with the MB coating is the one in the poorest condition since its coating layer has partially detached.

2.3. Optical Microscopy Characterization

Optical microscopy provides information about the surface of the samples both before and after the corrosion test. Interestingly, significant structural changes occur in each sample as identified in Figure 3. Figure 4 gives an overview of the micrographs obtained where the modifications at the microstructural level are shown.

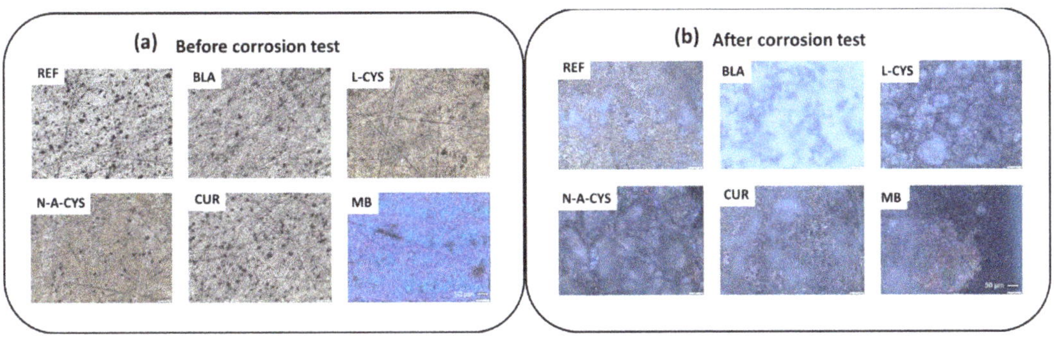

Figure 4. Optical micrographs of the samples: (**a**) before; and (**b**) after the corrosion test.

Figure 4a shows the surface of the different samples initially. The traces of the polishing prior to the application of the coating can be seen, which is carried out to remove any remaining oxide or dirt, but additionally the process improves the adhesion of the coating to the alloy surface. After the corrosion test in Figure 4b, the surface of all samples shows white crystalline deposits. Besides, in sample MB, the lifted coating is visible as can be observed in its respective micrograph by contrasting the brightness of each zone. In the L-CYS sample, cracks generated by the etching of the grain boundaries leading to grain detachment are observed along the entire surface.

2.4. Weight Variation

The samples were weighed before and after the corrosion test to examine weight variations. For accuracy, the samples were washed with distilled water after the test and left for 1 week in a desiccator to remove the humidity caused by the washing.

As a general remark, relatively high differences are noticed in the weight of the samples both before and after the corrosion test (Figure 5). In the MB sample, a negative weight variation of −1.5 mg is observed, which could be explained by a partial detachment of the sol–gel coating from the metal surface. In turn, the metal surface reacts with the ions of the medium, causing its dissolution. However, for the rest of the coated samples, the weight is slightly increased, probably due to the formation of corrosion products in the form of metal oxides, hydroxides or chlorides on the surface when the metal ions themselves react with the OH- ions of the saline medium. The uncoated sample (REF) has an excess weight of more than 35 mg (36.9 mg) and the CUR sample displays an increase of more than 20 mg (22.0 mg). In a second group, we can find the samples L-CYS and BLA showing an increase between 7 and 11 mg and finally the sample N-A-CYS, which barely modifies its weight (3.8 mg) (Figure 5).

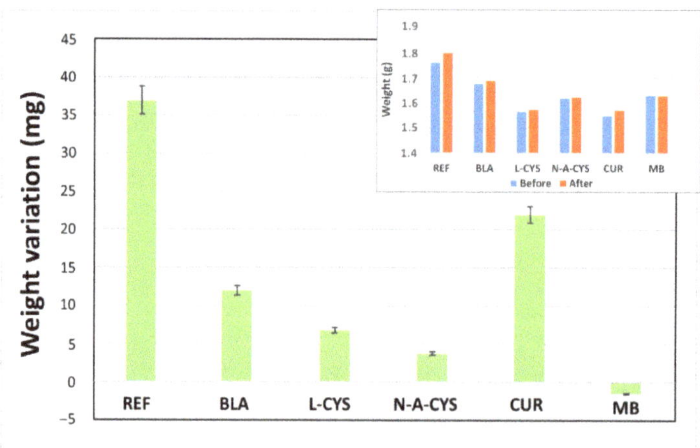

Figure 5. Weight variation of the samples before and after corrosion test. Inset shows absolute values of the weight before and after the test.

Results show how a thin layer of a sol–gel coating of less than 300 nm (BLA) is sufficient to significantly reduce corrosion on the alloy surface without coating (REF). Although it is logical to think that the thicker the coating is, the greater is its protective effect on the alloy, it is not always the case. Sol–gel coatings have an inherent nanoporosity in their structure and do not exert a barrier effect like other organic polymers (i.e., epoxy resins or polyurethanes). In the sol–gel coatings including organic inhibitors, the corrosion inhibiting effect of the chosen molecules must also be taken into account. Some previous studies mention that organic compounds bearing heteroatoms with high electron density such as phosphor, sulphur, nitrogen, oxygen or those containing multiple bonds are considered as adsorption centers, and thus are effective as corrosion inhibitors [21]. Compounds containing both nitrogen and sulphur in their molecular structure have exhibited greater inhibition compared with those containing only one of these atoms. The inhibitor seems to be chemically or physically adsorbed via both the thiol S atom and the carbonyl O atom to the metal and thus results in a higher experimental inhibition efficiency [22]. The non-bonding electron pair of N, O and S are known to adsorb onto metallic surfaces by electrostatic and electronic interactions with the vacant d orbital of iron. Due to these interactions, they can create a barrier layer on the metal, which displaces water and blocks the attack of aggressive species [23]. It seems like the sol–gel coating thickness improves the corrosion resistance as a barrier effect and the addition of organic inhibitor molecules improves the anti-corrosion effect.

2.5. SEM Characterization

The microstructure and surface morphologies were investigated by scanning electron microscopy (SEM), as in Figure 6. In addition, energy dispersive X-ray (EDX) analyses were carried out to reveal the compositions of the materials formed, which are listed in Table 3. The first sample analyzed is sample A, which corresponds to the AZ61 alloy without a coating or corrosion test. The SEM image shows a clean surface, with no traces of corrosion products or impurities. EDX analyses of the surface area of the image in Figure 6 A show the occurrence of Mg (92.5 wt.%) and Al (7.5 wt.%). Therefore, it indicates that no oxidation occurs on the surface. Regarding the REF sample (uncoated but submitted to the corrosion test), it is possible to observe flake-shaped crystals and certain grains with a rounded shape and more whitish color. The chemical analysis of the grains in the inset of Figure 6 REF shows a composition with Mg (66.7 wt.%), some Al (3.7 wt.%) and O (29.7 wt.%). Interestingly, the REF sample has a high amount of O, suggesting that it does possess a high amount of corrosion products on its surface. As for the whitish grains found in this sample, a spot analysis shows that they have a lower amount of O, implying that they are not as strongly oxidized as the background of the sample. Finally, the flaky crystals show a similar O composition to the background, so they also experienced a high oxidation. The BLA sample indicates a composition with Mg (39.5 wt.%), O (57.8 wt.%) and a small amount of Cl (2.7 wt.%) in its surface analysis. The high concentration of O along with Cl, is probably due to remnants of the NaCl solution that have not been washed out after the post-test wash. In turn, triangular-shaped crystals composed by different layers display a similar composition. As the percentage of Cl is residual, the formation of Mg hydroxide is probable.

Figure 6. Representative SEM images of the surface of samples A, REF, BLA, L-CYS, MB, CUR, and N-A-CYS.

Table 3. EDX compositional analysis of the microstructure of the samples.

Sample	C (wt.%)	O (wt.%)	Mg (wt.%)	Al (wt.%)	Si (wt.%)	Cl (wt.%)
A-background	-	-	92.5 ± 1.7	7.5 ± 1.7	-	-
REF-background	-	57.3 ± 2.4	42.7 ± 2.4	-	-	-
REF-white goop	-	29.7 ± 2.7	66.7 ± 2.7	3.7 ± 1.2	-	-
REF-crystal	-	57.9 ± 2.4	42.1 ± 2.4	-	-	-
BLA-background	-	57.8 ± 2.5	39.5 ± 2.4	-	-	2.7 ± 0.8
BLA-triangle	-	57.4 ± 2.2	42.6 ± 2.2	-	-	-
L-CYS-background	-	23.0 ± 2.6	77.0 ± 2.6	-	-	-
L-CYS-smooth area	-	14.7 ± 2.5	78.3 ± 2.7	5.8 ± 1.4	1.2 ± 1.0	-
L-CYS-stick	-	51.7 ± 2.0	46.1 ± 2.0	2.3 ± 0.7	-	-
MB-background	-	38.1 ± 1.3	50.1 ± 1.2	2.2 ± 0.5	9.6 ± 0.7	-
CUR-smooth area	19.2 ± 2.4	17.7 ± 1.1	60.3 ± 2.0	2.8 ± 0.3	-	-
CUR-crust area	-	57.0 ± 1.1	43.0 ± 1.1	-	-	-
N-A-CYS-background	9.0 ± 2.1	43.1 ± 1.4	43.1 ± 1.3	1.1 ± 0.3	2.7 ± 0.3	1.0 ± 0.2

In the L-CYS sample, surface area analysis reveals a predominant presence of Mg (77.0 wt.%) and O (23.0 wt.%), suggesting a relatively low concentration of O and indicative of less aggressive corrosion. Additionally, a smooth surface featuring small whitish rods is observed, with a rod composition made by a spot analysis comprising Mg (46.1 wt.%), O (51.7 wt.%), and a minor proportion of Al (2.3 wt.%). The structural background analysis reveals the presence of Si (1.2 wt.%), originating from the sol–gel coating. The surface composition analysis of the MB sample reveals a composition of Mg (50.1 wt.%), O (38.1 wt.%), Al (2.2 wt.%), and Si (9.6 wt.%). The presence of cracks on the surface indicates that the coating persists and, as in the L-CYS sample, a small proportion of Si was also detected. While the observed O content exceeds that of the L-CYS sample, it does not reach the levels identified in the REF sample; instead, it falls within an intermediate range.

Interestingly, two analyses are conducted on the CUR sample: one on the smoother and darker surface (left area), and the other one on the so-called crust zone (lighter right area). In the left area, C (19.2 wt.%) and Al (2.8 wt.%) appear together with O (17.7 wt.%) and Mg (60.3 wt.%), while in the lighter right area, there are only Mg (43.0 wt.%) and O (57.0 wt.%). In the separation region between the two areas (interface) can be seen rounded flaky crystals in the crust area. Moreover, a large flake-like crystallization is visible. The sample N-A-CYS presents all types of the constituents after surface analysis. It contains Mg (43.1 wt.%), O (43.1 wt.%), Al (1.1 wt.%), C (9.0 wt.%), Si (2.7 wt.%) and Cl (1.0 wt.%). Cracks were also found, indicating that the coating is present on the surface together with small triangular-shaped crystals with a different contrast. In the CUR sample, the crust area analysis displays the presence of O and Mg, suggesting the deposition of the corresponding hydroxide. In the smooth area, low values of O and Al are observed, accompanied by some C (19.2% by weight), possibly attributed to the thin layer of organic–inorganic coating. Finally, the N-A-CYS sample exhibits the presence of all the aforementioned elements. The concentration of Si indicates the presence of the coating, the percentage of Cl suggests minimal traces of marine dissolution, and the ratio of O to Mg indicates that the oxidation is not overly abrupt.

The SEM images from the cross-section of the samples in Figure 7 show sharp interfaces corresponding to the alloy (on the left side of the image) and the right side, corresponding to the epoxy resin of the embedding material for the non-corroded samples A and REF. The coatings of samples BLA, L-CYS, N-A-CYS and CUR, together with the corrosion products formed after the corrosion test, are marked in the SEM images for each of them with a blue double arrow. Their thicknesses are around ~1–2 μm and their composition indicates the presence of a large amount of Mg (40.0 wt.%), O (58.6 wt.%) and a small amount of Si (1.4 wt.%). The initial thickness of the coating seems to have increased due to oxidation of the magnesium at the alloy-coating interface. The adhesion of the coatings is, however, good due to the formation of Si-O-Mg chemical bonds and therefore no flaking is observed.

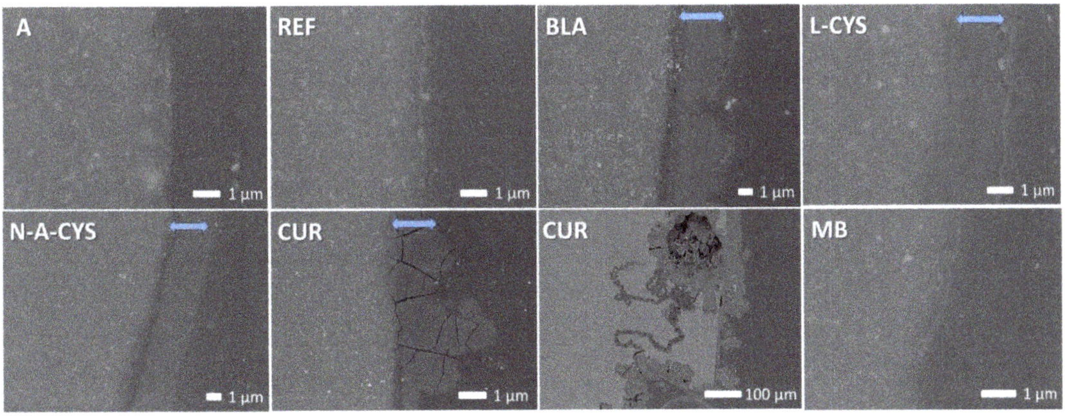

Figure 7. Representative SEM images from the cross-section of the samples A (non-corroded), REF, BLA, L-CYS, MB, CUR, and N-A-CYS (after corrosion test). The double blue arrow indicates the coating together with the corrosion products formed after the corrosion test.

In the SEM image of the CUR sample, the swollen coating is completely cracked, and in a lower magnification image, corrosion penetration lines can be seen, while the layer is peeling off in some areas. Finally, in the MB sample, an irregular interface can be seen where no coating or corrosion layer is visible. This is in agreement with the superficial analysis performed, which showed that the coating had peeled off in a large part of this sample. In contrast to the surface SEM images (Figure 6), no crystals are visible from the cross-section in Figure 7, which have probably been dissolved during polishing.

2.6. *Confocal Raman Microscopy Characterization*

Confocal Raman microscopy is a very versatile technique, and it can be highly useful in this type of corrosion study since it provides accurate compositional information at a local microscale level [24–27]. Specifically to this study, it enables us to know about the compounds formed considering the chemical strategies developed in the protective sol–gel coating. These compounds formed could hold the key to understanding the ability to withstand corrosion, in this case, in a saline environment. For this purpose, an uncoated (REF), coated (BLA) and coated with one type of inhibitor sample (N-A-CYS) were chosen. Optical images corresponding to representative zones are displayed in Figure 8a–c, respectively. The REF and BLA samples exhibit large homogeneous flat areas and no particle aggregates or evidence of a deposited coating on the BLA sample, as shown in the SEM images in Figure 6. In contrast, the N-A-CYS sample exhibits heterogeneity in its microstructure, alternating both flatter zones and particle aggregates, thus evidencing a high degree of surface roughness. In addition, there are easily found areas that form optical iridescence, characteristic of a silica-based sol–gel coating [28].

Raman mappings were performed on representative areas indicated with a red square in Figure 8a–c and different Mg-based compounds were identified. The integration of their main Raman bands was carried out and the in-plane Raman image was calculated (Figure 8d–f). Subsequently, the average Raman spectra of the compounds formed are retrieved and plotted in Figure 8g–i. Starting with the REF sample (Figure 8d,g), we notice the formation of bulk crystals of MgO with no contribution of hydroxyl groups (green color), agglomerates of $Mg(OH)_2$ nanocrystals (red color), and a continuous layer composed mainly of MgO and $Mg(OH)_2$ (turquoise color). The Raman spectrum of MgO is very characteristic since its rocksalt structure means that Raman modes are forbidden for bulk crystals [29]. Therefore, we are dealing with a flat Raman spectrum without observable Raman mode features.

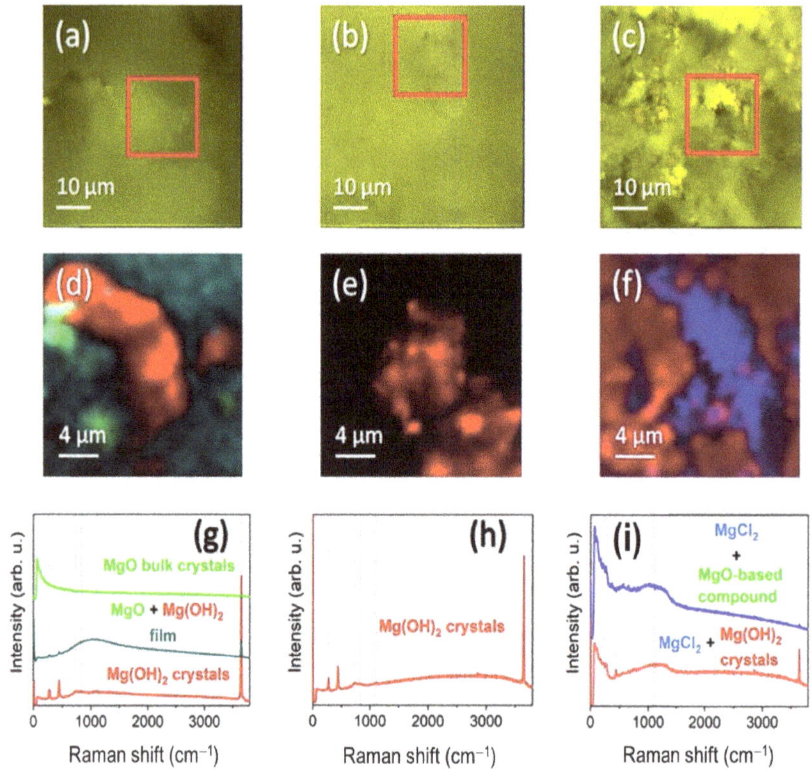

Figure 8. (**a**–**c**) Optical images (red squares indicate the representative areas where Raman mapping was performed); (**d**–**f**) superficial integrated Raman intensity images, (**g**–**i**) averaged Raman spectra obtained for the REF, BLA, and N-A-CYS samples.

It is true that this Raman spectrum could properly belong to the AZ61 metal alloy itself. However, given the high weight gain (Figure 5) and the high degree of oxidation detected (Table 3), we consider that the alloy surface has mostly oxidized after the corrosion test. Interestingly, several main Raman modes at 280, 446, 1088, and 1120 cm^{-1} are reported when the particle size is between 30 and 3000 nm [30,31]. Such Raman modes are clearly identified in the Raman spectrum obtained in the REF sample and they are indicated by dotted lines. Further modes also appear within the first-order Raman scattering such as the one around 730 cm^{-1} [31]. It is worth mentioning that the 1088 and 1120 cm^{-1} modes are fairly wide, spanning a long range between 800 and 1500 cm^{-1}. They correspond to surface phonon modes within the TO-LO phonon gap [29]. In contrast, $Mg(OH)_2$ shows great similarities with respect to MgO, except for the band located around 3600 cm^{-1}, corresponding to OH bonds [32]. Moreover, the formation of Mg hydroxides is very likely since MgO is highly hygroscopic and interacts with physisorbed water to form $Mg(OH)_2$ [33].

The BLA sample (with coating and no inhibitor), is compositionally homogeneous and the formation of $Mg(OH)_2$ is noted both in the flat zones and in the agglomerates detected (Figure 8e,h). However, the N-A-CYS sample with the inhibitor shows an additional formation of $MgCl_2$ with no presence of hydroxyl groups (blue color in Figure 8f,i). The presence of $MgCl_2$ is predominant in all regions studied and appears mixed with $Mg(OH)_2$ (red color in Figure 8f), although it is also observed mixed with a non-indexed compound, probably based on MgO. We assume that the blue Raman spectrum is an MgO-based compound since it possesses the same Raman characteristics as MgO, but the Raman bands are shifted towards lower wavenumbers. In addition, this non-indexed compound shows a low crystallinity as the Raman bandwidth is wide and the relative intensity of the Raman modes is really low. Coming back to the Raman features of $MgCl_2$, this compound features its Raman allowed modes below 300 cm^{-1} and can be identified mainly by the mode located around 254 cm^{-1} indicated with an X in Figure 8i [34]. For clarity, we compare the Raman spectra of MgO bulk with the mixture with $MgCl_2$ in Figure 9, indicating with a yellow color the Raman spectral zone characteristic of $MgCl_2$ [34,35].

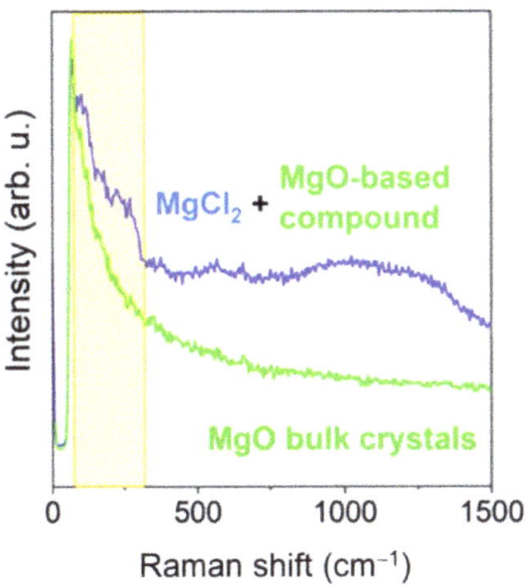

Figure 9. Raman comparison between MgO bulk crystals and $MgCl_2$ + MgO-based compound.

3. Conclusions

The investigation focused on the corrosion behavior of eco-friendly sol–gel coatings on the AZ61 (Mg-Al) alloy reveals valuable insights. Notably, two distinct groups based on coating thickness emerged: one with thicknesses below 200 nm (BLA and CUR) and another with thicknesses close to 300 nm (L-CYS, N-A-CYS, and MB).

The consistent composition of the coatings, coupled with simultaneous fabrication, indicated that samples containing organic inhibitors exhibit greater thickness. This observation suggests that the inhibitors play a catalytic role in accelerating polycondensation, facilitating their adhesion to the alloy surfaces. Noteworthy variations in contact angle values, with the CUR sample approaching the 90^0 threshold, underscored the predominantly hydrophilic character of all coatings. Through a comprehensive corrosion test conducted in a concentrated saline medium, the study aimed to assess the impact of different inhibitors on the alloy's corrosion resistance. Weight variation assessments highlighted the pivotal role of coatings in corrosion prevention. In this sense, all samples except MB increase in weight, indicating that the coating is the key in preventing or slowing down corrosion by the formation of hydroxides and chlorides that could adhere to the surface of the samples. Intriguingly, thicker coatings (L-CYS and N-A-CYS) demonstrated superior corrosion resistance.

Comparative EDX analyses consistently reveal lower corrosion in the presence of coatings, a trend aligned with SEM image observations. This leads to the inference that organic inhibitors, particularly those with S, N heteroatoms, and conjugated double bonds, stand out as promising candidates for effectively retarding and inhibiting corrosion in light

Mg-Al alloys. Raman experiments corroborate the oxidation and superficial modifications in coatings identifying several Mg-based compounds.

Notably, the environmentally friendly characteristics of these inhibitors enhance their potential for widespread industrial use inside sol–gel coatings on Mg alloys, making them economically viable. These coatings are easily applied by dip-coating or spray, since their densification is carried out at room temperature. Their properties are maintained over time, so that a possible industrial scale-up would be possible and easily implementable. Consequently, this research not only contributes to refining corrosion mitigation strategies for such alloys but also paves the way for future developments across diverse sectors and applications.

4. Materials and Methods

4.1. AZ61 Alloy

A light magnesium–aluminum alloy, commercially known as AZ61 (0.3% by mass of Mn, 0.7% by mass of Zn, 6% by mass of Al, and the remaining portion of Mg up to 100% by mass) was used as a substrate for the samples. Subsequently, samples with dimensions $2.0 \times 1.4 \times 0.3$ cm were polished to eliminate any traces of oxide layers and/or surface impurities that may have been present prior to the application of the coating.

4.2. Synthesis of the Sol–Gel Coatings for Inhibitory Performances

The alkoxides tetraethylorthosilicate (TEOS, Sigma-Aldrich, Madrid, Spain, 99%) and 3-(trimethoxysilyl)propyl methacrylate (MAP/MEMO, Sigma-Aldrich, 98%) served as precursors for the sol–gel coatings. Ethanol (EtOH, Sigma-Aldrich, 96%) was utilized as a solvent. Distilled water was carefully added drop by drop for hydrolysis until the reaction reached completion. Finally, acetic acid was introduced as a catalyst to facilitate polycondensation. The molar ratios of the components were 1:1:4:8:0.01, respectively. Subsequently, environmentally friendly organic corrosion inhibitors were added to this formulation. Four different organic molecules that are environmentally friendly, non-toxic, inexpensive and contain S, N heteroatoms, O and/or OH groups and/or conjugated double bonds were selected: L-cysteine (CYS), N-acetyl-cysteine (N-A-CYS), curcumin (CUR), and methylene blue (MB) in a ratio of 0.8 wt.%. Figure 10 illustrates the chemical structure of each compound:

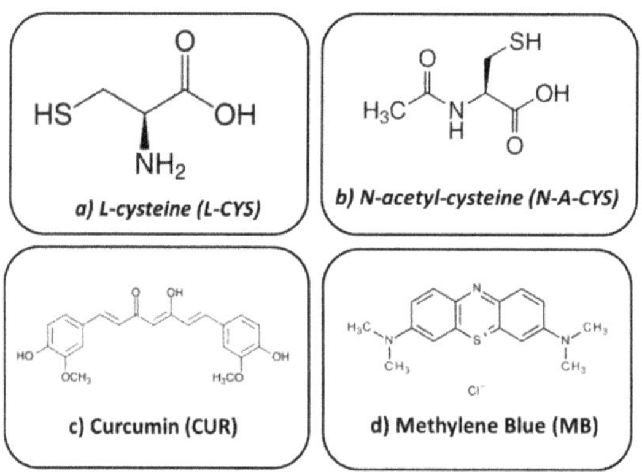

Figure 10. Chemical structure of corrosion inhibitors added to the sol–gel coating: (**a**) L-cysteine, (**b**) N-acetyl-cysteine, (**c**) curcumin and (**d**) methylene blue.

After mixing all the components, the sol was stirred with magnetic agitation for 24 h at 100 rpm to carry out the second stage of the sol–gel process, the polycondensation. Likewise, each solution includes the corresponding inhibitor, except for one called BLA, which does not include any inhibitor, and which will be subjected to the subsequent corrosion test.

Afterward, the dip-coating method was used for the application of the coatings since it provides good surface finishes and achieves a uniform coating layer. The extraction on the protective solution was carried out at a constant speed of 8 mm/s. After coating has been carried out on all respective samples, they are heat-treated in a muffle at 80 °C to carry out the densification step. Figure 11 illustrates the described sol–gel synthesis steps:

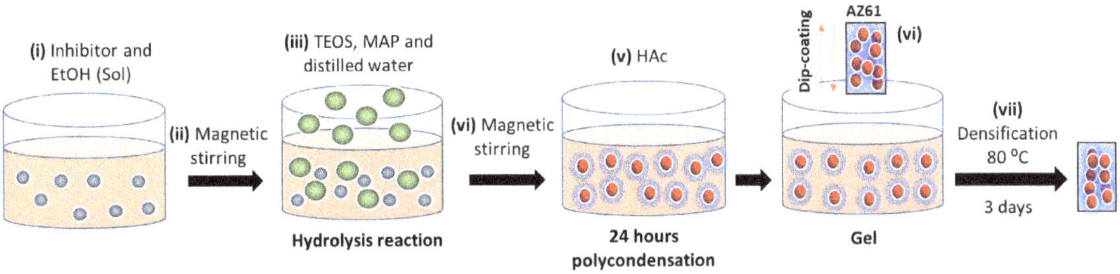

Figure 11. Sol–gel synthesis steps for coatings: (i) The corresponding inhibitor is dissolved in EtOH to form the sol. (ii) Homogenization of the sol by magnetic stirring. (iii) TEOS, MAP and distilled water are added dropwise where the hydrolysis process starts. (iv) Homogenization of the resulting solution. (v) Acetic acid (HAc) is added as catalyst where it is kept for 24 h under magnetic stirring. Then, (vi) the protective layer is deposited on the AZ61 alloy by dip-coating method. Finally, (vii) the densification treatment is carried out at 80 °C in air for 3 days.

Author Contributions: Ó.R.d.l.F., V.B., J.C.G. and N.C. conceived the work and coordinated the research. J.D.-M., F.G.-G., A.S. and J.L.-S. prepared the samples and carried out the weathering experiment. J.D.-M., J.L.-S., Ó.R.d.l.F., A.S. and N.C. carried out the structural characterization. J.C.G. and V.B. were responsible for the funding. J.D.-M., J.L.-S. and N.C. wrote the original draft. All authors have read and agreed to the published version of the manuscript.

Funding: This research was funded by the Ministry of Science and Innovation (MCINN, Spain) through the Project PID2019-104717RB-I00.

Institutional Review Board Statement: Not applicable.

Informed Consent Statement: Not applicable.

Data Availability Statement: The data presented in this study are openly available in article.

Acknowledgments: J.L.-S. acknowledges financial support from Spanish Ministry of Economic Affairs and Digital Transformation (MINECO) through the project PID2021-122980OB-C51MCIN and from grant RYC2022-035912-I funded by MCIU/AEI/10.13039/501100011033 and by the European Social Fund Plus (ESF+). A.S. acknowledges financial support from grant RYC2021-031236-I funded by MCIN/AEI/10.13039/501100011033 and by the "European Union NextGenerationEU/PRTR". The authors acknowledge the CNME Unit of the UCM for SEM/EDX measurements.

Conflicts of Interest: The authors declare no conflicts of interest.

References

1. Blanco, D.; Rubio, E.M.; Sáenz-Nuño, M.A.; Lorente-Pedreille, R.M. Comparison of sustainable cooling systems used in the drilling repair of Mg-Al and Mg-Ti multi-material parts in the aeronautical industry. *Tribol. Int.* **2022**, *175*, 107804. [CrossRef]
2. Akbarzadeh, S.; Santos, L.S.; Vitry, V.; Paint, Y.; Olivier, M. Improvement of the corrosion performance of AA2024 alloy by a duplex PEO/clay modified sol-gel nanocomposite coating. *Surf. Coat. Technol.* **2022**, *434*, 128168. [CrossRef]
3. Zhao, M.-C.; Liu, M.; Song, G.-L.; Atrens, A. Influence of Homogenization Annealing of AZ91 on MechanicalProperties and Corrosion Behavior. *Adv. Eng. Mater.* **2008**, *10*, 93–103. [CrossRef]
4. Kumar, R.; Bansal, R.K.; Sharma, V.; Minhas, N.; Thakur, A. Review: Corrosion behavior of friction stir welded magnesium alloys. *Mater. Today Proc.* **2023**, *80*, 272–277. [CrossRef]
5. Feliu, S., Jr.; Llorente, I. Corrosion product layers on magnesium alloys AZ31 and AZ61: Surface chemistry and protective ability. *Appl. Surf. Sci.* **2015**, *347*, 736–746. [CrossRef]

6. Kumar, P.P.; Bharat, A.R.; Sai, B.S.; Sarath, R.J.P.; Akhil, P.; Reddy, G.K.; Kondaiah, V.V.; Sunil, B.R. Role of microstructure and secondary phase on corrosion behavior of heat treated AZ series magnesium alloys. *Mater. Today Proc.* **2019**, *18*, 175–181 [CrossRef]
7. Dislich, H.; Hinz, P. History and principles of the sol-gel process, and some new multicomponent oxide coatings. *J. Non-Cryst. Solids* **1982**, *48*, 11–16. [CrossRef]
8. Wang, D.; Bierwagen, G.P. Sol-gel coatings on metals for corrosion protection. *Prog. Org. Coat.* **2009**, *64*, 327–338. [CrossRef]
9. Senani, S.; Campazzi, E.; Villatte, M.; Druez, C. Potentiality of UV-cured Hybrid sol-gel coatings for aeronautical metallic substrate protection. *Surf. Coat. Technol.* **2013**, *227*, 32–37. [CrossRef]
10. Zheng, X.; Liu, Q.; Ma, H.; Das, S.; Gu, Y.; Zhang, L. Probing local corrosion performance of sol-gel/MAO composite coating on Mg alloy. *Surf. Coat. Technol.* **2018**, *347*, 286–296. [CrossRef]
11. Murillo-Gutiérrez, N.V.; Ansart, F.; Bonino, J.-P.; Menu, M.-J.; Gressier, M. Protection against corrosion of magnesium alloys with both conversion layer and sol-gel coating. *Surf. Coat. Technol.* **2013**, *232*, 606–615. [CrossRef]
12. Barranco, V.; Carmona, N.; Galván, J.C.; Grobelny, M.; Kwiatkowski, L.; Villegas, M.A. Electrochemical study of tailored sol-gel thin films as pre-treatment prior to Organic coating for AZ91 magnesium alloy. *Prog. Org. Coat.* **2010**, *68*, 347–355. [CrossRef]
13. Rodríguez-Alonso, L.; López-Sánchez, J.; Serrano, A.; Rodríguez de la Fuente, O.; Galván, J.C.; Carmona, N. Hybrid sol-gel coatings doped with non-toxic corrosion inhibitors for corrosion protection on AZ61 magnesium alloy. *Gels* **2022**, *8*, 34. [CrossRef] [PubMed]
14. Yang, X.; Pan, F.; Zhang, D. A study on corrosion inhibitor for magnesium alloy. *Mater. Sci. Forum* **2009**, *610–613*, 920–926. [CrossRef]
15. Umoren, S.A.; Abdullahi, M.T.; Solomon, M.M. An overview on the use of corrosion inhibitors for the corrosion control of Mg and its alloys in diverse media. *J. Mater. Res. Technol.* **2022**, *20*, 2060–2093. [CrossRef]
16. Ashassi-Sorkhabi, H.; Moradi-Alavian, S.; Jafari, R.; Kazempour, A. Effect of amino acids and montmorillonite nanoparticles on improving the corrosion protection characteristics of Hybrid sol-gel coating Applied on AZ91 Mg alloy. *Mater. Chem. Phys.* **2019**, *225*, 298–308. [CrossRef]
17. BinSabt, M.; Abditon, M.; Galal, A. Highly efficient nanocomposite-containing coatings for the protection of Mg alloy against corrosion in chloride containing electrolytes. *Appl. Phys. A* **2022**, *128*, 151. [CrossRef]
18. Li, J.; Li, T.; Zeng, Y.; Chen, C.; Guo, H.; Lei, B.; Zhang, P.; Feng, Z.; Meng, G. A novel sol-gel coating via catechol/lysine polymerization for long-lasting corrosion protection of Mg alloy AZ31. *Colloids Surf. A Physicochem. Eng. Asp.* **2023**, *656*, 130361. [CrossRef]
19. Passos, M.L.C.; Saraiva, M.L.M.F.S. Detection in UV-visible spectrophotometry: Detectors, detection systems, and detection strategies. *Measurement* **2019**, *135*, 896–904. [CrossRef]
20. Montero, E.F.; García, M.A.; Villegas Broncano, M.A.; Llopis, J. Estudio de las propiedades ópticas de recubrimientos Sol-Gel dopados con fluoresceína en función de la concentración y del pH. *Boletín Soc. Española Cerámica Vidr.* **2004**, *43*, 8–11. Available online: http://boletines.secv.es/upload/2007013193009.43[1]8-11.pdf (accessed on 22 September 2022). [CrossRef]
21. *ISO Standard 11130*; Corrosion of Metals and Alloys—Alternate Immersion Test in Salt Solution. 2017. Available online: http://www.iso.org (accessed on 17 April 2023).
22. Solmaz, R. Investigation of the inhibition effect of 5-((E)-4-phenylbuta-1,3-dienylideneamino)-1,3,4-thiadiazole-2-thiol Schiff base on mild steel corrosion in hydrochloric acid, Corrosion. *Science* **2010**, *52*, 3321–3333. [CrossRef]
23. Oguzie, E.E.; Li, Y.; Wang, S.G.; Wang, F. Understanding corrosion inhibition mechanisms—Experimental and theoretical approach. *RSC Adv.* **2011**, *1*, 866–873. [CrossRef]
24. Udabe, E.; Somers, A.; Forsyth, M.; Mecerreyes, D. Design of polymeric corrosion inhibitors based on ionic coumarate groups. *Appl. Polym. Mater.* **2021**, *3*, 1739–1746. [CrossRef]
25. *ASTM Standard G31-72*; Standard Practice for Laboratory Immersion Corrosion Testing of Metals (Reapproved 1990). Annual Book of ASTM Standards; ASTM International: West Conshohocken, PA, USA, 2004; p. 302.
26. Aparicio, M.; Mosa, J.; Rodríguez, G.; Guzman, J.; Picard, Q.; Klein, L.C.; Jitianu, A. Consolidated Melting Gel Coatings on AZ31 Magnesium Alloy with Excellent Corrosion Resistance in NaCl Solutions: An Interface Study. *ACS Appl. Mater. Interfaces* **2019**, *11*, 3493–3505. [CrossRef] [PubMed]
27. López-Sánchez, J.; Serrano, A.; Del Campo, A.; Abuín, M.; Rodríguez de la Fuente, O.; Carmona, N. Sol–Gel Synthesis and Micro-Raman Characterization of ε-Fe_2O_3 Micro- and Nanoparticles. *Chem. Mater.* **2016**, *28*, 511–518. [CrossRef]
28. López-Sánchez, J.; Serrano, A.; del Campo, A.; Abuín, M.; Salas-Colera, E.; Muñoz-Noval, A.; Castro, G.R.; de la Figuera, J.; Marco, J.F.; Marín, P.; et al. Self-assembly of iron oxide precursor micelles driven by magnetic stirring time in sol–gel coatings. *RSC Adv.* **2019**, *9*, 17571. [CrossRef] [PubMed]
29. López-Sánchez, J.; Serrano, A.; del Campo, A.; Muñoz-Noval, Á.; Salas-Colera, E.; Cabero, M.; Varela, M.; Abuín, M.; Castro, G.R.; Rubio-Zuazo, J.; et al. A combined micro-Raman, X-ray absorption and magnetic study to follow the glycerol-assisted growth of epsilon-iron oxide sol-gel coatings. *J. Alloys Compd.* **2022**, *892*, 162061. [CrossRef]
30. Dekermenjian, M.; Ruediger, A.P.; Merlen, A. Raman spectroscopy investigation of magnesium oxide nanoparticles. *RSC Adv.* **2023**, *13*, 26683. [CrossRef]
31. Ishikawa, K.; Fujima, N.; Komura, H. First-order Raman scattering in MgO microcrystals. *J. Appl. Phys.* **1985**, *57*, 973. [CrossRef]

2. Sangster, M.J.L.; Peckham, G.; Saunderson, D.H. Lattice dynamics of magnesium oxide. *J. Phys. C Solid State Phys.* **1970**, *3*, 1026. [CrossRef]
3. Dawson, P.; Hadfield, C.D.; Wilkinson, G.R. The polarized infra-red and Raman spectra of $Mg(OH)_2$ and $Ca(OH)_2$. *J. Phys. Chem. Solids* **1973**, *34*, 1217–1225. [CrossRef]
4. Kumari, L.; Li, W.Z.; Vannoy, C.H.; Leblanc, R.M.; Wang, D.Z. Synthesis, characterization and optical properties of $Mg(OH)_2$ micro-/nanostructure and its conversion to MgO. *Ceram Int.* **2009**, *35*, 3355–3364. [CrossRef]
5. Sano, H.; Miyaoka, H.; Kuze, T.; Mori, H.; Mizutani, G.; Otsuka, N.; Terano, M. Raman Spectra of Magnesium Chloride Thin Films. *Surf. Sci.* **2002**, *502–503*, 70–74. [CrossRef]

gels

MDPI

Review

Green and Low-Cost Modified Pisha Sandstone Geopolymer Gel Materials for Ecological Restoration: A Phase Review

Changming Li [1,2,3,*], Yubing Fu [2], Haifeng Cheng [2], Yaozong Wang [2], Dongyang Jia [2] and Hui Liu [1]

1 Key Laboratory of Ecological Environment Protection and Restoration in the Yellow River Basin of Henan Province, Zhengzhou 450045, China; tuzi951@163.com
2 School of Civil Engineering and Transportation, North China University of Water Resource and Electric Power, Zhengzhou 450045, China; dryubingfu@163.com (Y.F.); chf6077@163.com (H.C.); wangyaozong2022@163.com (Y.W.); dongyangjia2022@163.com (D.J.)
3 International Joint Research Lab for Eco-Building Materials and Engineering of Henan, North China University of Water Resources and Electric Power, Zhengzhou 450045, China
* Correspondence: lichangmingh@163.com

Abstract: Pisha sandstone (PS) is a special interbedded rock in the middle reaches of the Yellow River that experiences severe weathering and is loose and broken. Due to severe multiple erosion events, the Pisha sandstone region is called "the most severe water loss and soil erosion in the world" and "the ecological cancer of the earth". As a special pozzolanic mineral, PS has the potential to be used as precursors for the synthesis of green and low-carbon geopolymer gel materials and applied in ecological restoration. This paper aims to undertake a phase review of the precursors for geopolymer gel materials. The genesis and distribution, physical and chemical characterization, erosion characteristics, and advances in the ecological restoration of PS are all summarized. Furthermore, current advances in the use of PS for the synthesis of geopolymer gel materials in terms of mechanical properties and durability are discussed. The production of Pisha sandstone geopolymer gels through the binder jetting technique and 3D printing techniques is prospected. Meanwhile, the prospects for the resource application of PS in mine rehabilitation and sustainable ecology are discussed. In the future, multifactor-driven comprehensive measures should be further investigated in order to achieve ecological restoration of the Pisha sandstone region and promote high-quality development of the Yellow River Basin.

Keywords: Pisha sandstone; geopolymer gels; physical and chemical characterization; ecological restoration; resource applications

1. Introduction

The Yellow River watershed is an important ecological barrier in China. However, the ecological environment of the Yellow River watershed is in a severe situation [1,2]. Soil erosion has become one of the most serious and complex global issues, significantly affecting soil nutrients, land productivity, and carbon cycling in terrestrial ecosystems, and further generating unfavorable environmental and economic impacts [3,4]. At large spatial scales, soil erosion can increase sediment deposition in river channels and downstream reservoirs, thereby increasing the risk of flooding and potential water pollution [5,6].

Located in the Yellow River Basin of North China and covering almost 454,000 km^2, the Loess Plateau is an important ecological barrier and fragile representative region. As the most serious region of soil erosion in China, the Loess Plateau accounts for 98% of the total region of soil erosion in the Yellow River Basin. Furthermore, it contributes 97% of sediment to the Yellow River, which seriously restricts the sustainable development of the Yellow River Basin [7]. The junction region of Shanxi, Shaanxi, and Mongolia in the middle reaches of the Yellow River is the most characteristic erosion region in the Loess Plateau, covering an area of 1.17×10^4 to 3.20×10^4 km^2 [8–12]. There is a special interbedded rock named Pisha sandstone (PS), which is severely weathered, loose, and broken. PS is as hard as stone when it is dry but rapidly disintegrates into sediment when it is exposed to water. It is the primary source of coarse sediment in the Yellow River. PS mainly refers to the interbedded rock composed of a thick layer of sandstone, mudstone, siltstone, and arenaceous shale in the Paleozoic Permian, Mesozoic Triassic, Jurassic, and Cretaceous. The Pisha sandstone region is characterized by fragmented topography, crisscrossing gullies, low soil nutrients, and low ecological carrying capacity. The climate is characterized by dramatic regional temperature variations, wind and sand, and frequent rainstorms. There are various types of erosion in this region, exhibiting a characteristic of interaction. The soil erosion modulus can reach approximately 30,000–40,000 $t/(km^2 \cdot a)$ [13]. The average annual sediment transport in the Pisha sandstone region is around 200 million t, accounting for about 30% of the total sediment imported in the middle and upper reaches of the Yellow River in the Loess Plateau region [14,15]. The coarse sediment ($d \geq 0.05$ mm) imported into the Yellow River in the Pisha sandstone region accounts for about 62% of the total coarse sediment imported into the Yellow River [16]. Additionally, around a quarter of the sediment is deposited in the lower reaches, which is one of the 'culprits' of frequent floods caused by riverbed uplift [17]. Furthermore, the coexistence of multiple hazards of desertification and soil erosion in the region leads to severe ecological degradation and a decline in carbon sequestration, exacerbating climate warming. Scholars call it "the most soil erosion in the world" and "the ecological cancer of the earth".

Great efforts have focused on ecological restoration since the 1960s. Significant and encouraging progress has been achieved in the management practice of the Pisha sandstone region through a series of projects such as the National Key Project for Soil and Water Conservation, the Soil and Water Conservation Project on the Loess Plateau supported by a World Bank loan, and the Key Project for the Management of Small Watersheds. The ecological restoration in the Pisha sandstone region is facing new opportunities and challenges, with the proposal of the national strategy of "Ecological Protection and High-Quality Development in the Yellow River Basin" [18]. PS has good pozzolanic activity due to its high contents of silica and alumina. There is an excellent opportunity to develop PS as a green, low-carbon geopolymer gel material to control soil erosion and restore ecology. At the same time, based on Pisha sandstone geopolymer gel materials, the material preparation process can be developed to meet different engineering purposes. Combined with the potential ecological restoration function of Pisha sandstone, the development of comprehensive measures for ecological restoration is a future research direction.

Therefore, it is necessary to carry out a phased review to facilitate a comprehensive understanding of Pisha sandstone as well as the ecological restoration techniques, and to demonstrate the main problems and gaps faced by ecological restoration in the Pisha sandstone region. This paper reviews the current studies related to the physical and chemical properties, erosion characteristics, ecological restoration advances, and potential applications of PS, with particular emphasis on the properties of modified Pisha sandstone geopolymer gel materials. Additionally, prospects for the potential applications of PS in mining area restoration are also reviewed. Finally, the future direction of ecological restoration is prospected with new insights to establish a multifactor-driven comprehensive technical model.

2. Distribution of Pisha Sandstone

In the Paleozoic before the Carboniferous, the Pisha sandstone region was a sea. During the Permian period, the crust was uplifted and the sea disappeared, turning into an inland basin where red mudstone and clastic rock of continental facies were deposited. In the early and middle Triassic, the basin sank rapidly, and red mudstone and clastic rock of continental facies continued to be deposited. Subsequently, the basin was uplifted under the influence of Indonesian tectonic movement. Weathering erosion caused false conformity and angular unconformity contact between strata. In the late period, the crust slowly descended, causing parallel and angular unconformity contacts between strata.

Meanwhile, the red stratum gradually decreased, and organic substances such as gray-green clastic rocks and coal-bed lenses began to be deposited, which was characterized by lacustrine deposition. At the end of the period, due to the Indochinese tectonic movement, the basin was fully uplifted and the strata were generally eroded. In the early Jurassic period, the crust gradually sank. The climate changed from a dry and hot oxidation environment to a warm and humid reduction environment. Meanwhile, gray-green and gray coal-bearing fine sand shale strata began to be deposited. In the middle and late Jurassic, due to the descent of the crust, the climate transitioned from warm and humid to dry and hot. Meanwhile, red mudstone deposits appeared. Under the influence of the late Yanshan tectonic movement, the crust continued to rise. The basin was uplifted for a long time and suffered from weathering and stripping. Meanwhile, the Lower Cretaceous was unconformity over the Middle Jurassic. In the Cretaceous, the crust turned to sink, and mudstone and coarse debris were deposited in large amounts, which were characterized by lacustrine deposition. At the end of the Cretaceous, the crust was gradually lifted by the late Yanshan tectonic movement, and the uplifted basin suffered wind erosion. In the Pliocene period of the Tertiary, the crust began to sink, and the red mudstone clay layer was deposited, which was characterized by lacustrine deposition. At the end of the last century, a plateau was formed during the upward and downward movements of the crustal structure. In the Quaternary period, the loess became sand, ancient humans appeared, and the modern landscape was formed [19–24].

At present, PS is concentrated in the junction region of Shanxi, Shaanxi, and Mongolia, ranging from the Yellow River in the east, reaching Maobula River in Hangjin Banner of Inner Mongolia in the west, Shenmu County in Shaanxi Province in the south, and the southern edge of Hobq Desert in the north. Erdos Dongsheng District, Junger Banner, Ejin Horo Banner, Dalad Banner, and Hangjin Banner are the main distribution region of PS. There are scattered distributions in Shenmu County and Fugu County in Shaanxi, Hequ County and Baode County in Shanxi, and Qingshui River County in Inner Mongolia [8]. The distribution of Pisha sandstone is shown in Figure 1. Depending on the surface cover, the rocky region can be divided into three types of regions: bare PS region, sand-covered PS region, and soil-covered PS region. The distribution region of PS is shown in Table 1. It can be seen that the distribution region of PS ranges from 1.17×10^4 to 3.2×10^4 km^2. The defined regions may vary due to different research methods and definitions of the type of regions, as well as the fact that the bare PS near the desert may evolve into sand-covered PS over time.

(a)

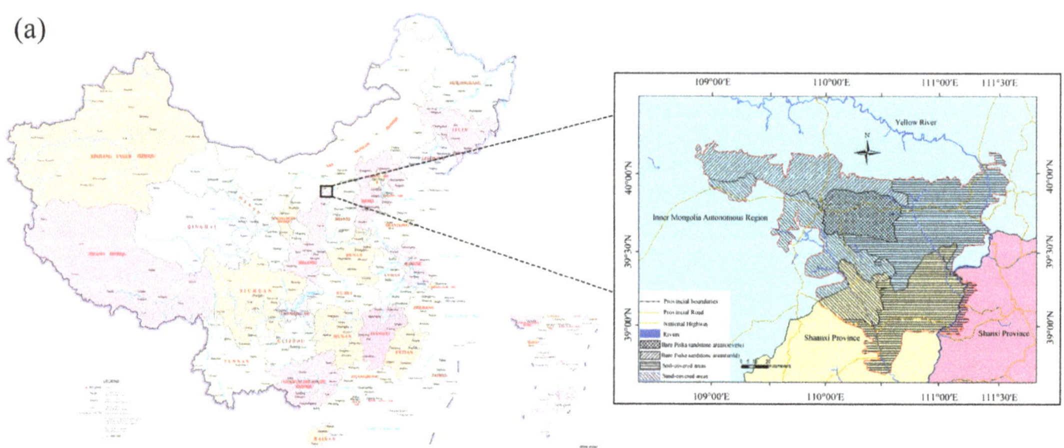

Figure 1. *Cont.*

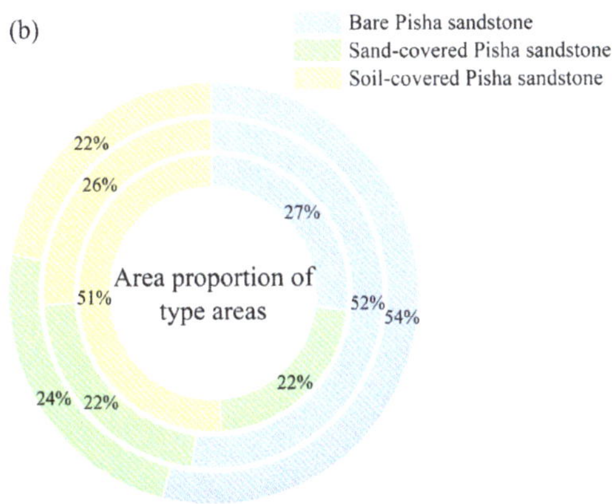

Figure 1. The distribution of Pisha sandstone. (**a**) The distribution map of Pisha sandstone in China [8]. (**b**) Proportional area of Pisha sandstone-type regions [8–10].

Table 1. Distribution regions of PS unit: 10^4 km^2.

Reference	Bare PS	Sand-Covered PS	Soil-Covered PS	Total
[8]	0.45	0.37	0.84	1.66
[9]	0.63	0.26	0.31	1.20
[10]	0.63	0.28	0.26	1.17
[11]	/	/	/	3.20
[12]	/	/	/	1.97 ①

Note: ①—The particle size of the sediment yield is greater than or equal to 0.05 mm and the sediment transport modulus is greater than 5000 t/(km^2·a).

3. Physical and Chemical Characterization of Precursors for Pisha Sandstone Geopolymer Gels

3.1. Physical and Chemical Properties

PS shows a diversity of colors due to its different strata and environment. The main colors are pink, purple-red, gray-white, gray-green, and yellow-green [16,25–27]. The mineral composition of PS of different colors and regions is shown in Figure 2 [11,28–33]. It can be seen that PS is mainly composed of quartz (17.10% to 88.80%), calcium montmorillonite (2.51% to 51.33%), potassium feldspar (3.70% to 23.00%), plagioclase (0.20% to 30.20%), calcite (1.00% to 52.70%), illite (0.22% to 22.00%), kaolinite (0.00% to 44.00%), etc. In addition, some PS contains small amounts of hematite, with a content of about 1%. In addition, mica minerals are also detected in PS [33]. The contents of plagioclase, calcite, illite, and kaolinite vary considerably for different regions of PS. The contents of quartz, calcium montmorillonite, plagioclase, and calcite vary greatly among different colors of PS. However, there is little difference in other mineral contents. The large amount of quartz in the Pisha sandstone can provide particle support for geopolymer gels. Meanwhile, clay minerals such as calcium montmorillonite, kaolinite, and illite offer the possibility of Pisha sandstone as a reactive precursor for geopolymer gels.

The chemical component of PS is shown in Figure 3 [9,11,26,27,30,32,34–40]. It can be seen that the main chemical components of PS are SiO_2 (51.20~78.25%), Al_2O_3 (9.57~15.37%), and Fe_2O_3 (0.05~9.28%). The total contents of alkaline oxides Na_2O, K_2O, and CaO are 0.90~20.76%. The contents of other chemical components are lower, within 0.00~1.00%. The presence of alkaline oxides makes the rocks alkaline, with pH values ranging from 7.66 to 10.02. The difference in the content of the main chemical components of PS of different colors is small. The chemical components with a low content, such as CaO and Fe_2O_3, have greater differences in content. The content of CaO in gray-white PS is the highest, exceeding 14.00%. The content of Fe_2O_3 in red PS is about 9.00%. The difference in Fe_2O_3 content is the main reason for the red and white color of PS. In addition, some PS also has small amounts of SO_3, CO_2, MnO, and P_2O_5. The abundance of silica–aluminum–calcium chemicals in Pisha sandstone provides a source for the formation of Pisha sandstone geopolymer gels. However, the activity of these chemicals determines the amount of geopolymer gel produced. The method of excitation of the activity needs to be further investigated.

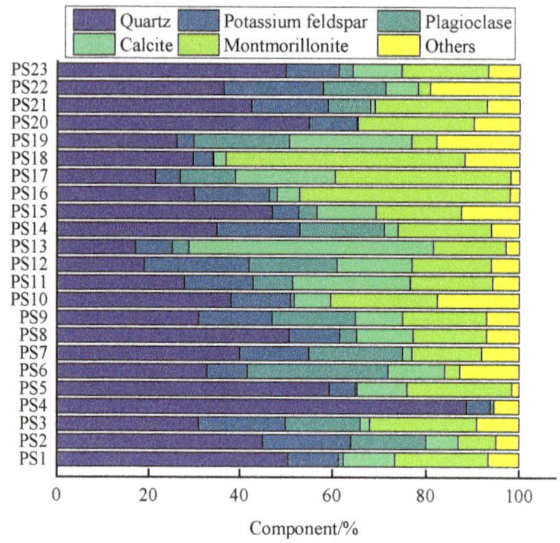

Figure 2. Mineral composition of PS [11,28–33]. Note: PS1–7 are gray-white, PS8–11 are purple-red, PS12–15 are pink, PS16–19 are yellow, PS20 is gray-green, and PS21 is gray-white and purple-red stripes. PS2, PS3, PS9, and PS12 are cited from the literature [11]; PS1, PS8, and PS21 are cited from the literature [28]; PS4-6, PS10, PS11, PS13, and PS16–20 are cited from the literature [29]; PS23 is cited from the literature [30]; PS7 and PS14 are cited from the literature [31]; PS15 is cited from the literature [32]; and PS22 is cited from the literature [33].

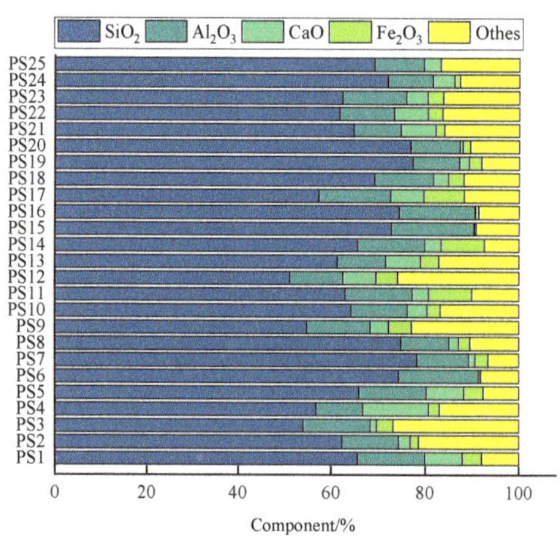

Figure 3. Chemical component of Pisha sandstone [9,11,26,27,30,32,34–40]. Note: PS1–7 are gray-white, PS8 and PS9 are purple, PS10–15 are pink, and PS16 is yellow. PS22 is cited from the literature [9]; PS2, PS3, PS8, and PS11 are cited from the literature [11]; PS4, PS9, and PS12 are cited from the literature [27]; PS6 and PS14 are cited from the literature [35]; PS5 and PS13 are cited from the literature [36]; PS17 is cited from the literature [30]; PS18 is cited from the literature [32]; PS1 and PS10 are cited from the literature [38]; PS19–21 are cited from the literature [34]; and PS7, PS15, and PS16 are cited from the literature [39].

In general, the bulk density of PS is 1.56–2.50 g/cm^3, and the particle density is 2.62–2.84 g/cm^3. The porosity is 6.02~35.15%. The permeability coefficient is 5.2×10^{-3} mm/s. The liquid limit W_L and plastic limit W_p are 29.30% and 19.60%, respectively. The plasticity index I_p is 9.40 [38,41–43]. This indicates that the Pisha sandstone has large pores and good permeability, and that the rock structure is loose. The cumulative distribution curves of the particle size of PS are shown in Figure 4. It can be seen that the rock particles mainly consist of 10.73~57.60% of medium and fine sand (0.10~0.25 mm) and 1.40~44.76% of medium and coarse sand (0.25~0.50 mm). Most of the particles are concentrated in the range of 0.10~0.50 mm, accounting for 50.00% of the total. The percentage of clay particles with $d \leq 0.005$ mm is 1.40~10.40%. The coefficient of nonuniformity of PS ranged from 2.55% to 23.90%, which is poorly graded sand. The curvature coefficient is 0.51~5.54%. The content of sand and powder particles is large, and the content of clay particles is relatively small. Therefore, PS is a typical sandy soil and has a non-equigranularity structure [26,28,30,31,35,38,40–47].

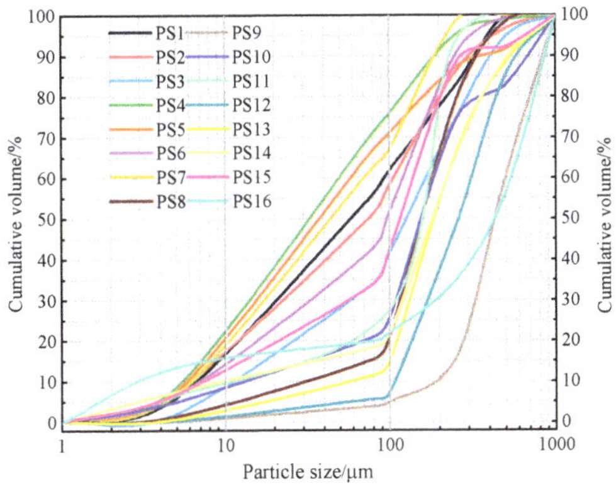

Figure 4. Cumulative distribution curves of PS. Note: PS 1–6 are cited from the literature [26], PS 7 and PS 8 are cited from the literature [31], PS 9 is cited from the literature [35], PS 10 and PS 11 are cited from the literature [40], PS 12 is cited from the literature [41], PS 13 is cited from the literature [45], PS 14 and PS 15 are cited from the literature [46], and PS 16 is cited from the literature [47].

The internal pores of PS are numerous and poorly connected, indicating that the distribution is non-homogeneous. Mineral particles are mainly connected by cementation. The types of cementations are divided into pressure-embedded cementation, substrate-porous cementation, and crystal stock-contact cementation. Carbonate and clay are the main cement. Pressure-embedded cementation structures are very tightly cemented between particles, with fewer pores and fissures between particles, which are mainly filled and wrapped by muddy cement and flaky cement. The particles of substrate-porous cementation are mainly composed of quartz and feldspar, with fewer inter-particle pores but relatively more cracks. At the same time, there is more cement between the particles and on the surface of the particles. The particles of the crystal stock-contact cementation structure are mostly adhered by muddy cement, and the degree of cementation between particles is relatively weak [28,38,48]. In conclusion, the cement between the particles of the Pisha sandstone is mainly clay minerals, which can help with the preparation of geopolymer gels. However, the clay mineral content, morphology, and distribution characteristics vary between different-colored Pisha sandstone. It is necessary to combine the respective properties for the synthesis of geopolymer gels from Pisha sandstone.

The mechanical properties of PS are shown in Table 2. The mechanical strength of Pisha sandstone varies according to color. There are also significant differences in the strength of the rock in different states. The compressive strength in the natural state is 0.40 to 39.70 MPa. The compressive strength in the dry state can be increased to 25.30 MPa.

The difference in shear strength between the dry and saturated states is small, ranging from 0 to 4.16 MPa. The tensile strength ranges from 0.01 to 0.74 MPa. The compressive strength, tensile strength, and shear strength decrease significantly after encountering water, and the softening coefficient ranges from 0.05 to 0.42. The low water-saturated shear strength and dry shear strength determine the weak resistance of PS to runoff erosion and collapse [28,30,32,38,40,42,49]. Therefore, water resistance is a factor that must be considered for the application of Pisha sandstone geopolymer materials for ecological restoration.

Table 2. Mechanical properties of PS.

Reference	Sample Collection Site	Type	Softening Coefficient	Compressive Strength/MPa	Shear Strength/MPa	Tensile Strength/MPa
[28,32]	Jungar Banner, Ordos	Gray-white, purple-red, gray-white with purple-red stripes	0.05~0.42 0.20 (AVG)	5.19~39.70 (dry) 18.72 (AVG) 1.00~8.69 (water-saturated) 3.73 (AVG)	0.09~4.15 1.16 (AVG)	0.06~0.74 0.21 (AVG)
[30]	Boldonggou, Shagedu town, Ordos	Gray-white, purple-red, gray-white with purple-red stripes	-	-	0.090~0.470 (dry)	-
[38]	Erlaohuo gully, Shagedu town, Ordos	White	-	0.40~1.20	-	-
		Red	-	0.90~2.80	-	-
[40]	Jungar Banner, Ordos	Brown red, light red	-	-	0.072 (cohesive force)	-
[42]	Jungar Banner, Ordos	Gray-white with purple-red stripes	-	7.30~10.90	0.0021~0.0082	0.0100~0.0430
		Gray-white	-	13.70~20.10	0.0122~0.0210	0.0527~0.0896
		Purple-red	-	23.20~27.60	0.0275~0.0393	0.1010~0.1696
		Pink	-	30.30~30.60	0.0464~0.0674	0.2275~0.2627
[49]	Jungar Banner, Ordos	-	-	1.33~3.11	-	0.01~0.16

3.2. Erosion Characteristics

The Pisha sandstone region is characterized by the interaction of hydraulic erosion, wind erosion, freeze–thaw erosion, gravity erosion, and anthropogenic erosion. Bi et al. [50] classified rock erosion into "seasonal rainfall runoff erosion" and "perennial non-runoff erosion" according to the type of dynamics. Hydraulic erosion occurs in the form of raindrop splashes, slope, and gully erosion. In particular, gully erosion plays an important role in sediment transport, which accounts for more than 70% of the sediment in the basin. The wind direction in the Pisha sandstone region is mostly perpendicular to the gully or crosses the gully at a large angle, which leads to frequent wind erosion. It is characterized by wind accumulation and wind erosion [51]. The high-intensity wind erosion regions are formed between high landforms or windward slopes and both sides. The weak wind erosion regions are formed on leeward slopes or in relatively flat regions. Sediment accumulation regions form on leeward slopes, depressions, and gullies [51,52]. Gravity erosion forms mainly in gullies, caused by weathering and other factors [44]. According to statistics, the sediment yield of gravity erosion accounts for about one third of the total sediment yield in the basin. The main manifestations of gravity erosion are debris slides and landslides. Additionally, the accumulation of various kinds of sliding sandstone blocks or granules caused by slumps and avalanches at the foot of the slope is also an important manifestation of gravity erosion. Furthermore, sliding sandstone blocks would be scoured and disintegrated into coarse sediment in the gradual convergence of the water flow, which indirectly provides a sand source [51]. Freeze–thaw erosion is caused by mechanical damage such as severe temperature fluctuation, solid–liquid circulation transformation of water, and the expansion, contraction, and fracture of rocks. Eventually, the rock moves and is lost under the action of hydraulic and gravity erosion. Furthermore, the bonded water between the particles is transformed into solid water during the freezing process, leading to soil expansion. Subsequently, the melting process of solid water between particles leads to changes in the soil structure, resulting in gravity erosion of marginal soil [53,54]. The multiple dynamic erosion cycles on an annual scale are presented in Figure 5.

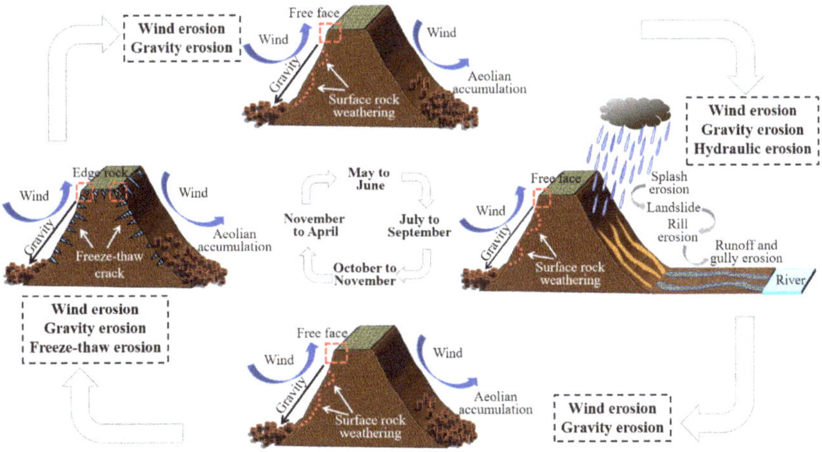

Figure 5. Multiple dynamic erosion cycles on an annual scale.

The erosion of different types of regions is shown in Table 3 [25,55]. It can be seen that gravity erosion exists in the multiple erosion of all three types of regions. Moreover, freeze–thaw erosion occurs wherever PS is bare. Meanwhile, multiple erosion patterns in the Pisha sandstone region are coupled in space and alternated in time. Sediment loss mainly occurs in summer and autumn. In winter and spring, it is mainly freeze–thaw erosion and hydraulic erosion caused by gravity erosion and wind erosion. The transition period of spring and summer is the high incidence of freeze–thaw erosion and wind erosion. April and May have the high incidence of wind erosion, and May and June have the high incidence of gravity erosion. July to September have the high incidence of hydraulic erosion. May to September are the coupling period of gravity erosion and hydraulic erosion. Other periods are characterized by interactive coupling of wind, gravity, and freeze–thaw erosion [53,56].

Table 3. Erosion modulus of different Pisha sandstone-type regions.

Pisha Sandstone-Type Region	Distribution Region	Main Erosion Types	Main Composite Erosion Types	Average Erosion Modulus/[t/(km^2·a)]
Bare regions	Kuye River, Huangfuchuan Watershed, Ten Tributaries, etc.	Hydraulic erosion	Hydraulic erosion, wind erosion, freeze–thaw erosion	2.1
Soil-covered regions	Kuye River, Gushanchuan Watershed, Qingshui River, Huangfuchuan Watershed, Hun River, Ten Tributaries, etc.	Hydraulic erosion	Hydraulic erosion, wind erosion, gravity erosion, freeze–thaw erosion	1.5
Sand-covered regions	Kuye River, Gushanchuan Watershed, Ten Tributaries, etc.	Wind erosion	Hydraulic erosion, wind erosion	0.8

4. Advances in the Ecological Restoration of Pisha Sandstone Region

In the early 1990s, China began to conduct systematic research on the erosion of PS, and the Yellow River Conservancy Commission of the Ministry of Water Resources completed the "Distribution range of PS and erosion type region in the junction region of Shanxi, Shaanxi and Mongolia" and the "General report of the experimental research stage of plant 'flexible dam' in Pisha sandstone region (1995–1998)" [57]. Subsequently, the biological measures mainly based on sea buckthorn were explored through the "National Key Management Project of Soil and Water Conservation", "Soil and Water Conservation Ditch Control Backbone Project", "Sea Buckthorn Ecological Project in Pisha Sandstone of Shanxi, Shaanxi and Mongolia", "Comprehensive Management Project of Small Watersheds", and "Small Watershed Comprehensive Management Project" [58–60]. Good ecological benefits have been achieved. Based on the harness of seabuckthorn, Bi [61] proposed the idea of "a seabuckthorn flexible dam". Many scholars have successively researched the use of seabuckthorn to construct flexible dams in gully channels to intercept sediment, and many results have been achieved [62–64]. However, seabuckthorn showed a phenomenon of withering and death and could only be planted in gullies. Moreover, erosion occurring on slopes and gully slopes was difficult to prevent.

The Institute of Geographical Sciences and Natural Resources Research, CAS, joined enterprises and universities to develop PS and sand compounded into soil technology. Based on the complementary structural characteristics, PS and sand were compounded into a new type of soil, which could help with the resource utilization of PS [65,66]. However, this technology was mainly used for sand transformation, and it was difficult to solve the problem of soil erosion in the Pisha sandstone region. Wen et al. [67,68] found that PS had outstanding electrostatic adsorption and ion exchange capacity and had good adsorption properties, which could realize the adsorption of heavy metals to restore contaminated soil. It was a good natural adsorption material and was economical and environmentally friendly. Zhen et al. [69] selected three types of typical soils in mining areas, namely, PS, loess, and sandy soil, and designed the soil structure to simulate the infiltration process. It showed that PS could reduce the infiltration capacity of soil and could be applied to the ecological restoration of mining areas. Su et al. [70,71] carried out experiments on the effect of EN-1 curing agent on the mechanical properties of PS. The hydrodynamic properties under different factors were investigated by simulating the slope runoff scour. The optimum curing agent dosage, compaction, moisture content, and curing age were obtained. However, the current EN-1 curing agent could not be directly used as a modified material, because it could not meet the comprehensive performance requirements of consolidation, erosion resistance, water storage, and vegetation promotion. Meanwhile, it could not cope with the environment of multiple erosion coupling and interaction. It was difficult to solve the erosion problem of high slopes. Based on the project "Integration and Demonstration of Anti-erosion and Vegetation-promoting Technology in Pisha Sandstone Region in the Middle Reach of the Yellow River", the Yellow River Institute of Hydraulic Research has considered the limitations of traditional engineering measures and biological measures and invented anti-erosion and vegetation-promoting materials and modified dam-building technology. Meanwhile, soil and water conservation material, engineering biological measures, and a two-dimensional slope–gully integrated management mode system were proposed, and the management effect was remarkable [72–75]. Meanwhile, the Dalian University of Technology carried out research on modified dam construction technology and prepared modified dam construction materials with good performance [35,36,76].

Figure 6 presents the trend of sediment delivery from the Yellow River over the last 100 years [77]. The water and soil conservation work in the middle reaches of the Yellow River has been effective in recent years. It can be seen that the measured annual sediment transport at Tongguan station in 2021 was 171 million t, which showed a significant decrease compared to the average annual sediment transport of 1.592 billion t from 1919 to 1959. However, the contradiction of the incongruent water–sediment relationship was still prominent. Focusing on the Pisha sandstone region, there were deficiencies in the comprehensiveness, uniformity, and continuity of the implementation of traditional harness measures. Moreover, there was a lack of integrated and comprehensive region-wide and mechanism-wide harnesses. The trend of ecological deterioration in the Pisha sandstone region has not been fundamentally curbed.

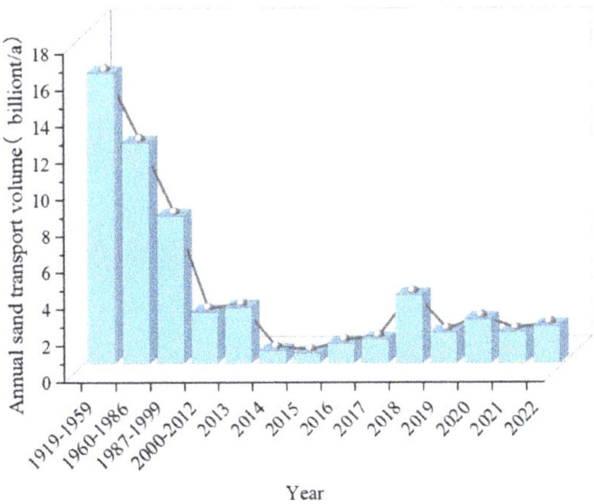

Figure 6. Annual sediment transport in the Yellow River [77].

5. The Synthesis of Pisha Sandstone Geopolymer Gels

As a special pozzolanic mineral, Pisha sandstone contains large amounts of clay minerals such as calcium montmorillonite, kaolinite, and illite. Alkali activation and thermal activation can turn clay minerals into a highly reactive precursor, providing a source for the synthesis of geopolymer gels. Thermal activation causes dehydroxylation of the clay minerals, resulting in the formation of high-energy chemical bonds that make it very readily available for reaction. Alkali activation dissolves the silicate monomers in the clay mineral, providing a source of reactants for the polycondensation reaction [78,79]. The effects of saturated lime water and different concentrations of sodium hydroxide solution on the dissolution of SiO_2 and Al_2O_3 in PS were also studied. It was concluded that the dissolution of SiO_2 and Al_2O_3 in Pisha sandstone increased significantly with the concentration of NaOH solution. The dissolved amount reached above 30% in 2 mol/L NaOH solution, showing high activity. At the same time, the activity of PS was studied by a mortar strength test. The compressive strength results of the collodion specimens showed that the activity index of arsenopyrite was close to 60%, and the flexural strength test results showed that the activity index of arsenopyrite was around 65% [35].

Considering that the mechanical properties of the material were not excellent when only PS was used as a precursor, additional sources of precursors were combined to generate stronger geopolymer gels [80,81]. Pisha sandstone combined with mineral additives was used as a raw material, and alkaline modifiers were used to stimulate mineral activity to prepare geopolymer gel materials. Specimens were prepared by using a compression molding method for ambient- or high-temperature curing. The mechanical and microscopic properties of the materials were investigated by compressive strength tests and microscopic tests (XRD, TG, FTIR, SEM, EDS). Combined with the results of microscopic tests, the significant improvement in the mechanical and durability properties of the material was attributed to the generation of a large number of geopolymer gels that bound the particles to form a stable skeletal structure. PS–fly ash composites were prepared with a maximum compressive strength of 20.3 MPa under optimal conditions. It was concluded that the main factors affecting the mechanical properties of the material are the fineness of the PS, type of activator, fly ash amount, alkali content, and curing temperature. PS was also calcined to prepare PS–fly ash composites. The results showed that the thermal activation treatment improved the reaction level and enhanced the mechanical properties of the materials. The compressive strength was up to 21.9 MPa. In addition, the water resistance and permeability of the PS–fly ash geopolymer materials were also studied. With the increase in the amount of fly ash, the total porosity of the materials decreased, the water permeability resistance increased significantly, and the softening coefficient reached 0.86 [82–85]. The application potential of blast furnace slag–PS composites, steel slag–PS composites, and silica fume–PS composites was also studied. The results showed that the compressive strength of blast furnace slag–PS composites and steel slag–PS composites reached up to 56.2 MPa and 46 MPa, respectively, under the optimum level of mineral replacement (40%) and optimum activator (1.5% NaOH). The compressive strength of silica fume–PS composites reached 44.8 MPa with the addition of 10% silica fume and 3% sodium hydroxide. Meanwhile, blast furnace slag–PS composites and silica fume–PS composites had excellent water resistance with softening coefficients of 0.95 and 0.98, respectively [86–88].

PS-based cement gel materials have also been widely studied. Dong et al. [36] studied the feasibility of PS as a concrete admixture. The results showed that the flexural strength and durability of the material increased significantly when the amount of PS was lower than 20%. The initial setting time and compressive strength were not affected. However, the performance of the material decreased significantly when the amount exceeded 30%. Geng et al. [89] studied the effect of cement amount on the properties of PS. The results showed that with the increase in cement incorporation, the porosity of PS gradually decreased, the cohesion tended to increase, the material tended to be brittle, the pore deterioration of PS under freeze–thaw action was reduced, and the frost resistance was significantly improved. It was concluded that the optimum cement amount was 20%. Dong et al. [90] selected sodium hexametaphosphate and calcium lignosulfonate as plasticizers. The mechanical properties, rheological properties, dry shrinkage properties, and hydration properties of PS–cement composites were investigated. It was concluded that the compressive strength of the material reached 52.7 MPa after curing at 80 °C under the optimal mixing ratio. Yang et al. [91,92] studied the effects of fly ash and basalt fiber on the mechanical properties of PS–cement materials. It was concluded that a small amount of fly ash improved the compactness and deformation control of the material. Basalt fibers with a content of less than 0.2% had a reinforcement effect on the material, which inhibited the deformation of the material. Zhang et al. [93,94] tested the performance of cement mortar with PS as a substitute for river sand. The results showed that the substitution rate of red PS below 10% had little effect on the dry shrinkage of specimens at various ages. Concrete of different strength grades met the mechanical property requirements at a 20% substitution rate. However, the water–cement ratio of cement mortar increased significantly at a substitution rate of more than 50%. This led to a decrease in strength and an increase in dry shrinkage, resulting in poor durability. Therefore, the addition of a water reducer to control the water–cement ratio was proposed to improve the substitution rate of PS and the mechanical properties of cement mortar. The feasibility of replacing river sand with PS in the preparation of building materials was verified.

6. Ecological Restoration Applications of Pisha Sandstone Geopolymer Gels

6.1. Engineering Composite Materials

The most common and conventional approaches are engineering measures to control water and soil loss. Engineering measures include slope engineering (horizontal ditch, fish-scale pit, ditch edge ridge, intercepting ditch, etc.), ditch head protection engineering, and ditch engineering (check dam, pond dam, small retaining engineering, etc.). However, Pisha sandstone cannot be used as an engineering material directly because of its poor cementation and poor anti-water erosion ability. Industrial wastes such as PS, fly ash, blast furnace slag, silica fume, and steel slag are rich in silicate minerals. These minerals are more reactive under alkali activation and can produce more compact hydration products, resulting in a more stable backbone structure and producing engineering composite materials with even better properties [95,96]. Figure 7 presents the mechanical properties of different types of engineering composite materials for Pisha sandstone. The mechanical properties of the composite materials prepared from Pisha sandstone combined with solid waste (e.g., blast furnace slag, fly ash, silica fume, steel slag) meet the engineering requirements. Moreover, the composite materials have excellent resistance to water erosion [82–84]. This can be attributed to the fact that the generated geopolymer gels fill the pores of the material and the particles are more tightly bonded to form a stable skeleton structure [80]. Based on the above research, materials for dam engineering were prepared and applied in the Erlaohuo gully dam project of Ordos Jungar Banner to verify the feasibility of PS-modified materials in the engineering of check dams in Pisha sandstone regions [72,97].

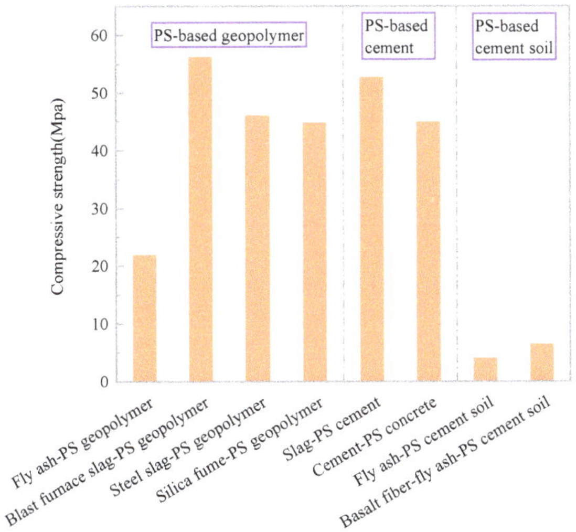

Figure 7. Compressive strength of different types of engineering composite materials for Pisha sandstone.

On the other hand, PS-based cement materials used in the construction of check dams and slope engineering have also been widely studied [36,89,93,94]. However, even though concrete has desirable mechanical strength, there are limitations to its durability performance. Due to its poor deformation control, Pisha sandstone expands and contracts after the freeze–thaw process, resulting in cracking and damage to the concrete. Fly ash and basalt fibers have a reinforcing effect on the material and can improve deformability [92,98]. Nevertheless, further research is necessary on the durability properties of PS-based cement materials. In addition, Zhu et al. [91,99] prepared compound soil by mixing fly ash with PS cement soil through alkali excitation technology. It was concluded that fly ash significantly improved the mechanical properties and microstructure of PS cement soil. The strength of the compound soil reached 6.85 MPa, which met the design requirements of highway structural soil.

However, Pisha sandstone geopolymer gel materials have limitations in practical applications. The preparation process of the material is unable to cope with different functional requirements for ecological restoration. There is an increasing trend in 3D concrete printing technology. PS-based geopolymer material and PS-based cement material are improved to be binder jetting materials applied to slope protection projects and in ground dam projects, which will greatly improve the operability and application of ecological restoration projects. At the same time, the material preparation process is upgraded to 3D printing technology, which makes the material both producible and economical [100–102]. This is a potential research direction for the future. Ensuring the strength of the slurry while controlling the fluidity of the sprayed slurry is the key technical issue. Meanwhile, the curing process of printed blocks is also an issue worthy of in-depth study in the future.

6.2. Mine Restoration Green Materials

The Pisha sandstone region is located in the core region of China's coal energy base. Energy exploitation has brought great economic benefits, but it has also seriously polluted the local ecological environment. In particular, the open-pit coal mine would cause irreversible damage to the atmosphere, soil, vegetation, water source, and topography [103]. Therefore, the ecosystems of coal mine dumps are in urgent need of restoration.

The structure of PS determines its strong impermeability and adsorption [104]. PS could effectively reduce the infiltration capacity of the soil and improve the water-holding capacity. Meanwhile, it could reduce soil toxicity and improve soil physicochemical properties, which have a significant effect on the improvement of polluted soil. Compared to other materials, PS is widely available and can be locally sourced, which has the potential for soil ecological remediation. Ma et al. [105] mixed PS with sandy soil, irrigated silt soil, and loess in different proportions to study the process of water infiltration into the soil. The results showed that the PS significantly reduced the infiltration rate and saturated hydraulic conductivity of the soil. The sandy soil mixed with PS at a ratio of 1:3 had the best water-holding capacity. Zhen et al. [69] selected three types of typical soils in the mine region, namely, PS, loess, and sandy soil, and simulated the infiltration process of single configuration, layered configuration, and soil-rock mixed configuration. Wen et al. [67,68] studied the adsorption effect of three different colors (red, white, and grey) of PS on Pb and analyzed the effects of adsorbent dosage, pH, and ionic strength. The effect of the amount of PS on the morphology distribution and toxic leaching content of Pb was studied through simulation experiments. The changes in soil porosity, soil water-holding capacity, soil enzyme activity, and pH were also studied. The results showed that PS significantly reduced the content of weak acid extractable fraction and toxic leaching in soil. The adsorption effect was mainly related to the Pb concentration, adsorbent dosage, pH, and ionic strength, which was proportional to the adsorbent dosage and pH and inversely proportional to the ionic concentration. In addition, the adsorption rates of different colors of PS were different, showing that gray > white > red.

However, relying only on the impermeability and adsorption properties of the Pisha sandstone protolith to repair the soil is limiting for the restoration of the mine site and cannot deal with the multiple ecological problems in the mine site. The application of the adsorption of materials to solve the problem of heavy metal pollution of industrial wastewater from mining areas has been widely discussed [106,107]. The preparation of binder jetting materials by combining Pisha sandstone with mine wastes to solve the problem of heavy metal ions infiltration underground from mine effluent may be a very interesting research direction. However, the pH has a large influence on the adsorption property because it is closely related to the mechanism involved in the adsorption process. How to control the pH of the material so that Pisha sandstone maintains good adsorption properties is a technical problem that needs to be further studied.

7. Conclusions and Prospects

Although many studies and practices have been carried out on ecological restoration in the Pisha sandstone region, further research and development of more innovative and practical technologies should be undertaken. The preparation process of Pisha sandstone geopolymer materials needs to be further improved. Combined with 3D printing technology, preparation as slurry material or extrusion-based block material is a potential future research direction. PS is used as a precursor to synthesize a green and low-loss geopolymer gel material, reducing its negative environmental effects while enhancing ecological sustainability. Additionally, the multi-dimensional harness mode combining modified Pisha sandstone geopolymer gel materials with anti-erosion and vegetation-promoting materials and vegetation measures has good effects on sediment interception, soil erosion retardation, and vegetation restoration, which is the new direction of ecological restoration in the Pisha sandstone region in the future. Furthermore, based on the study of physical and chemical properties, the development of PS into mine restoration materials is also a practical method of resource application. A multifactor-driven synergistic mechanism between resource applications and ecological security should be established to deal with soil erosion, forming a technical model for resource applications, ecological restoration, and security assurance. In the future, multifactor-driven comprehensive technical measures should be further investigated in order to achieve ecological restoration of the Pisha sandstone region and promote high-quality development of the Yellow River Basin.

Author Contributions: C.L.: Writing—original draft, Software, Resources, Methodology, Investigation, Formal analysis, Data curation, Conceptualization. Y.F.: Writing—review & editing, Writing—original draft, Visualization, Validation, Formal analysis, Data curation. H.C.: Validation, Investigation. Y.W. and D.J.: Validation, Investigation. H.L.: Validation, Investigation. All authors have read and agreed to the published version of the manuscript.

Funding: This research was funded by Research Programs of the Key Laboratory of Ecological Environment Protection and Restoration in the Yellow River Basin of Henan (LYBEPR202104), the National Natural Science Foundation of China (51708216), the Henan Colleges and Universities Key Re-search Programs (No. 24A570004), the Training Plan for Young Backbone Teachers in Colleges and Universities in Henan (2023GGJS078), the Technology Development Project of the Power Construction Corporation of China, and He-nan Wanshan Green Building Materials Co., Ltd. (ZDJWS-2022-004).

Institutional Review Board Statement: Not applicable.

Informed Consent Statement: Not applicable.

Data Availability Statement: Data are contained within the article.

Conflicts of Interest: The authors declare no conflicts of interest.

References

1. Wen, X.; Deng, X. Current soil erosion assessment in the Loess Plateau of China: A mini-review. *J. Clean. Prod.* **2020**, *276*, 123091. [CrossRef]
2. Xiao, J.; Lv, G.; Chai, N.; Hu, J.; Jin, Z. Hydrochemistry and source apportionment of boron, sulfate, and nitrate in the Fen River, a typical loess covered area in the eastern Chinese Loess Plateau. *Environ. Res.* **2022**, *206*, 112570. [CrossRef] [PubMed]
3. Yan, R.; Zhang, X.; Yan, S.; Chen, H. Estimating soil erosion response to land use/cover change in a catchment of the Loess Plateau, China. *Int. Soil Water Conserv.* **2018**, *6*, 13–22. [CrossRef]
4. Nearing, M.A. Modeling response of soil erosion and run off to changes in precipitation and cover. *Catena* **2005**, *61*, 131–154. [CrossRef]
5. Pandey, S.; Kumar, P.; Zlatic, M.; Nautiyal, R.; Panwar, V.P. Recent advances in assessment of soil erosion vulnerability in a watershed. *Int. Soil Water Conserv.* **2021**, *9*, 305–318. [CrossRef]
6. Andualem, T.G.; Hewa, G.A.; Myers, B.R.; Peters, S.; Boland, J. Erosion and Sediment Transport Modeling: A Systematic Review. *Land* **2023**, *12*, 1396. [CrossRef]
7. Hu, C.; Zhang, X.M. Loess Plateau soil erosion governance and runoff-sediment variation of Yellow River. *Water. Resour. Hydropower Eng.* **2020**, *51*, 1–11. (In Chinese) [CrossRef]
8. Wang, Y.; Wu, Y.; Li, M. *Research on Soil and Water Loss and Its Control Measures in the Pisha Sandstone Area*; The Yellow River Water Conservancy Press: Zhengzhou, China, 2007; ISBN 9787807342120. (In Chinese)
9. Jin, Z. *Research on Soil and Water Conservation and Development of Agriculture and Animal Husbandry in Pisha Sandstone Area*; The Yellow River Water Conservancy Press: Zhengzhou, China, 2003; ISBN 9787806215463. (In Chinese)
10. Yang, J.; Fang, D.; Bi, C.; Qiao, W.; Li, G.; Li, W. Study on the model of freeze-thaw and weathering erosion of smaller watershed in soft rock area. *Chin. J. Geol. Hazard Control* **2003**, *2*, 89–95. (In Chinese) [CrossRef]
11. Wu, Y.; Yang, Z.; Liu, H.; Cai, H.; Wei, M. Effect of Composition on Nutrient of Pisha Sandstone. *Yellow River* **2016**, *38*, 18–21+25. (In Chinese) [CrossRef]
12. Xu, J.; Wu, C.; Lin, Y.; Gao, Y.; Zuo, Z.; Yu, Q. Definition on Source Area of Centralized Coarse Sediment in Middle Yellow River. *J. Soil Water Conserv.* **2006**, *8*, 19–21+52. (In Chinese) [CrossRef]
13. Liang, Z.; Wu, Z.; Yao, W.; Noori, M.; Yang, C.; Xiao, P.; Leng, Y.; Deng, L. Pisha sandstone: Causes, processes and erosion options for its control and prospects. *Int. Soil Water Conserv.* **2019**, *7*, 1–8. [CrossRef]
14. Zhang, P.; Yao, W.; Liu, G.; Xiao, P.; Sun, W. Experimental study of sediment transport processes and size selectivity of eroded sediment on steep Pisha sandstone slopes. *Geomorphology* **2020**, *363*, 107211. [CrossRef]
15. Yang, F.; Cao, M.; Li, H.; Wang, X.; Bi, C. Simulation of sediment retention effects of the single seabuckthorn flexible dam in the Pisha Sandstone area. *Ecol. Eng.* **2013**, *52*, 228–237. [CrossRef]
16. Wang, L.; Li, C.; Dong, J. Study on Distribution and Lithologic Characters of Feldspathic Sandstone. *Yellow River* **2013**, *35*, 91–93. (In Chinese) [CrossRef]

17. Zhang, P.; Xiao, P.; Yao, W.; Liu, G.; Sun, W. Characteristics of sediment particle size distribution in the Pisha sandstone area of the middle Yellow River under the effects of vegetation and complex erosion. *Earth Surf. Process. Landf.* **2023**, *48*, 1281–1289 [CrossRef]
18. Xi, J. Speech at the symposium on ecological protection and quality development in the Yellow River Basin. *China Water Resour.* **2019**, *20*, 1–3. (In Chinese) [CrossRef]
19. Verburg, K.; Baveye, P. Hysteresis in the binary exchange of cations on 2:1 clay minerals; a critical review. *Clays Clay Miner.* **1994**, *42*, 207–220. [CrossRef]
20. Zhao, Y. Study on Palaeo-Rock-Solution of Carbonate Rockof Ordovician and Its Reservoir. Ph.D. Thesis, Southwest Petroleum University, Chengdu, China, 2006. (In Chinese)
21. Dang, B. The Tectonic and Sedimentary Evolution and its Relationship to Gas Accumulation of Lower Paleozoic in Ordos Basin Ph.D. Thesis, Northwest University, Xi'an, China, 2003. (In Chinese) [CrossRef]
22. Yang, J. *Techonic Evolution and Oil-Gas Reservoirs Distribution in Ordos Basin*; Petroleum Industry Press: Beijing, China, 2002, ISBN 7-5021-3801-3. (In Chinese)
23. Liu, X. *Basin Analyses and Simulation of Up-Paleozoicin Ordos Basin*; Southwest Petroleum University: Chengdu, China, 2005 (In Chinese) [CrossRef]
24. Liang, J. *Research on Sedimentary System and Sequence Stratigraphy of Jurassic of the Ordos Basin*; Southwest Petroleum University: Chengdu, China, 2007. (In Chinese) [CrossRef]
25. Wang, Y.; Wu, Y.; Kou, Q.; Min, D.; Chang, Y.; Zhang, R. Definition of arsenic rock zone borderline and its classification. *Sci. Soil Water Conserv.* **2007**, *5*, 14–18. [CrossRef]
26. Liu, H.; Shi, M.; Cai, H.; Wu, Y. Experimental Study of the Physical Characteristics of the Pisha Sandstone on Its Erosion. *Yellow River* **2016**, *38*, 8–10. (In Chinese) [CrossRef]
27. Leng, Y.; Yao, W. Features and utilization of arsenic sandstone. *China Water Resour.* **2015**, *8*, 15–17. (In Chinese) [CrossRef]
28. Ye, H.; Shi, J.; Li, X.; Hou, H.; Shi, Y.; Cheng, Y. The Effect of Soft Rock Lithology upon Its Anti-erodibility. *Acta Geosci. Sin.* **2006**, *27*, 145–150. (In Chinese) [CrossRef]
29. Peng, B.; Sun, Y.; Sun, X. Mineral composition of soft rock in Mu Us sandy land in China. *Land Dev. Eng. Res.* **2018**, *3*, 29–34 (In Chinese)
30. Shi, Y.; Ye, H.; Hou, H.; Bi, Z. The Internal Cause of the Erosion in "Pisha" Sandstone Area, Southern Inner Mongolia. *Acta Geosci. Sin.* **2004**, *25*, 659–664. (In Chinese) [CrossRef]
31. Wu, Y.; Cai, H.; Liu, H.; Yang, Z. Relationship between shear strength and structural characteristics of Pisha sandstone. *Adv. Sci. Technol. Water Resour.* **2019**, *39*, 21–26. (In Chinese) [CrossRef]
32. Ye, H.; Shi, J.; Wang, G.; Hou, H.; Shi, Y.; Cheng, Y. Effect of chemical compositions of Pisha sandstone on the gravity erosion *Hydrogeol. Eng. Geol.* **2006**, *6*, 5–8. (In Chinese) [CrossRef]
33. Wang, Q.; Sun, X.; Liu, Y.; Zhang, G.; Shi, J.; Ye, H. Indoor Modeling the Effect of Water-Rock Interaction on the Weathering and Erosion of Pi-Sandstone. *Yellow River* **2013**, *35*, 45–47+68. (In Chinese) [CrossRef]
34. Shi, J.; Ye, H.; Wang, Q.; Sun, Y. Effect of Water-rock Interaction on the Weathering and Erosion of Pi-sandstone, Southern Inner Mongolia, China. *Geoscience* **2009**, *23*, 171–177. (In Chinese) [CrossRef]
35. Li, C.; Zhang, T.; Wang, L. Pozzolanic Activity of Pisha Sandstone and Mechanical Properties of Alkali-activated Pisha Sandstone Materials. *J. Chin. Ceram. Soc.* **2015**, *43*, 1090–1098. (In Chinese) [CrossRef]
36. Dong, J.; Zhang, T.; Wang, L.; Li, C. Properties of Pisha Sandstone in Inner Mongolia and Its Influence to Mechanical Properties of Cement. *Soil Water Conserv. China* **2015**, 46–49+73. (In Chinese) [CrossRef]
37. Song, T.; Liu, L.; Wang, Y.; Liu, N.; Yu, M. Characteristics and genesis of the bleached Pisha sandstone in Ordos Basin. *Oil Gas Geol.* **2014**, *35*, 679–684. (In Chinese)
38. Li, C.; Song, L.; Wang, L. Mineral composition and anti-erodibility of Pisha sandstone. *Sci. Soil Water Conserv.* **2015**, *13*, 11–16. (In Chinese) [CrossRef]
39. Yao, W.; Liu, H.; Liang, Z. A New Approach to Ecological Restoration in Pisha Sandstone Area—An Environment-Friendly Technique of Anti-Erosion and Vegetation-Promoting. *Bull. Soil Water Conserv.* **2018**, *38*, 118–124. (In Chinese) [CrossRef]
40. Li, X.; Su, Y.; Qi, X.; Liang, W. The Experimental Analysis Study of Soft Sandstone Soil Properties in the Plateau Hilly Region. *J. Inn. Mong. Agric. Univ.* **2011**, *32*, 315–318. (In Chinese)
41. Li, X.; Yu, J.; Liu, L.; Zhai, T. The experimental research on mechanical properties of Pisha–Sandstone of Erdos. *J. Arid Land Resour. Environ.* **2016**, *30*, 118–123. (In Chinese) [CrossRef]
42. Guo, J.; Ye, H.; Shi, Y.; Wu, L.; Sun, Y. *Indicator System Establishment of Sandstone Corrosion Resistance Inner Mongolia*; Forum on Urban Geoenvironment and Sustainable Development: Shanghai, China, 2010; pp. 186–191. (In Chinese)
43. Li, X.; Li, M.; Xie, W. Experimental research on mechanical properties and constitutive model of Pisha-sandstone remolded soil *J. Drain. Irrig. Mach. Eng.* **2019**, *37*, 909–914. (In Chinese) [CrossRef]
44. Ye, H.; Shi, J.; Hou, H.; Shi, Y.; Cheng, Y.; Guo, J. Effect of the Lithologic Characters of Pisha Sandstone on Gravity Erosion in South Inner Mongolia. *Arid Zone Res.* **2008**, *3*, 402–405. (In Chinese) [CrossRef]
45. Li, X.; Zhai, T.; Zhang, Q. Experiment on mechanical properties of Pisha-sandstone at recurrent shear. *Trans. Chin. Soc. Agric. Eng.* **2015**, *31*, 154–159. (In Chinese) [CrossRef]

46. Jin, Z.; Shi, P. A Study on the Soil Erosion Processses in the Semiarid Region of Inner Mongolia. *J. Arid Land Resour. Emviron.* **1987**, *2*, 55–66. (In Chinese)
47. Qin, F.; Yang, Z.; Li, L. Research progress on soil erosion mechanism and ecological restoration technology in feldspathic sandstone region. *J. Beijing For. Univ.* **2020**, *42*, 142–150. (In Chinese) [CrossRef]
48. Ge, L.; Li, J.; Peng, B. Study on microstructure characteristies of soft rocks in Ordos. *Land Dev. Eng. Res.* **2019**, *4*, 36–42. (In Chinese)
49. Wu, L.; Li, X.; Shi, J.; Ye, H.; Guo, J.; Cheng, Y.; Wang, Q. Quantitative characteristics of microstructure of Pisha-sandstone. *Acta Geosci. Sin.* **2007**, *6*, 597–602. (In Chinese) [CrossRef]
50. Bi, C.; Wang, F. Research on soil erosion mechanism in soft rock regions. *J. Sediment. Res.* **2008**, *1*, 70–73. (In Chinese) [CrossRef]
51. Zhang, P.; Liu, Y.; Zhang, Z. The Features of Sediment Production and the Analysis of Genesis by Erosion in Huangfuchuan Watershed. *J. Soil Water Conserv.* **1992**, *2*, 15–24. (In Chinese)
52. Zhao, G. Research on the Laws of Soil and Water Loss in Sand Rock Region, Inner Mongolia. *Res. Soil Water Conserv.* **2001**, *4*, 158–160. (In Chinese) [CrossRef]
53. Zhang, P.; Yao, W.; Liu, G.; Xiao, P. Dynamic characteristics of complex erosion in a typical small watershed of soft sandstone area. *J. Hydraul. Eng.* **2019**, *50*, 1384–1391. (In Chinese) [CrossRef]
54. Chen, S.; Li, X.; Li, Q.; Li, M. Deformation characteristics of red pisha-sandstone during freezing-thawing cycles in Ordos. *Sci. Soil Water Conserv.* **2016**, *14*, 34–41. (In Chinese) [CrossRef]
55. Liu, Y.; Niu, G. Analysis on the influencing factors of soil erosion in the Pisha sandstone area. *Inn. Mong. Water Resour.* **2012**, *3*, 55–56. (In Chinese)
56. Tang, Z.; Cai, Q.; Li, Z.; Zhao, H. Study on Interaction Among Wind Erosion, Hydraulic Erosion and Gravity Erosion in Sediment-Rock Region of Inner Mongolia. *J. Soil Water Conserv.* **2001**, *2*, 25–29. (In Chinese) [CrossRef]
57. Yao, W.; Li, C.; Zhang, P.; Wang, Y. Prospect and Research on the Erosion Mechanism of Pisha Sandstone. *Yellow River* **2018**, *40*, 1–7+65. (In Chinese) [CrossRef]
58. Yin, L.; Liang, Y. Water-holding Capability of Litter Layers of Artificial Seabuckthorn Stands in Arsenic Sandstone Area. *Water Resour. Dev. Manag.* **2007**, *3*, 10–14. (In Chinese) [CrossRef]
59. Li, X.; Li, H. Creating a new model for managing World Bank financed Project-Brief introduction of Inner Mongolia Autonomous Region of Loess Plateau Watershed Rehabilitation Project. *China Water Resour.* **2005**, *12*, 45–46. (In Chinese) [CrossRef]
60. Wang, J. Analysis on the Sediment Reduction Effect of Seabuckthorn Ecological Project in the Concentrated Source Area of Yellow River Coarse Sediment. *Soil Water Conserve. China* **2016**, *9*, 86–88. (In Chinese) [CrossRef]
61. Bi, C.; Li, G. Special functions of seabuckthorn in controlling soil erosion in arsenic sand areas. *Express Water Resour. Hydropower Inf.* **1998**, *18*, 2–4. (In Chinese) [CrossRef]
62. Cao, Q. Study on Constructing Approach of Forest and Grass in Pisha Rock Aera. *Bull. Soil Water Conserv.* **1997**, *7*, 2–5. (In Chinese)
63. Wang, J.; Xue, S.; Gao, F. Effect, Experience, and Development Strategy of Seabuckthorn Control in Bare Arsenic Sandstone Areas. *Hippophae* **2002**, *1*, 1–4. (In Chinese)
64. Guo, J.; He, J.; Yin, L.; Zhao, X.; Xing, E.; Liu, T. Analysis on Community Structure and Plant Diversity of Artificial *Hippophae rhamnoides* in Soft Rock Region. *Water Resour. Dev. Manag.* **2011**, *9*, 24–29+33. (In Chinese) [CrossRef]
65. Han, J.; Liu, Y.; Luo, L. Research on the Core Technology of Remixing Soil by Soft Rock and Sand in the Maowusu Sand Land Region. *China Land Sci.* **2012**, *26*, 87–94. (In Chinese) [CrossRef]
66. Fu, P.; Wang, H.; Luo, L.; Han, J.; Ma, Z.; Tong, W.; Cheng, J. Farmland-building Technology of Turning Arsenic Rock and Sand into Soil. *Bull. Soil Water Conserv.* **2013**, *33*, 242–246. (In Chinese) [CrossRef]
67. Wen, J.; Zhu, Y.; Yin, X.; Zhang, X. Pb (II) adsorption property of Pisha sandstone. *Acta Sci. Circumst.* **2014**, *34*, 2491–2499. (In Chinese) [CrossRef]
68. Wen, J.; Zhu, Y.; Zhang, X.; Pan, H.; She, X.; Wang, L. Remediation of lead contaminated soil by adding Pisha sandstone in resource-rich region of Shanxi, Shaanxi and Inner Mongolia. *Acta Sci. Circumst.* **2015**, *35*, 873–879. (In Chinese) [CrossRef]
69. Zhen, Q.; She, X.; Zhang, Y.; Han, F.; Zhang, X. Simulation of Infiltration Characteristics with Various Soil Configurations in Shanxi-Shaanxi-Inner Mongolia Energy Zone. *Trans. Chin. Soc. Agric. Mach.* **2015**, *46*, 90–96. (In Chinese) [CrossRef]
70. Su, T.; Zhang, X.; Wang, R. Effect of EN-1 on Shear Strength Characteristics of Pisha Sandstone Solidified Soil. *Trans. Chin. Soc. Agric. Mach.* **2013**, *44*, 86–90+67. (In Chinese) [CrossRef]
71. Su, T.; Zhang, X. Effects of EN-1 Soil Stabilizer on Slope Runoff Hydraulic Characteristics of Pisha Sandstone Stabilized Soil. *Trans. Chin. Soc. Agric. Mach.* **2011**, *42*, 68–75. (In Chinese)
72. Li, C.; Yao, W.; Wang, L.; Shen, Z.; Liu, H.; Leng, Y. Research on properties of modified Yellow River Pisha sandstone dam construction materials. *Adv. Sci. Technol. Water Resour.* **2019**, *39*, 16–20+26. (In Chinese) [CrossRef]
73. Liang, Z.; Wu, Z.; Yang, C.; Yao, W.; Leng, Y. Mechanism of erosion resistance and vegetation promotion by W-OH in Pisha sandstone. *J. Hydraul. Eng.* **2016**, *47*, 1160–1166. (In Chinese) [CrossRef]
74. Yao, W.; Shi, M.; Wu, Z.; Wang, L.; Yang, C. Experimental Study of the Physical Characteristics of the Pisha Sandstone on Its Erosion. *Yellow River* **2016**, *38*, 1–7+10. (In Chinese) [CrossRef]
75. Yao, W.; Xiao, P.; Wang, Y.; Shen, Z. Research progress on soil erosion control technologies in Pisha sandstone areas. *Adv. Sci. Technol. Water Resour.* **2019**, *39*, 1–9+15. [CrossRef]

76. Li, C.; Wang, L.; Zhang, T.; Yang, D.; Song, L. Preparation of Geopolymeric Materials by Pisha Sandstone. *J. Build. Mater.* **2016**, 19, 373–378. (In Chinese) [CrossRef]
77. Wang, G.; Zhong, D.; Wu, B. Future trend of Yellow River sediment changes. *China Water Resour.* **2020**, *1*, 9–12+32. (In Chinese)
78. Buchwald, A.; Hohmann, M.; Posern, K.; Brendler, E. The suitability of thermally activated illite/smectite clay as raw material for geopolymer binders. *Appl. Clay Sci.* **2009**, *46*, 300–304. [CrossRef]
79. Marsh, A.; Heath, A.; Patureau, P.; Evernden, M.; Walker, P. Alkali activation behaviour of un-calcined montmorillonite and illite clay minerals. *Appl. Clay Sci.* **2018**, *166*, 250–261. [CrossRef]
80. Provis, J.; Bernal, S. Geopolymers and Related Alkali-Activated Materials. *Annu. Rev. Mater. Sci.* **2014**, *44*, 299–327. [CrossRef]
81. Bondar, D.; Lynsdale, C.J.; Milestone, N.B.; Hassani, N.; Ramezanianpour, A.A. Effect of adding mineral additives to alkali activated natural pozzolan paste. *Constr. Build. Mater.* **2011**, *25*, 2906–2910. [CrossRef]
82. Li, C.; Zhang, T.; Wang, L. Mechanical properties and microstructure of alkali activated Pisha sandstone geopolymer composites. *Constr. Build. Mater.* **2014**, *68*, 233–239. [CrossRef]
83. Li, C.; Dong, J.; Zhao, S.; Liu, H.; Yao, W.; Wang, L. Development of low cost supplementary cementitious materials utilizing thermally activated Pisha sandstone. *Constr. Build. Mater.* **2018**, *174*, 484–495. [CrossRef]
84. Li, C.; Wang, L.; Zhang, T.; Dong, J. Development of building material utilizing a low pozzolanic activity mineral. *Constr. Build. Mater.* **2016**, *121*, 300–309. [CrossRef]
85. Li, C.; Zhang, T.; Wang, L. Effect of Dosage of Fly Ash and NaOH on Properties of Pisha Sandstone-Based Mortar. *ACI Mater. J.* **2016**, *113*, 173–183. [CrossRef]
86. Dong, J.; Li, C.; Liu, H.; Zhang, L.; Liu, J. Investigating the mechanical property and reaction mechanism of geopolymers cement with red Pisha Sandstone. *Constr. Build. Mater* **2019**, *201*, 641–650. [CrossRef]
87. Dong, J.; Wang, L.; Zhang, T. Study on the strength development, hydration process and carbonation process of NaOH-activated Pisha Sandstone. *Constr. Build. Mater.* **2014**, *66*, 154–162. [CrossRef]
88. Dong, J.; Zhang, T.; Liu, H.; Wang, K.; Liu, J.; Wang, Z. Silica fume-reinforced alkali-activated uncalcined Pisha Sandstone-based geopolymer cement. *Constr. Build. Mater.* **2021**, *269*, 121296. [CrossRef]
89. Geng, K.; Chai, J.; Qin, Y.; Li, X.; Zhou, H. Collapse inhibition mechanism analysis and durability properties of cement-stabilized Pisha sandstone. *Bull. Eng. Geol. Environ.* **2022**, *81*, 144. [CrossRef]
90. Dong, J.; Zhang, T.; Zhu, Y.; Zhang, L.; Jiang, Y. Synthesis of alkali-activated uncalcined Pisha sandstone cement composites. *Compos. Eng.* **2021**, *225*, 109311. [CrossRef]
91. Yang, J.; Li, X.; Wang, H.; Geng, K. Indoor experimental study on mechanical properties of composite fly ash and Pisha-sandstone cement soil. *J. Drain. Irrig. Mach. Eng.* **2021**, *39*, 1230–1236. (In Chinese) [CrossRef]
92. Wang, H.; Li, X.; Yang, J.; Geng, K. Experimental study on mechanical properties of basalt fiber-Pisha sandstone cement soil. *J. Drain. Irrig. Mach. Eng.* **2021**, *39*, 1264–1269+1290. (In Chinese) [CrossRef]
93. Zhang, M.; Luo, Z.; Yin, H.; Zheng, Y.; Lu, B.; Yang, J. Study of properties of cement mortars prepared of red pisha sandstone fine aggregate. *Concrete* **2015**, *6*, 112–114. (In Chinese) [CrossRef]
94. Zhang, M.; Luo, Z.; Yin, H.; Lu, B.; Zheng, Y.; Yang, J. Study of erosion-wear of concrete prepared of Pisha sandstone fine aggregate. *Concrete* **2016**, *6*, 158–160. (In Chinese) [CrossRef]
95. Duan, P.; Yan, C.; Zhou, W. Compressive strength and microstructure of fly ash based geopolymer blended with silica fume under thermal cycle. *Cem. Concr. Compos.* **2017**, *78*, 108–119. [CrossRef]
96. Zhang, D.; Wang, X.; Kang, S.; Cheng, G.; Wu, W. The effect of slag and fly ash content on the properties of electric furnace nickel slag-based geopolymer used for repair materials. *Case Stud. Constr. Mater.* **2023**, *19*, e02284. [CrossRef]
97. Xiao, Q.; Yao, W.; Shen, Z.; Jiao, P. Resistance to Erosion and Growth Promoting Technique of Pisha Sandstone Region in the Middle Yellow River. *Soil Water Conserv. China* **2016**, *9*, 73–75. (In Chinese) [CrossRef]
98. Yang, J.; Li, X.; Wang, H.; Geng, K. A Study of the Structural Evolution and Strength Damage Mechanisms of Pisha Sandstone Cement Soil Modified with Fly Ash. *J. Renew. Mater.* **2021**, *9*, 2241–2260. [CrossRef]
99. Zhu, J.; Li, X. Strength and microscopic mechanism analysis of CFBCA-based geopolymer–Pisha sandstone cement composite soil. Environ. *Sci. Pollut. Res.* **2022**, *29*, 65197–65210. [CrossRef] [PubMed]
100. Shakor, P.; Chu, S.H.; Puzatova, A.; Dini, E. Review of binder jetting 3D printing in the construction industry. *Prog. Addit. Manuf.* **2022**, *7*, 643–669. [CrossRef]
101. Valente, M.; Sambucci, M.; Chougan, M.; Ghaffar, S.H. Composite alkali-activated materials with waste tire rubber designed for additive manufacturing: An eco-sustainable and energy saving approach. *J. Mater. Res. Technol.* **2023**, *24*, 3098–3117. [CrossRef]
102. Khan, S.A.; Jassim, M.; Ilcan, H.; Sahin, O.; Bayer, İ.R.; Sahmaran, M.; Koc, M. 3D printing of circular materials: Comparative environmental analysis of materials and construction techniques. *Case Stud. Constr. Mater.* **2023**, *18*, e02059. [CrossRef]
103. Lei, K.; Pan, H.; Lin, C. A landscape approach towards ecological restoration and sustainable development of mining areas. *Ecol. Eng.* **2016**, *90*, 320–325. [CrossRef]
104. Feng, Z.; Li, X. Microbially induced calcite precipitation and synergistic mineralization cementation mechanism of Pisha sandstone components. *Sci. Total Environ.* **2023**, *866*, 161348. [CrossRef] [PubMed]
105. Ma, W.; Zhang, X. Effect of Pisha sandstone on water infiltration of different soils on the Chinese Loess Plateau. *J. Arid Land* **2016**, *8*, 331–340. [CrossRef]

106. Godwin, J.; Njimou, J.R.; Abdus-Salam, N.; Panda, P.K.; Tripathy, B.C.; Ghosh, M.K.; Basu, S. Nanoscale ZnO-adsorbent carefully designed for the kinetic and thermodynamic studies of Rhodamine B. *Inorg. Chem. Commun.* **2022**, *138*, 109287. [CrossRef]
107. Grba, N.; Baldermann, A.; Dietzel, M. Novel green technology for wastewater treatment: Geo-material/geopolymer applications for heavy metal removal from aquatic media. *Int. J. Sediment Res.* **2023**, *38*, 33–48. [CrossRef]

gels

Article

Development of Environmentally Friendly and Economical Flood-Prevention Stones Based on the Sediments of the Yellow River

Ying Liu [1], Hao Xiao [2], Yongxiang Jia [3], Yajun Lv [4], Li Dai [2,5,*] and Chen Yang [2,*]

1 College of Water Resources, North China University of Water Resources and Electric Power, Zhengzhou 450046, China
2 Collaborative Innovation Center for Efficient Utilization of Water Resources, North China University of Water Resources and Electric Power, Zhengzhou 450046, China
3 Henan ZhongGong Design & Research Group Co., Ltd., Zhengzhou 450046, China
4 College of Architecture, North China University of Water Resources and Electric Power, Zhengzhou 450046, China
5 School of Environmental and Municipal Engineering, North China University of Water Resources and Electric Power, Zhengzhou 450046, China
* Correspondence: daili@ncwu.edu.cn (L.D.); yangchen@ncwu.edu.cn (C.Y.)

Abstract: The deposition of Yellow River sediment in the middle and lower reaches is a significant factor in the siltation of reservoirs and the occurrence of serious flooding along the river. The efficient and valuable utilization of Yellow River sediment has already become a key research topic in this field. In this study, we have employed Yellow River sediment as the primary material, in conjunction with commercially available slag, fly ash, and quicklime as the binder, to develop a novel type of artificial flood-prevention stone. Following a 28-day standard curing procedure, the highest compressive strength of the prepared artificial stone was recorded at 4.29 MPa, with a value exceeding 0.7 MPa under wet conditions. The results demonstrated that the prepared artificial stone met the specifications for artificial flood-prevention stones. The curing mechanism, as evidenced by analyses from SEM and XRD testing, indicated that the alkali excitation process in the binder, which produced C-A-S-H gel, was the key factor in enhancing the compressive strength of the specimens. Notably, an evaluation of the amount of CO_2 emissions and the cost of the artificial stone concluded that the preparation process was both environmentally friendly and cost-effective.

Keywords: Yellow River sediment; flood-prevention stones; alkali excitation; C-A-S-H gel; environmentally friendliness

1. Introduction

The Yellow River is a prototypical sediment-laden river, exhibiting the highest sediment load in the world [1,2]. As the river slows down in the middle and lower reaches, it gradually accumulates sediment, depositing more than 400 million cubic meters of sediment each year, making it vulnerable to various hydraulic disasters [3–6]. The occurrence of floods and other hydraulic disasters in the middle and lower reaches of the Yellow River has resulted in significant damage and has constituted a major constraint on China's socio-economic development [7–10]. In order to mitigate the impact of these disasters, management strategies have been implemented for the Yellow River. As one of the key management strategies, significant quantities of flood-prevention stones, characterized by their specific volume and weight, have been strategically positioned in crucial locations such as perilous bends, checkpoints, and culverts along the middle and lower sections of the river [11–15]. It is estimated that the current annual consumption of stone for flood prevention is 600,000 cubic meters. As China's economic development model has shifted to prioritize environmental protection, the extraction of non-renewable natural stone for flood prevention has declined [16,17]. Therefore, facile preparation methods of artificial flood-prevention stones would effectively address this deficit.

The utilization of alkali activation for converting silt into cementitious materials has been well documented by numerous research teams. This process leverages the catalytic effect of alkaline activators to expedite the hydration process and enhance the hydration reaction rate [18–20]. This applied technology is presently employed across various engineering applications, including the resource utilization of Yellow River sediment [21–23]. For instance, Deng and co-workers, in 2022, presented a preparation scheme for ultra-high-performance concrete (UHPC) from ultra-fine sand sourced from the Yellow River [22]. The scheme was developed using the alkali activation principle and discussed the probability of transporting sediment from the bottom of the reservoir to the shore and building an embankment on both sides of the reservoir.

In comparison, the number of studies conducted on the utilization of Yellow River sediment for the preparation of artificial flood-prevention stones is relatively limited. Wang, in 2017, used the Yellow River sediment alkali activation method to produce a kind of artificial flood-prevention stone, and studied the effects of the amount of alkali, slag content, and curing time on the compressive strength of artificial flood-prevention stones, and the test block made with the optimal ratio could meet the requirements of flood-prevention stones [24]. Zheng and co-worker modified Yellow River sediment by alkali excitation, and added granulated blast furnace slag powder, red coal slime, black coal slime, and other admixtures to prepare flood-prevention stones with a 90d compressive strength greater than 10 MPa [25].

While instances of preparing artificial flood-prevention stones have been previously mentioned, it is imperative to acknowledge the limitations of previous studies. Firstly, the alkaline activation method typically requires the addition of alkaline compounds, such as $Ca(OH)_2$ or NaOH, as activators in a mass ratio of at least 5% to produce specimens with high compressive strength. However, the presence of these alkaline substances in the stones placed in water can significantly increase the pH of the water, which can have irreversible negative impacts on the environment and its corresponding organisms. Secondly, the financial implications of the preparation of artificial flood-prevention stones have been omitted from the aforementioned works. We hypothesize that the use of specific industrial by-products in place of the alkaline activators mentioned above has the potential to significantly reduce the cost of stone preparation, thereby increasing the viability of producing and marketing artificial flood-prevention stones.

In light of the aforementioned research, and the shortcomings in the previous studies, we herein present the synthesis of a novel artificial flood-prevention stone. The optimal amount of each component of the binder was preliminarily determined through the use of an orthogonal experimental design and simulation calculations by Design-Expert. Moreover, the mechanical properties of the prepared specimens and the pH of the leaching solution were evaluated, and it was determined that the prepared specimens satisfied the criteria for utilization as artificial flood-prevention stones. The microstructure of the samples was analyzed, and it was concluded that the hydrogel reaction between calcium oxide in slag and reactive silicon dioxide and aluminum oxide in fly ash to form C-A-S-H gel was the dominant reason for the increase in the compressive strength of the specimens. It is noteworthy that the calculation of CO_2 emissions from the preparation process and the cost of preparation indicates that the production process of the designed stones is environmentally friendly. The lower production cost is favorable for the further promotion of the stones on the market.

2. Materials and Methods

2.1. Materials

The sediment from the Yellow River, used in this work, was sourced from the dredged deposits of the Sanmenxia Reservoir in the Henan Province. Prior to the curing procedure, the sediment was subjected to drying in an oven maintained at a temperature of 105 °C for a period of 24 h. The slag employed in the treatment of the sediment was the by-product of ore crushing and processing, procured in Gongyi City. The fly ash was produced by the Shandong Rongchangsheng Environmental Protection Materials Factory. The visual characteristics and the outcomes of the particle size distribution tests for the raw materials are illustrated in Figures 1 and 2, respectively.

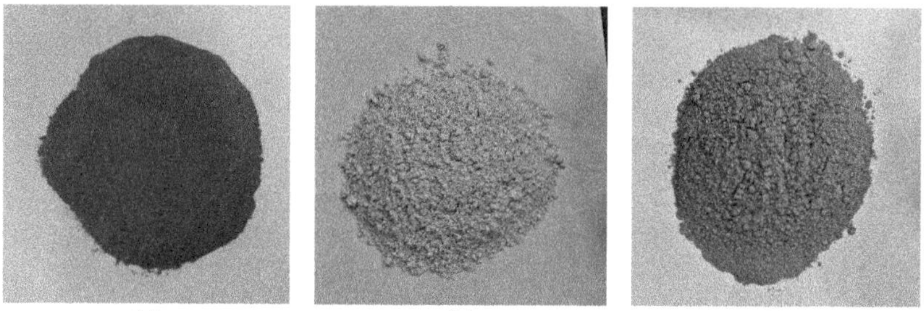

Figure 1. Appearance of Yellow River sediment (**a**), slag (**b**), and fly ash (**c**).

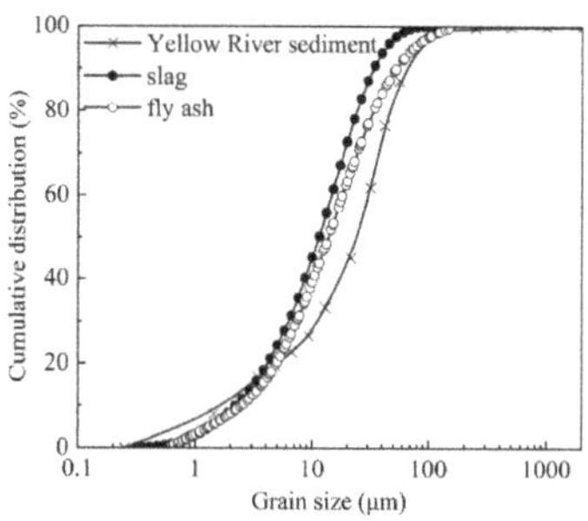

Figure 2. Gradation curves of Yellow River sediment, slag, and fly ash in this work.

Samples for XRF analysis were prepared according to the following procedure: The selected blocks were dried at 105 °C until a constant weight was reached. The materials were then crushed, and the resulting powder was sieved through an 80 μm sieve. The mass of the samples tested was at least 3 g. The aforementioned materials were subjected to a composition analysis through the utilization of X-ray fluorescence spectrometry (XRF) (Panalytical Axios, PANalytical B.V., Almelo, Netherlands), with the findings presented in Table 1. Furthermore, the quicklime that was employed as one of the selected binders was identified as dolomitic lime, with a calcium oxide (CaO) content of 70.42% and a magnesium oxide (MgO) content of 10.51%.

Table 1. Chemical composition of Yellow River sediment, slag and fly ash.

Chemical Composition (%)	Yellow River Sediment	Slag	Fly Ash
CaO	11.09	49.42	7.20
SiO_2	57.02	25.57	46.44
Fe_2O_3	6.47	0.31	3.12
Al_2O_3	15.88	13.58	38.01
TiO_2	0.90	2.15	0.61
Na_2O	1.17	0.44	0.33
SO_3	0.15	2.36	0.69
K_2O	3.31	0.33	0.88
MgO	3.30	5.32	0.23
Other	0.71	0.52	2.19

2.2. Sample Preparation

The dried Yellow River sediment and designed binder mixture were placed in an electrical agitator, followed by the addition of water with a mass fraction of 25%. Subsequently, the mixture was compacted into 40 mm × 40 mm × 40 mm specimens by compression molding. In order to ensure a compacted density of 1.78 g/cm^3, a minimum pressure of 2.33 MPa was applied during the molding process. Following a 24 h period of room temperature curing, the specimens were transferred to a standard curing box (20 ± 2 °C, 95% RH) for a further period of curing. The curing durations were set at 7 and 28 days.

The binder employed in this study was a combination of slag, fly ash, and quicklime. In order to ascertain the optimal binder formulation, a multifactorial test was designed, which incorporated the aforementioned three materials. In the binders, the proportions of slag and fly ash were varied at three levels: 2%, 5%, and 8%, with quicklime serving as the complementary component. The specific quantities of each substance in the binder are delineated in Table 2. It is notable that the dosage of the binder constitutes 20% of the combined mass of the specimen.

Table 2. Dosage of each substance in the binder [a].

Entry	Slag (%)	Fly Ash (%)	Quicklime (%)
1	2	2	16
2	2	5	13
3	2	8	10
4	5	2	13
5	5	5	10
6	5	8	7
7	8	2	10
8	8	5	7
9	8	8	4

[a] The Yellow River sediment accounts for 80% of the total mass of the specimens.

Recently, Song investigated the protocol for the production of concrete by alkali excitation using fly ash and sodium silicate [26]. The results showed that an appropriate amount of quicklime promoted the formation of N-A-S-H and C-S-A gels, thereby increasing the compressive strength of the specimens. In order to guarantee the progression of alkali activation in our research, the minimum dosage of quicklime in the binder was established at 20%, representing 4% of the total specimen mass. To ensure the optimal mix ratio of the binder, a single-factor influence experiment was conducted, wherein the quantity of slag was systematically varied to replace the fly ash continuously. The experimental design is outlined in Table 3.

Table 3. Design of slag replacement fly ash in single-factor influence experiment [a].

Name	Slag (%)	Fly Ash (%)	Quicklime (%)
S4FA12	4	12	4
S6FA10	6	10	4
S8FA8	8	8	4
S10FA6	10	6	4
S12FA4	12	4	4

[a] The Yellow River sediment accounts for 80% of the total mass of the specimens.

2.3. General Testing Methods

To characterize the strength of the specimens for Yellow River sediment-based flood-prevention stones, unconfined compressive strength tests (UCS) were carried out in accordance with ASTM D4219. For the purpose of pH testing the leaching solution, the specimens were crushed and then sieved through a square sieve with a mesh size of 0.08 mm. Subsequently, the specimens were combined with water in a 1:10 ratio for the purpose of pH testing, in accordance with standard ASTM D4972-01. The freeze–thaw resistance of the prepared specimens, which were cured for 28 days, was examined through a series of freeze–thaw cycles, with intervals of 2, 5, 10, and 20 days [27]. The softening coefficient of the specimens, designated as flood-prevention stone and cured for a duration of 28 days [28], was calculated with the below equation.

$$\text{softening coefficent} = \frac{\text{wetting strength (MPa)}}{\text{drying strength (Mpa)}}$$

Microscopic morphology testing of samples under different ratios was conducted using a scanning electron microscope (SEM) (MIRA LMS, Tescan, Brno-Kohoutovice, Czech Republic) at resolutions set to 3 nm. Testing samples were magnified at 500 and 5000 times (20 μm and 2 μm). The acceleration voltage ranged from 0.5 kV to 30 kV, with low vacuum pressure set between 1 Pa and 270 Pa. Additionally, the samples cured for 7 and 28 days under different mix ratios were tested using an X-ray diffractometer (XRD) (Rigaku Ultima IV X, Rigaku, Japan). The scanning angle range was set from 5° to 80° at a scanning rate of 20°/min, with a wavelength of 1.5418, voltage of 40 kV, and current of 40 mA. Powder from specimens passing through an 80 μm sieve was used for testing.

3. Results and Discussion

3.1. Analysis of Multi-Factor Compressive Strength Test Results

The outcomes of the compressive strength assessments of the prepared specimens under the conditions of varying binder formulations (Table 2) are illustrated in Figure 3. In general, the compressive strength index of the specimens exhibited a notable enhancement following a 28-day curing period, in comparison to the 7-day curing period. The formulae for the compressive strength test results are presented as follows:

$$\text{Compressive strength (MPa, 7d)} = 3.35 \times A + 1.73 \times B + 1.31 \times C$$

$$\text{Compressive strength (MPa, 28d)} = 4.82 \times A + 0.13 \times B + 1.52 \times C + 7.04 \times AB - 3.04 \times AC + 2.3 \times BC$$

A: dosage of slag (%); *B*: dosage of fly ash (%); C: dosage of quicklime (%). The regression coefficients were calculated by a linear regression analysis with interaction using Design-Expert.

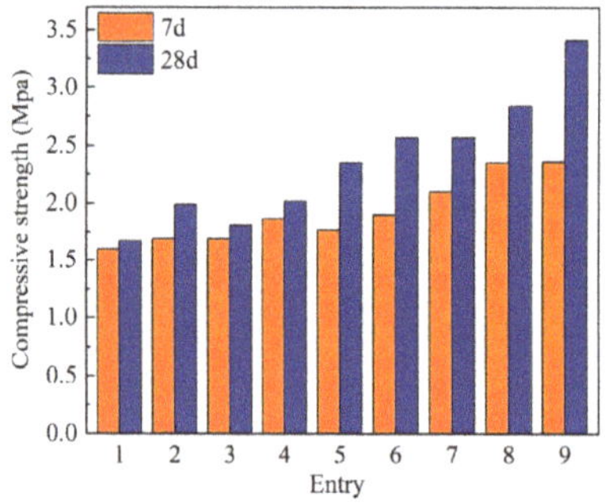

Figure 3. Compressive strength test results of multi-factor experiments.

Notably, the compressive strength of the specimens derived from the various binder formulations exhibited significant variation. The maximum compressive strengths of 2.36 MPa and 3.41 MPa were achieved at 7 and 28 days of curing, respectively, using a binder comprising 40% slag, 40% fly ash, and 20% quicklime (Table 2, entry 9). Furthermore, the data presented in Figure 3 suggest that reducing the quantity of quicklime and increasing the proportion of slag resulted in an improvement in the compressive strength of the specimens.

To accurately illustrate the influence of the binder components on the compressive strength test outcomes, a simulation analysis was conducted using Design-Expert (Figure 4). The compressive strength of Yellow River sediment, cured for seven days, demonstrated a nearly linear relationship with variations in the slag, fly ash, and quicklime (Figure 4a). The highest compressive strength was observed when the slag content was at 100%, indicating that the strength of the cured Yellow River sediment at 7 days was predominantly influenced by the slag content. Figure 4b of the compressive strength analysis reveals that the compressive strength of the cured Yellow River sediment at 7 days increased with an increase in the slag content but decreased with an increase in the fly ash and quicklime content. Indeed, an increase in the proportion of slag in the binder resulted in a corresponding rise in the levels of reactive silicon dioxide (SiO_2) and aluminum oxide (Al_2O_3), which in turn led to an enhancement in the strength of the cured Yellow River sediment. Consequently, the slag within the binder was identified as the primary component responsible for conferring early strength to the solidified Yellow River sediment.

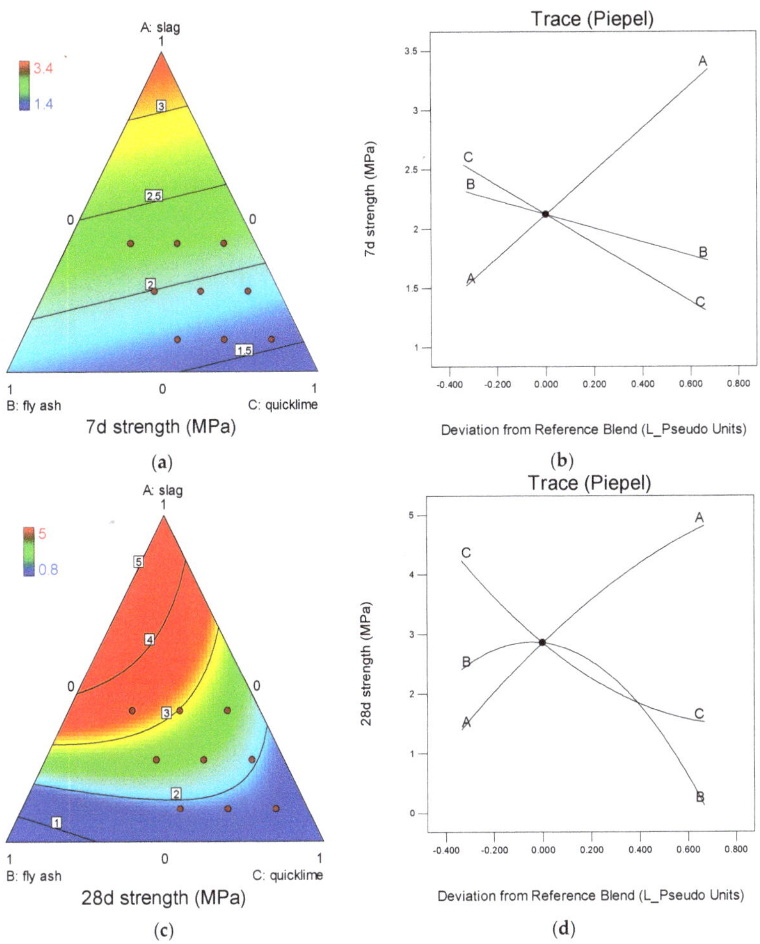

Figure 4. Analysis of the compressive strength of the prepared flood−prevention stone. (**a**) Ternary plot of 7−day compressive strength; (**b**) single−factor influence analysis of 7−day compressive strength; (**c**) ternary plot of 28−day compressive strength; (**d**) single−factor influence analysis of 28−day compressive strength.

The compressive strength results after a 28-day curing period exhibited a distinct trend (Figure 4c) when compared to the observations made after a 7-day curing period. The analysis indicated that the optimal compressive strength of the specimen, at 4.31 MPa, was achieved with a dosing composition of 68% slag and 32% fly ash, in the absence of quicklime. An initial increase in the compressive strength was observed as slag replaced fly ash, followed by a subsequent decrease (Figure 4d). Meanwhile, the incorporation of quicklime had a negative effect on the compressive strength during the specified curing duration.

The results of the aforementioned analyses indicated that the primary contribution of slag to the compressive strength of Yellow River sediment was during its initial curing process. During the extended curing period, the inclusion of fly ash in addition to the compressive strength provided by the slag led to a subsequent enhancement in the compressive strength of the specimens. However, it is noteworthy that while the inclusion of quicklime slightly decreased the compressive strength of the specimens, a 20% quicklime supplement was incorporated as a binder in the following curing experiments to ensure the initiation of the alkaline excitation process.

It is notable from the data that the minimum compressive strength achieved after a curing period of 28 days was merely 0.8 MPa, as depicted in Figure 4c. This value is considerably lower than the minimum compressive strength of 1.4 MPa observed after just 7 days of curing, as shown in Figure 4a. The discrepancy in these results can be directly attributed to the compositional differences in the binder used. Specifically, the binder utilized in the specimens that underwent 28 days of curing consisted entirely of fly ash. Referring to Table 2, it becomes evident that the chosen fly ash composition possesses an exceptionally low content of reactive CaO. This characteristic impedes its ability to effectively react with reactive SiO_2 and Al_2O_3. Consequently, under such conditions, the primary contributor to the compressive strength of the specimens is the internal consolidation process.

3.2. *The Influence of Slag Content on Macroscopic Performance*

3.2.1. Compressive Strength Testing

Upon establishing that 4% of the total weight comprised quicklime, an investigation was conducted into the impact of varying doses of slag and fly ash on the compressive strength of Yellow River sediment flood-prevention stones (Table 3). In accordance with the preceding research results, the compressive strength was determined for the conditions of 7 and 28 days of curing time, respectively, and the results are presented in Figure 5.

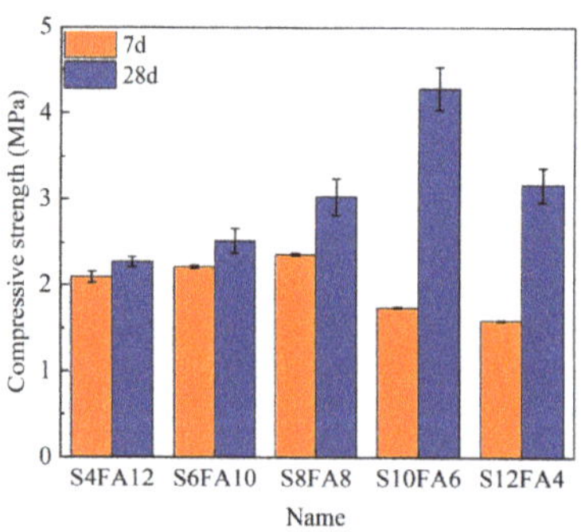

Figure 5. Compression strength results with stepwise slag and fly ash ratios.

In general, an increase in the curing duration resulted in a positive impact on the compressive strength of the specimens. The results presented in Figure 5 demonstrated that a prolonged curing time, in conjunction with an elevated slag content in the binder, resulted in a notable enhancement in the compressive strength. For example, the addition of 10% slag and 6% fly ash to the binder (S10FA6) resulted in a 146% increase in the compressive strength after 28 days of curing compared to the same curing conditions with a 7-day curing time.

A 7-day curing period resulted in the activation of SiO_2 and Al_2O_3 in the fly ash, which was evident from the formation of a C-A-S-H gel with enhanced compressive strength in the specimens. The highest compressive strength index was observed for specimens with a slag dosage of 8% (S8FA8). Nevertheless, an increase in the dosage of slag resulted in a reduction in the compressive strength of the specimens. It was hypothesized that the excess CaO in the slag was unable to form an effective C-A-S-H gel, leading to the downward trend.

Conversely, the specimens cured for 28 days with a composition of 10% slag and 6% fly ash exhibited the highest compressive strength, reaching 4.29 MPa. These findings were in accordance with the results of our prior analysis (Figure 4c). The observed enhancement in the compressive strength can be attributed to two primary factors. Firstly, the optimal ratio of CaO from slag and the activation of SiO_2 and Al_2O_3 from fly ash in the formation of C-A-S-H gel within the binder. Secondly, the extension of the curing period facilitated additional growth of the C-A-S-H gel, thereby bolstering the compressive strength of the specimens. Our presumption in this section was further proved by subsequent microscopic structure analyses.

3.2.2. Leaching Solution pH Testing

The components of Yellow River sediment flood-prevention stones, particularly CaO, which is the main constituent of slag and quicklime, exert a considerable influence on the pH of the water during long-term immersion. Herein, we examined the pH of the leachate in the case of different durations of immersion in water of the specimens after 28 days of curing, and the results are presented in Figure 6.

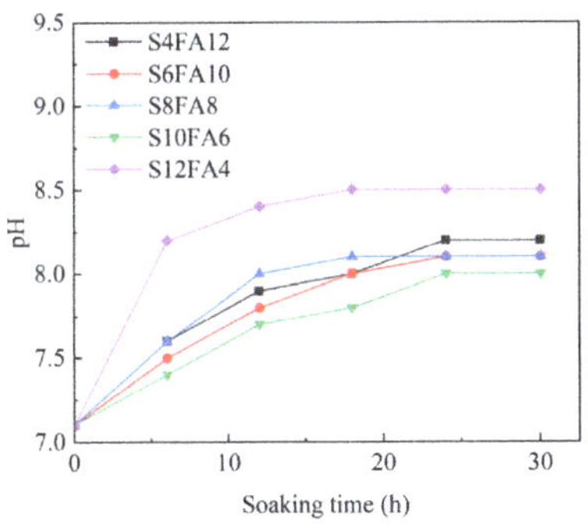

Figure 6. pH of leaching solution.

The test results demonstrated that specimens containing less than 10% slag content within the binder exhibited a leachate pH of below 8.2 after being immersed for 30 h. It was hypothesized that the presence of CaO in slag and quicklime facilitated an effective combination with active SiO_2 and Al_2O_3 in fly ash, resulting in the production of C-A-S-H gel, which was responsible for enhancing the compressive strength as previously discussed. It is notable that an increase in the slag content to 12% resulted in a reduction in the quantity of active substances presented in the fly ash, thereby enabling the excess CaO to interact with water and form alkaline $Ca(OH)_2$ in a process that resulted in a marked increase in the pH of the specimens during the initial period of immersion, with the final pH maintained at 8.5. Notably, compared to the leaching solution pH outcomes for biomass polymer lignin-stabilized silt (with a final pH reaching 10.0) [29] and cement-stabilized soils (exceeding a final pH of 12) [30], the artificial flood-prevention stone produced in our study appears to be environmentally benign.

3.2.3. Rapid Freeze–Thaw Cycling Testing

A rapid freeze–thaw cycle test was conducted to evaluate the influence of the freeze–thaw process on the prepared specimens, particularly in regard to their loss of compressive strength and mass. The corresponding test results are presented in Figure 7.

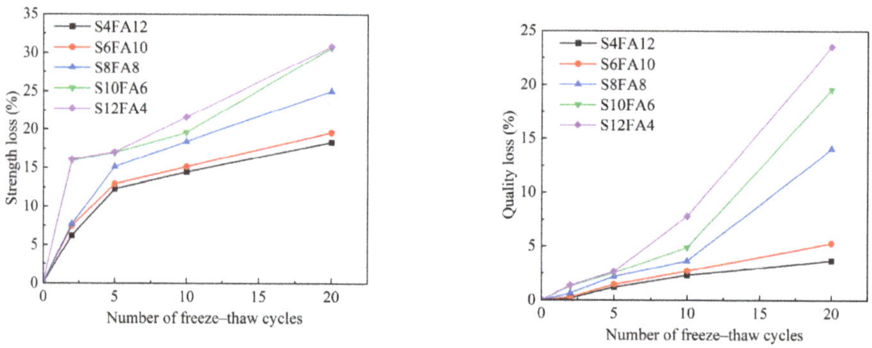

(**a**) (**b**)

Figure 7. Effect of rapid freeze–thaw cycle on the strength (**a**) and quality (**b**) of flood-prevention stones. (**a**) Compressive strength loss in rapid freeze–thaw cycling; (**b**) quality loss in rapid freeze–thaw cycling.

In the initial five freeze–thaw cycles, a notable decline in the compressive strength was observed for the specimens (Figure 7a). The rate of loss of compressive strength in the specimens exhibited a progressive deceleration with an increase in the number of freeze–thaw cycles. It is noteworthy that an increase in the proportion of fly ash in the binder resulted in a reduction in the rate of strength degradation and an improvement in the stone's durability. In the case study involving a 10% supplementation of fly ash to the test block (S6FA10), the reduction in compressive strength remained below 20% following 20 freeze–thaw cycles. It is pleasing to report that the prepared specimens have met the specified requirements for artificial flood-prevention stones.

Similarly, the quality of the prepared stone specimens was found to be adversely affected by rapid freeze–thaw cycling. As the frequency of freeze–thaw cycles increased, a reduction in the compressive strength was accompanied by a clear decrease in the quality of the specimens (Figure 7b). Fortunately, after 10 freeze–thaw cycles, the majority of specimen blocks exhibited a mass loss of less than 5%, which is lower than the similar work by Gutiérrez et al, in 2020, who declaim a 7.45% loss of mass [31], thereby satisfying the requisite specifications for artificial flood-prevention stones.

It is plausible to posit that the formation of a C-A-S-H gel plays a crucial role in preserving the compressive strength and mass of specimens during rapid freeze–thaw cycling. This is evident from the results depicted in Figure 7, which show that an increased dosage of fly ash in the binder leads to a negligible increase in the mass loss of specimens as the number of freeze–thaw cycles escalates. We hypothesized that augmenting the mass fraction of fly ash, rich in active SiO_2 and Al_2O_3 components, in the binder, along with an additional 4% mass fraction of quicklime, facilitated the generation of C-A-S-H gel. Conversely, in specimens with elevated slag content, such as S12A4 and S10A6, surplus active CaO failed to effectively form C-A-S-H gel due to limited active SiO_2 and Al_2O_3. We also anticipated that the production of $Ca(OH)_2$ and following $CaCO_3$ from excess CaO during rapid freeze–thaw cycling did not fulfill the requirements for the compressive strength and quality of the specimens.

3.2.4. Softening Coefficient Testing

In order to guarantee the maintenance of robust strength following an extended period of immersion, a series of experiments were conducted with the objective of determining the softening coefficient of the flood-prevention stones that were presented (Figure 8). In general, the compressive strength of the prepared specimens was analyzed under conditions of moisture. The highest compressive strength was observed in both dry and wet conditions for the test samples that contained 10% slag content (SA10F6).

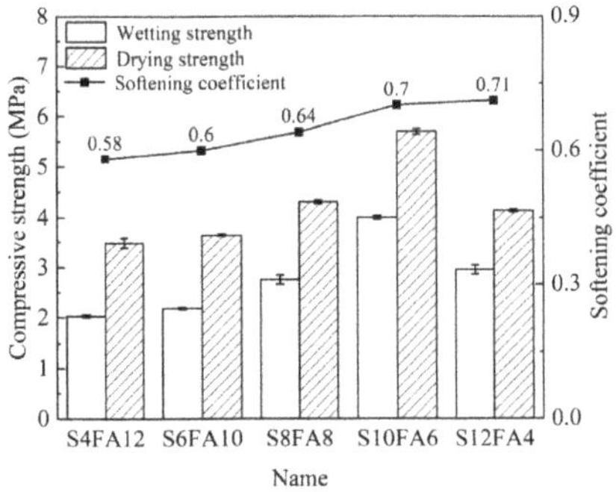

Figure 8. Softening coefficient of Yellow River sediment-based flood-prevention stones.

Notably, the specimens exhibiting a slag content in the binder in excess of 10% displayed a softening coefficient of at least 0.7. Prior analyses determined that an amount of C-A-S-H gel, which enhanced the compressive strength, was sufficient to be formed after 28 days of curing. The creation of the gel maintained the compressive strength of the specimen in wet conditions.

3.3. *Effect of Slag Content Variation on Microstructure*

3.3.1. Mechanism Analysis

Our findings indicate that the formation of a C-A-S-H gel, resulting from the reaction between CaO in slag and active SiO_2 and Al_2O_3 in fly ash, is a necessary condition for the specimens to exhibit the desired properties and to meet the specifications for artificial flood-prevention stones. The potential reaction equations are presented below:

Firstly, CaO reacted with water to form $Ca(OH)_2$, as shown in the following chemical equation:

$$CaO + H_2O \rightarrow Ca(OH)_2$$

The hydration process of slag entailed a pozzolanic reaction involving active SiO_2, Al_2O_3, and $Ca(OH)_2$ within the slag, leading to the creation of C-S-H and C-A-H gels, which were recognized as the primary source of compressive strength for the prepared Yellow River sediment flood-prevention stones in this work. The key reaction is demonstrated as follows:

$$\text{Active } SiO_2 + m_1Ca(OH)_2 + nH_2O \rightarrow m_1CaO{\cdot}SiO_2{\cdot}nH_2O$$

$$\text{Active } Al_2O_3 + m_2Ca(OH)_2 + nH_2O \rightarrow m_2CaO{\cdot}Al_2O_3{\cdot}nH_2O$$

3.3.2. Microscopic Structure

Figure 9 illustrates the scanning electron microscope (SEM) images of the Yellow River sediment-based flood-prevention stones under different binders. A comparison of the SEM images of samples cured for 7 and 28 days under identical binders revealed that the structural interlacing of the 28-day cured samples exhibited greater complexity. This was characterized by an increased presence of needle-like calcium aluminate silicate hydrate (C-A-S-H) gel, the same results as in previous reports [32,33], in comparison to the 7-day cured samples. This increase in structural complexity markedly enhances the mechanical properties of the Yellow River sediment-based flood-prevention stones after 28 days of standard curing, thereby corroborating our hypotheses presented in the preceding discussion.

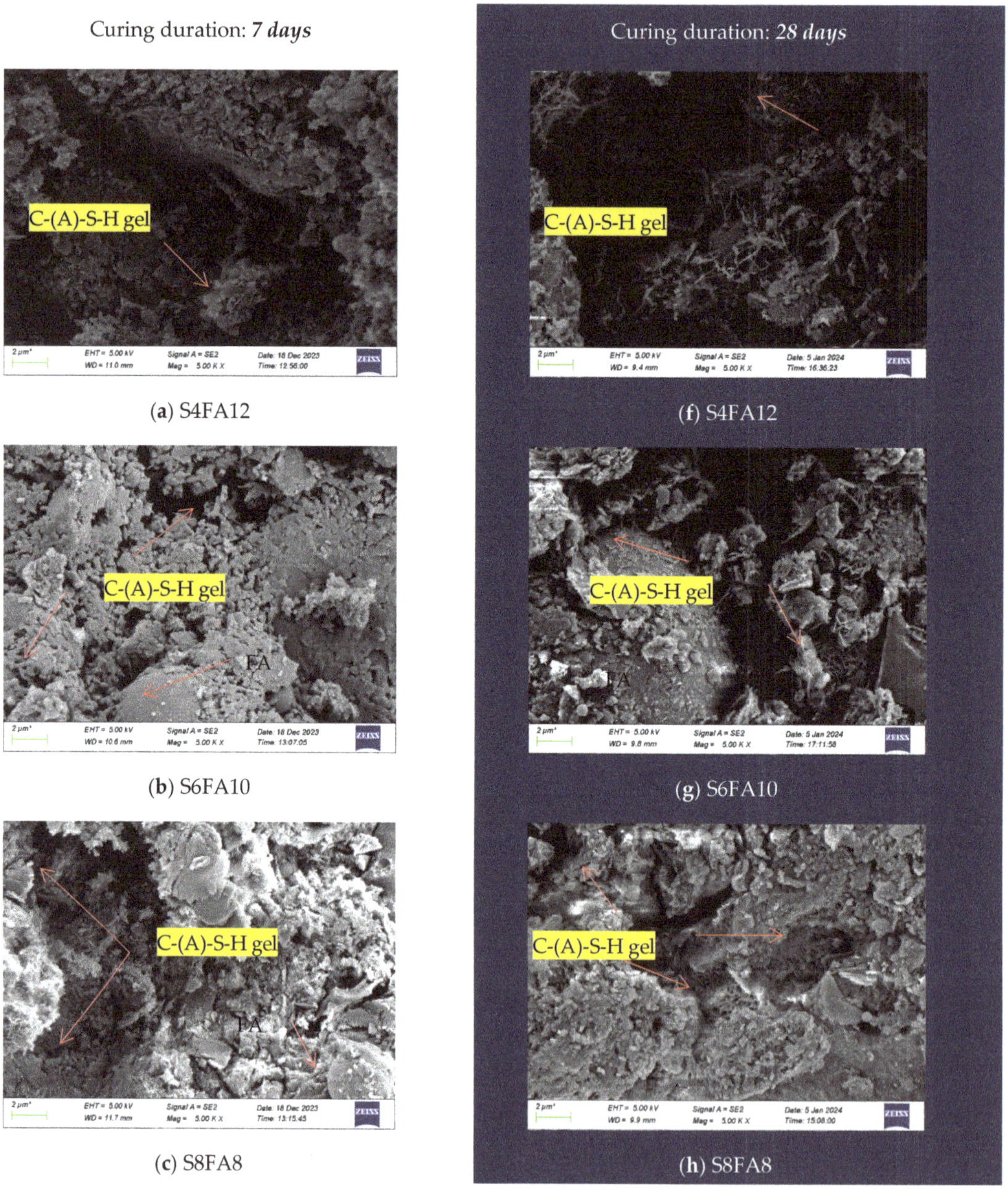

Figure 9. *Cont.*

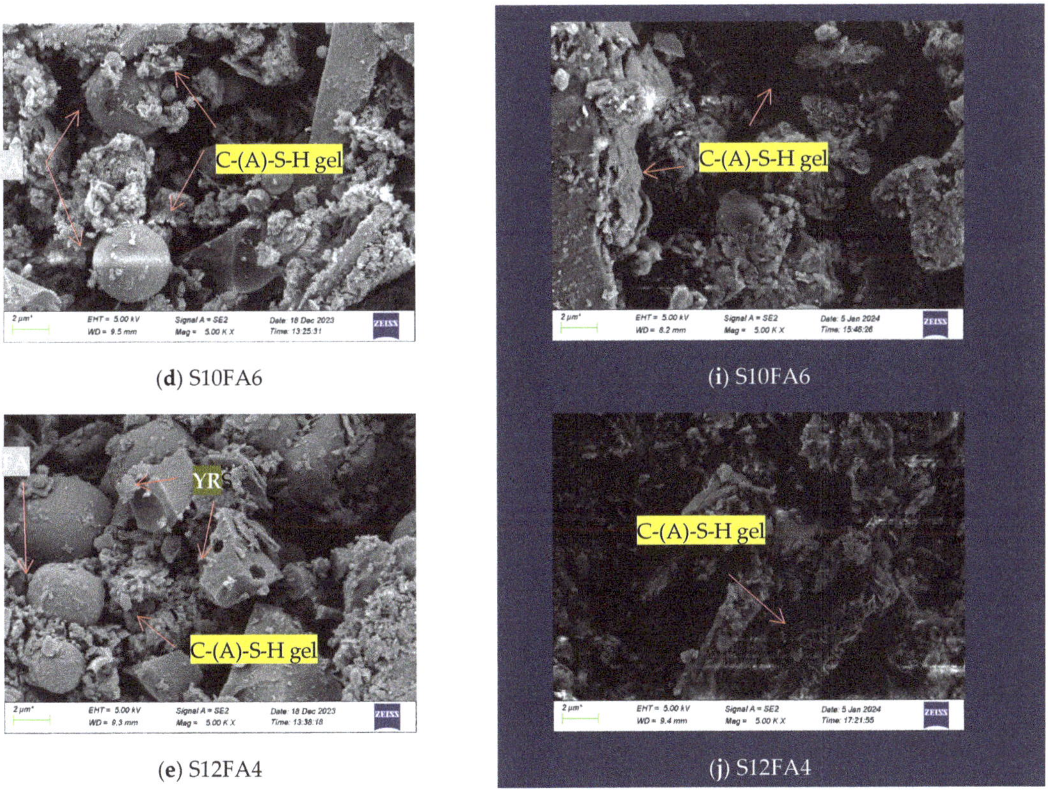

Figure 9. SEM diagram of prepared flood-prevention stones at 7 and 28 days of curing. FA: fly ash; YRS: Yellow River Sediment.

Figure 9a–e (left column) illustrate the SEM images of the Yellow River sediment-based flood-prevention stones that have undergone a standard 7-day curing process. The results demonstrated the formation of C-A-S-H gel within the specimens during a relatively short curing period. The structure of the C-A-S-H gel continued to expand as the slag content increased, particularly when the dosage of slag in the specimen was below 8%. However, when the slag content exceeded 8% of the total mass of the specimen, the structure of C-A-S-H gels in the specimen was significantly reduced, accompanied by a large number of unreacted spherical slag particles. This structural alteration led to a modification in the compressive strength of the specimen, which aligned with the results of our prior experimental work (Figure 5).

In contrast, Figure 9f–j (right column) show the SEM images of the 28-day cured stones. Following an extended curing period, the images revealed the formation of numerous needle-like C-A-S-H gels, which significantly contributed to the observed increase in the compressive strength of the specimens. When the slag content was increased above an 8% mass fraction, the SEM images show that the observed hydration products were intricately interwoven to form a stable structure. However, at lower slag contents (below 6%), the porous hydration products resulted in a decrease in the compressive properties of the specimens (Figure 9f,g). This observed trend also corroborated the results obtained during the compressive strength testing discussed in the previous section.

3.3.3. XRD Analysis for Composition of Flood-Prevention Stones

The X-ray diffraction (XRD) analysis results for the Yellow River sediment-based flood-prevention stones, which were cured for 7 and 28 days, are illustrated in Figure 10. The raw sediment from Sanmenxia was found to consist of calcite, montmorillonite, and a significant amount of quartz [34–36]. As depicted in Figure 10a, all specimens prepared from Yellow River sediment exhibited signals indicative of C-A-S-H gel formation during testing. Additionally, a notable reduction in the intensity of the quartz peak was observed in the XRD tests conducted under a curing condition of 7 days. The results indicated that the SiO_2 present in the raw material participated in the hydration reaction, contributing to the formation of a gel with enhanced compressive strength. It is noteworthy that the specimen labeled S8FA8, which displayed the most pronounced C-A-S-H gel signal in the XRD spectrum, exhibited superior performance during the compressive strength tests (Figure 5).

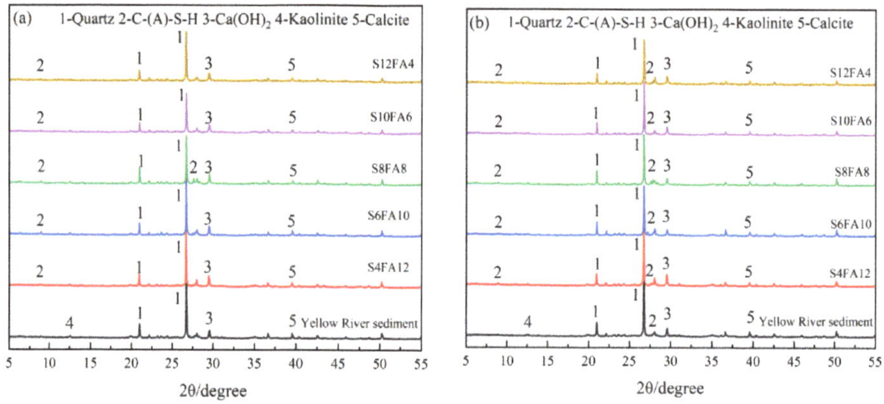

Figure 10. XRD testing of prepared flood-prevention stones after 7 and 28 days of curing. (**a**) XRD testing after 7 days of curing. (**b**) XRD testing after 28 days of curing.

Similarly, the specimens that underwent a curing time of 28 days yielded identical results (Figure 10b). The stronger C-A-S-H gel signal observed in the XRD test implied that the production of the gel should be promoted after prolonging the curing time, which indicated an enhancement in the compressive strength of the specimens. This conclusion was corroborated by the results of the preceding compressive strength tests and the SEM images.

3.4. Environmental and Economic Evaluations

The quantity of each component employed in the binder during the solidification of Yellow River sediment had a direct influence on the cost and CO_2 emissions associated with the manufacturing process of artificial flood-prevention stones. Indeed, these indices exerted a considerable influence on the practical utility of the prepared stones. In this study, we conducted an evaluation of the CO_2 emissions and manufacturing costs associated with the production of our testing specimens.

3.4.1. Environmental Evaluation

In evaluating the environmental impact of the artificial flood-prevention stones, we identified the ratio of CO_2 emissions to the corresponding intensity as a key evaluation index. The corresponding calculation data are presented in Figure 11. S10FA6 exhibited the lowest emissions of 0.18 kg (CO_2)/(kg·MPa) due to its satisfactory compression strength, demonstrating minimal environmental impact.

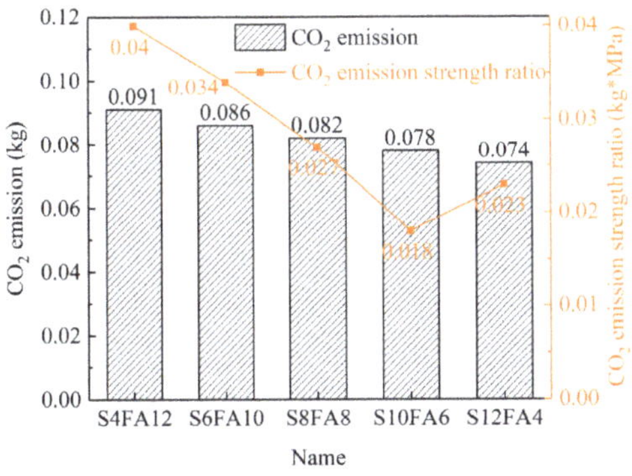

Figure 11. Carbon emissions calculation on Yellow River sediment-based flood-prevention stones.

3.4.2. Economic Analysis

In this section, we set out to examine the production costs of artificial flood-prevention stones through two scenarios. The total production costs of the prepared stones were evaluated (Figure 12). As the proportion of higher-priced slag in the binder increases, the cost of the artificial stones generally rises. The specimen labelled as S12F4 exhibited the highest production cost at 60.4 yuan/t (8.3 USD/t), which was significantly lower than the price of commercially available natural flood-prevention stones, which is approximately 80 yuan/t (11 USD/t).

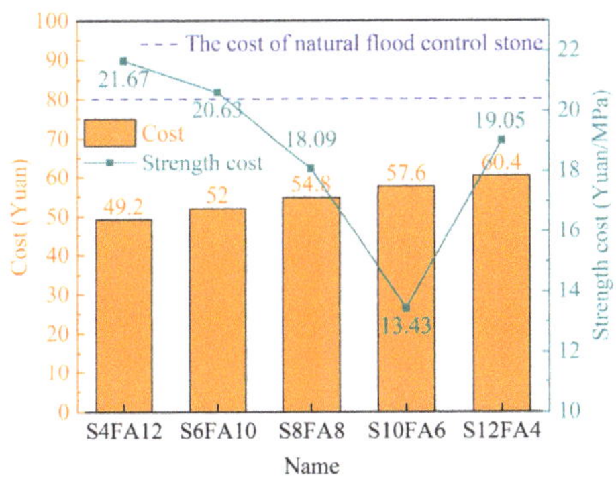

Figure 12. Cost of strength for Yellow River sediment-based flood-prevention stones.

Additionally, the ratio of cost to compressive strength was employed as an additional reference indicator. With the assistance of the higher compressive strength, the lowest calculated value, 13.4 yuan/t·MPa (1.4 USD/t·MPa), was obtained for the specimen designated as S10F6. These favorable outcomes instill confidence in the prospective utilization of the prepared artificial flood-prevention stones.

4. Summary and Conclusions

In this study, we have employed slag, fly ash, and quicklime as binders with a 20% additive concentration to prepare a novel artificial Yellow River sediment-based flood-prevention stone. The main work of our study is outlined below:

1. Cross-verifying the theoretical and experimental results on compressive strength. The results of an orthogonal experimental design, in conjunction with Design-Expert for simulation prediction, were employed to ensure the compressive strength of a prepared artificial Yellow River sediment-based flood-prevention stone with a high de-

gree of accuracy. Furthermore, the individual components of the binder that influence the compressive strength under different curing durations have been elucidated.

2. Clarification of the mechanism of high compressive strength of prepared specimens. Analyses conducted using scanning electron microscopy (SEM) and X-ray diffraction (XRD) revealed the formation of C-A-S-H gel and its growth at different curing periods.
3. Demonstration of the environmental friendliness and economy of artificial Yellow River sediment-based flood-prevention stones. The CO_2 emissions associated with the preparation of the artificial flood-prevention stones were calculated. In accordance with the optimal compressive strength conditions, the emission on preparation of the desired stones was determined to be 0.18 kg $(CO_2)/(kg \cdot MPa)$, which is considered to be typical of environmentally friendly production processes.

Based on the research findings, we have derived the following relevant conclusions:

1. Following the completion of the standard curing procedure over a 28-day period, the compressive strength of the prepared artificial stone was found to be 4.29 MPa. The pH of the identical specimen leachate was found to stabilize at 8.2 after 30 h of immersion. The softening coefficient of the S10FA6 specimen was tested with a value of over 0.7 under wet conditions. These important results confirmed that the prepared stones with the selected binder met the specifications for artificial flood-prevention stones.
2. The pozzolanic reaction between CaO in slag and active SiO_2 and Al_2O_3 in fly ash to form a C-A-S-H gel was proved as the main reason for the compressive strength enhancement of the specimens. The testing results on SEM and XRD of the tested specimens demonstrated a direct correlation between the alterations in the structure of the C-A-S-H gel and the compressive strength of the blocks.
3. It has been determined that the cost of the produced artificial flood-prevention stones, utilizing various binders, is notably lower than the market price for commercially available natural stones. The cost-effectiveness of the artificial stones will facilitate their promotion in the market.

The potential applications of solidified Yellow River sediment in other areas of construction and engineering are currently under investigation in our laboratory.

Author Contributions: Conceptualization, Y.L. (Ying Liu) and C.Y.; methodology, Y.J.; investigation, H.X. and C.Y.; resources, Y.L. (Ying Liu); writing—original draft preparation, Y.L. (Ying Liu), Y.J.; Y.L. (Yajun Lv) and L.D.; writing—review and editing, H.X., L.D. and C.Y.; funding acquisition, Y.L. (Ying Liu) and C.Y. All authors have read and agreed to the published version of the manuscript.

Funding: This research was funded by National Key R&D Program of China (2021YFC3200402), Program for Science & Technology Innovation Talents in Universities of Henan Province Science (21HASTIT010), Key R&D Program of Henan Province: (232102321107), Research Fund of Key Laboratory of Water Management and Security for Yellow River Basin, Ministry Water Resources (2022-SYSJJ-09), Key R&D and Promotion Projects (Scientific and Technological Research) of Henan Province (232102310367). The APC was funded by National Key R&D Program of China (2021YFC3200402) & Program for Science & Technology Innovation Talents in Universities of Henan Province Science (21HASTIT010).

Institutional Review Board Statement: Not applicable.

Informed Consent Statement: Not applicable.

Data Availability Statement: The original contributions presented in the study are included in the article, further inquiries can be directed to the corresponding authors.

Acknowledgments: We would like to thank Changming Li for his constructive comments during the revision of the manuscript.

Conflicts of Interest: The authors declare no conflict of interest. There are no potential commercial interests with the company in the affiliation.

References

1. Wang, Y.; Chen, H.; Song, Y.; Wang, Z.; Zhong, Y.; Zhang, B. Experimental study on the stability of soybean urease-induced calcium carbonate solidification in sediment water in the Yellow River. *J. Hydraul. Eng.* **2024**, *55*, 71–79. [CrossRef]
2. Wang, Z.J.; Ta, W.Q.; Zheng, J.; Zhang, K. Total sediment transport from an urbanizing watershed in the upper Yellow River. *Environ. Sci. Pollut. Res.* **2019**, *26*, 13702–13716. [CrossRef] [PubMed]
3. Zhang, J.L.; Shang, Y.Z.; Liu, J.Y.; Lu, J.; Wei, S.T.; Wan, Z.W.; Luo, Q.S.; Chen, C.X.; Tong, L.; Wang, Q.; et al. Optimisation of reservoir operation mode to improve sediment transport capacity of silt-laden rivers. *J. Hydrol.* **2021**, *594*, 125951. [CrossRef]
4. Nian, X.M.; Zhang, W.G.; Wang, X.T.; Wang, X.T.; Hutchinson, S.M.; Zhao, X.Q.; Liu, K.B. Multi-Centennial Variability of Yangtze Delta Growth Over the Last 2000 Years: Interplay of Climate and People. *Earths Future* **2022**, *10*, e2021EF002461. [CrossRef]
5. Liu, H.Q.; Liu, G.J.; Wang, J.; Yuan, Z.J.; Da, C.N. Fractional distribution and risk assessment of heavy metals in sediments collected from the Yellow River, China. *Environ. Sci. Pollut. Res.* **2016**, *23*, 11076–11084. [CrossRef]
6. Chen, C.X.; Fu, J.; Wu, M.X.; Gao, X.; Ma, L.M. High-efficiency Sediment Transport Requirements for operation of the Xiaolangdi Reservoir in the Lower Yellow River. *Water Supply* **2022**, *22*, 8572–8586. [CrossRef]
7. Wu, P.; Liu, D.; Ma, J.; Gu, L.; Yong, J. Reconstructing the man-made Yellow River flood of Kaifeng City in 1642 AD using documentary sources. *Int. J. Disaster Risk Reduct.* **2019**, *41*, 101289. [CrossRef]
8. Li, T.; Li, J.; Zhang, D.D. Yellow River flooding during the past two millennia from historical documents. *Prog. Phys. Geogr. Earth Environ.* **2020**, *44*, 661–678. [CrossRef]
9. He, H.; Tian, Y.; Mu, X.; Zhou, J.; Li, Z.; Chen, N.; Zhang, Q.; Keo, S.; Oeurng, C. Confluent flow impacts of flood extremes in the middle Yellow River. *Quat. Int.* **2015**, *380–381*, 382–390. [CrossRef]
10. Wang, Z.; Liu, C. Two-thousand years of debates and practices of Yellow River training strategies. *Int. J. Sediment Res.* **2019**, *34*, 73–83. [CrossRef]
11. Liu, G. On the geo-basis of river regulation in the lower reaches of the Yellow River. *Sci. China-Earth Sci.* **2012**, *55*, 530–544. [CrossRef]
12. Tian, F.; Ma, B.; Yuan, X.; Wang, X.; Yue, Z. Hazard Assessments of Riverbank Flooding and Backward Flows in Dike-Through Drainage Ditches during Moderate Frequent Flooding Events in the Ningxia Reach of the Upper Yellow River (NRYR). *Water* **2019**, *11*, 1477. [CrossRef]
13. Zhang, X.; Song, C.; Hu, D. Research on the periodic regularity of annual runoff in the Lower Yellow River. *J. Water Clim. Chang.* **2019**, *10*, 130–141. [CrossRef]
14. Manopiniwes, W.; Irohara, T. Optimization model for temporary depot problem in flood disaster response. *Nat. Hazards* **2020**, *105*, 1743–1763. [CrossRef]
15. Huang, L.; Yu, Z.; Wang, W.; Song, Y.; Wang, Y. Multiobjective programming model for a class of flood disaster emergency material allocation. *J. Flood Risk Manag.* **2023**, e12918. [CrossRef]
16. Ma, G.; Wang, J.; Yu, F.; Yang, W.; Ning, J.; Peng, F.; Zhou, X.; Zhou, Y.; Cao, D. Framework construction and application of China's Gross Economic-Ecological Product accounting. *J. Environ. Manag.* **2020**, *264*, 109852. [CrossRef]
17. Feng, C.; Cui, B.; Ge, H.; Huang, Y.; Zhang, W.; Zhu, J. Reinforcement of Recycled Aggregate by Microbial-Induced Mineralization and Deposition of Calcium Carbonate—Influencing Factors, Mechanism and Effect of Reinforcement. *Crystals* **2021**, *11*, 887. [CrossRef]
18. Li, M.H.; Lu, Y.; Yang, S.; Chu, J.; Liu, Y. Study on the Early Effect of Excitation Method on the Alkaline Steel Slag. *Sustainability* **2023**, *15*, 4714. [CrossRef]
19. Wang, Q.; Yang, J.W.; Yan, P.Y. Influence of initial alkalinity on the hydration of steel slag. *Sci. China-Technol. Sci.* **2012**, *55*, 3378–3387. [CrossRef]
20. Jiang, M.Y.; Qian, Y.F.; Sun, Q. Preparation of controlled low-strength materials from alkali-excited red mud-slag-iron tailings sand and a study of the reaction mechanism. *Environ. Sci. Pollut. Res.* **2023**, *30*, 22232–22248. [CrossRef]
21. Wang, B.; Wang, W.; Liang, X.; Liu, H.; Han, J.; Zhao, L.; Yang, X.; Yan, J. Development of Cementitious Materials Utilizing Alkali-activated Yellow River Silt. *J. Wuhan Univ. Technol. Mat. Sci. Edit.* **2021**, *36*, 364–373. [CrossRef]
22. Yan, J.; Zhong, S.; Chen, S.; Lv, Y.; Yang, L.; Peng, G.; Deng, A. Study on the Application of Sediment-Based Embankment Building and Ultra-High-Performance Concrete (UHPC) Preparation in the Resource Utilization of Yellow River Sediment. *Materials* **2022**, *15*, 5668. [CrossRef] [PubMed]
23. Jiang, S.; Xu, J.; Song, Y.; Xu, Y. Alkali-activated fly ash foam concrete with Yellow River silt: Physico-mechanical and structural properties. *Constr. Build. Mater.* **2023**, *373*, 130879. [CrossRef]
24. Wang, B.; Li, G.; Han, J.; Zheng, Y.; Liu, H.; Song, W. Study on the properties of artificial flood-prevention stone made by Yellow River silt. *Constr. Build. Mater.* **2017**, *144*, 484–492. [CrossRef]
25. Zheng, L. *Research on Consolidation and Gelling Technology of Flood Control Stone Made from Yellow River Sediment*; Dalian University of Technology: Dalian, China, 2017. Available online: https://kns.cnki.net/kcms2/article/abstract?v=aHgEko1xHjjAposLR8iSE5IgyHPQi4toMJKniypW0HRp1z8PzmetegHlnqgSnmgBXA5lClyHMcTN7DhaYoOoQ3lr-Z_WKDHX9Vwzjvt3cptGPiNMbgL9l2NNXBDGx_nxoNjXa3ZazhKNDQqdIDdB0w==&uniplatform=NZKPT&language=CHS (accessed on 23 June 2024).
26. Song, Y. *Experimental Study on the Mechanical Properties and Durability of Alkali-Activated Fly Ash Concrete under Freeze-Thaw-Wind Erosion Environment*; Inner Mongolia Agricultural University: Hohhot, China, 2023. [CrossRef]

27. *GB/T 9966.1-2020*; Test Methods for Natural Stone—Part 1: Determination of Dry, Wet and after Freezing Compressive Strength. National Standardization Administration: Beijing, China, 2020.
28. *GB/T 23561.7-2009*; Methods for Determining the Physical and Mechanical Properties of Coal and Rock—Part 7: Methods for Determining the Uniaxial Compressive Strength and Counting Softening Coefficient. National Standardization Administration: Beijing, China, 2009.
29. Zhang, T.; Cai, G.; Liu, S.; Puppala, A.J. Engineering properties and microstructural characteristics of foundation silt stabilized by lignin-based industrial by-product. *KSCE J. Civ. Eng.* **2016**, *20*, 2725–2736. [CrossRef]
30. Tremblay, H.; Duchesne, J.; Locat, J.; Leroueil, S. Influence of the nature of organic compounds on fine soil stabilization with cement. *Can. Geotech. J.* **2002**, *39*, 535–546. [CrossRef]
31. Rivera, J.F.; Orobio, A.; Cristelo, N.; de Gutiérrez, R.M. Fly ash-based geopolymer as A4 type soil stabiliser. *Transp. Geotech.* **2020**, *25*, 100409. [CrossRef]
32. Arulrajah, A.; Yaghoubi, M.; Disfani, M.M.; Horpibulsuk, S.; Bo, M.W.; Leong, M. Evaluation of fly ash- and slag-based geopolymers for the improvement of a soft marine clay by deep soil mixing. *Soils Found.* **2018**, *58*, 1358–1370. [CrossRef]
33. Li, C.; Dong, J.; Zhao, S.; Liu, H.; Yao, W.; Wang, L. Development of low cost supplementary cementitious materials utilizing thermally activated Pisha sandstone. *Constr. Build. Mater.* **2018**, *174*, 484–495. [CrossRef]
34. Sukmak, P.; Horpibulsuk, S.; Shen, S.-L. Strength development in clay–fly ash geopolymer. *Constr. Build. Mater.* **2013**, *40*, 566–574. [CrossRef]
35. Ortega-Zavala, D.E. Chemically bonded ceramic/cementitious materials of alkali activated metakaolin processed by cold pressing. *Constr. Build. Mater.* **2021**, *279*, 121365. [CrossRef]
36. Lv, Y.; Zhang, K.; Qu, J.; Yin, K.; Yang, C.; Zheng, H.; Xiang, T. Study on physicochemical and anti-corrosion performance of methylsiloxane-based hydrophobic powder modified mortar. *Constr. Build. Mater.* **2024**, *425*, 135999. [CrossRef]

gels

MDPI

Article

Sol–Gel Derived Alumina Particles for the Reinforcement of Copper Films on Brass Substrates

Samah Sasi Maoloud Mohamed [1], Marija M. Vuksanović [2,*], Dana G. Vasiljević-Radović [3], Ljiljana Janković Mandić [2], Radmila M. Jančić Heinneman [1], Aleksandar D. Marinković [1] and Ivana O. Mladenović [3,*]

1 Faculty of Technology and Metallurgy, University of Belgrade, Karnegijeva 4, 11000 Belgrade, Serbia; samahsasi1@gmail.com (S.S.M.M.); radica@tmf.bg.ac.rs (R.M.J.H.); marinko@tmf.bg.ac.rs (A.D.M.)
2 VINČA Institute of Nuclear Sciences—National Institute of the Republic of Serbia, University of Belgrade, 11000 Belgrade, Serbia; ljmandic@vin.bg.ac.rs
3 Institute of Chemistry, Technology and Metallurgy, University of Belgrade, Njegoševa 12, 11000 Belgrade, Serbia; dana@nanosys.ihtm.bg.ac.rs
* Correspondence: marija.vuksanovic@vin.bg.ac.rs (M.M.V.); ivana.mladenovic@ihtm.bg.ac.rs (I.O.M.); Tel.: +38-1116455654 (M.M.V.); +38-1112628587 (I.O.M.)

Abstract: The aim of this study is to provide tailored alumina particles suitable for reinforcing the metal matrix film. The sol–gel method was chosen to prepare particles of submicron size and to control crystal structure by calcination. In this study, copper-based metal matrix composite (MMC) films are developed on brass substrates with different electrodeposition times and alumina concentrations. Scanning electron microscopy (FE-SEM) with energy-dispersive spectroscopy (EDS), TEM, and X-ray diffraction (XRD) were used to characterize the reinforcing phase. The MMC Cu-Al_2O_3 films were synthesized electrochemically using the co-electrodeposition method. Microstructural and topographical analyses of pure (alumina-free) Cu films and the Cu films with incorporated Al_2O_3 particles were performed using FE-SEM/EDS and AFM, respectively. Hardness and adhesion resistance were investigated using the Vickers microindentation test and evaluated by applying the Chen–Gao (C-G) mathematical model. The sessile drop method was used for measuring contact angles for water. The microhardness and adhesion of the MMC Cu-Al_2O_3 films are improved when Al_2O_3 is added. The concentration of alumina particles in the electrolyte correlates with an increase in absolute film hardness in the way that 1.0 wt.% of alumina in electrolytes results in a 9.96% increase compared to the pure copper film, and the improvement is maximal in the film obtained from electrolytes containing 3.0 wt.% alumina giving the film 2.128 GPa, a 134% hardness value of that of the pure copper film. The surface roughness of the MMC film increased from 2.8 to 6.9 times compared to the Cu film without particles. The decrease in the water contact angle of Cu films with incorporated alumina particles relative to the pure Cu films was from 84.94° to 58.78°.

Keywords: alumina; sol–gel; copper films; co-electrodeposition; microhardness; adhesion

1. Introduction

Alumina and aluminum hydroxides are notable materials due to their structural and property characteristics, which can be precisely tailored through appropriate processing methods [1]. A variety of distinct structures can be selected [2] to achieve the desired particle morphology and size, enabling the material to exhibit properties designed to fit the requirements of the final application [3,4]. In certain low-temperature applications, surface modification, using silane and organic modifiers on the particles, can significantly improve their incorporation into a polymer matrix [5]. When the particles are prepared for use in the metal matrix, the organic molecules attached to the surface suffer during the preparation of the composite, and therefore, the chosen way to prepare suitable particles is to control their size and crystal structure [6,7]. The morphology of prepared particles could also influence the properties of the prepared films in several ways. The film hardness can be improved, the film can change surface energy, and adhesion to the substrate can be modified.

The sol–gel technique allows for an easy way to produce very fine alumina particles using relatively accessible laboratory conditions and to control the quality of the particles. This way of producing the particles enables the control of their chemical composition and tuning of their properties. The exact crystal structure of the material is controlled by the calcination process that follows the first step, maintaining the fine grain and crystallite size.

Copper is used as a material that has several good properties, such as an aesthetic one where the material is covered to have a specific colour and resistance to the environment. On the other hand, copper has some interesting properties in contact with some chemicals and environments [8] and can be used to carry the particles that could eventually exhibit some specific activity [9]. Copper itself has very low hardness and is not very resistant to wear and mechanically harsh environments. When it is meant to serve as the matrix of the composite, the intention is to obtain a material that is harder, stronger, and more resistant to environmental impact [9]. Also, if it is possible to control properties such as surface energy or water or oil affinity, it can change the possible use of this sort of material [10].

The incorporation of alumina particles into copper matrices via electrochemical methods has been extensively studied in recent decades, primarily using commercial particles [11–13]. The sol–gel technique involves using substances that are high in highly reactive chemicals as precursors. These precursors combine with solvents to produce reactions like alcoholysis and hydrolysis. A stable and transparent sol matrix forms within the solution as a result of these processes [14,15]. The use of sol–gel-prepared alumina to reinforce MMC films represents a novel approach. Additionally, the synergistic effects of film thickness and alumina concentration on the composite films' absolute hardness [16] and adhesion parameters [17,18] have not been previously investigated. The effects of alumina are most often observed by varying the electrochemical parameters [11]. Thiemig et al. [12] studied the impact of the pulse plating (PC) parameters on the co-electrodeposited γ-Al_2O_3 particles in Cu- and Ni-based MMC films and compared them to DC (direct current), PC, and PRC (pulse reverse current) plating, whereby pulsed deposition modes are favoured due to greater incorporation of alumina into the composite. Further information on the co-deposition mechanisms of inert, semi-conductive, and conductive particles in copper films on metal substrates was provided by [13]; the study utilized α-Al_2O_3 nanoparticles (300 nm) and demonstrated that particle content in films increased with current density. Additionally, the effects of varying potential differences during the co-electrodeposition of Cu-Al_2O_3 composites were examined [19]. The authors used α-Al_2O_3 nano-sized alumina ($\approx$40 nm) particles and indicated that alumina has not been deposited at lower magnitudes of potential differences. On the other hand, it is found that the surface wettability of Cu-Al_2O_3 composite films increases as the potential difference increases, and more hydrophilic films are obtained, and as such, they are suitable for the enhancement of heat transfer applications [19].

Numerous studies have explored the preparation of MMC films on material surfaces, making this a widely researched topic. The deposition conditions used in this study were previously optimized for copper films. Composite films are prepared using electrochemical deposition/co-deposition under uniform current density. Ceramic particles are maintained in suspension in electrolytes through continuous stirring and agitation. The size of the particles needs to be submicrometric. The thickness of the deposit is regulated by the duration of the deposition process [20] and verified by the mechanical comparator [18]. Materials produced this way should exhibit improved hardness, adhesion to the substrate, and water affinity, as the added particles are relatively polar. However, challenges include achieving uniform particle distribution both throughout the film's depth and on its surface. Additionally, the particles may alter the film's microstructure, affecting the shape and size of the forming crystals. Those properties are studied and correlated with other material properties. Roughness can be influenced by the properties of the reinforcement as well as by the conditions of film formation [16,21].

This paper will present the results of the preparation of alumina particles with the aim of obtaining them to be suitable for film deposition with appropriate properties, and these properties include surface hardness, surface wettability, and roughness. This study aimed to prepare Cu-Al_2O_3 nano-microscale composites, where alumina particles were synthesized via the sol–gel method and then co-deposited with the matrix using electrodeposition, and to study the effect of alumina on the film properties. The description of the synthesis of alumina nanoparticles according to sol–gel is given in Section 4.1. The study of grain refinement/non-refinement of copper and its composite, for the particles and phases present, in terms of roughness, wettability, hardness, and adhesion using different techniques, was applied. The alumina contents in the electrolyte were adjusted to be 0, 1.0, 3.0, and 5.0 wt.% in the sulphate electrolyte. Film thicknesses of 2, 22, and 52 μm were adjusted based on the deposition rate (electrodeposition time) for both cases for the deposited films without/with alumina. A comparative study with all the advantages and disadvantages of the synthesized material is presented to understand the possibility of applying different synthesis methods in a cheap, fast, economical, and eco-friendly way. The goal of this research is to enhance the composite film's surface topography and mechanical characteristics over pure copper film with minimal degradation.

2. Results and Discussion

This investigation includes two research directions: (1) synthesis and characterization of alumina particles and (2) synthesis and characterization of copper films without/with alumina particles with a characterization of the metallic substrate prior to deposition of the films.

2.1. Microstructural and Chemical Characterization of Alumina Nanoparticles

2.1.1. Microstructure of Al_2O_3-FE/SEM and TEM

An FE-SEM microscope was used to examine the alumina particles prepared in the sol–gel process. The morphology of the particles is shown in Figure 1a. The mean diameter of the particles was determined using FE-SEM images and Image Pro Plus 6.0 image analysis software (Media Cybernetics, Rockville, MD, USA) [22]. Figure 1b displays the particle size distribution. The histogram that was obtained indicates that most of the particles had a diameter of around 100 nm.

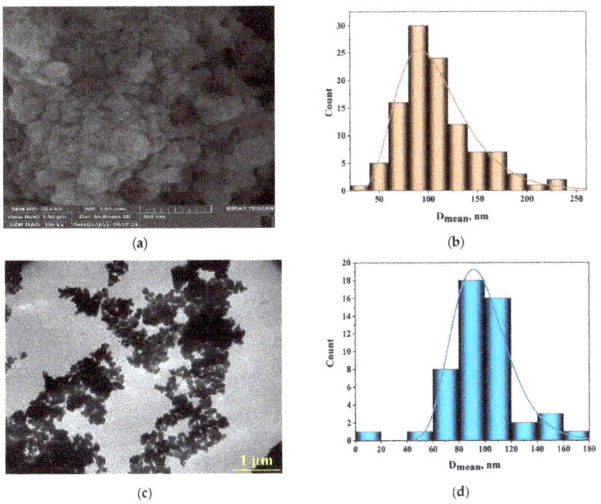

Figure 1. Particle morphology of alumina: (**a**) FE-SEM picture, (**b**) diameter size, (**c**) TEM micrographs, and (**d**) corresponding histogram of particles.

Micrograms obtained by transmission electron microscopy (TEM) show an inhomogeneous distribution of different sizes of submicrometre agglomerates and nano-sized particles of alumina powder (Figure 1c). Spherical–oval-shaped particles were observed in individual locations where no agglomerates were formed and their size is 100–120 nm, which agrees with SEM analysis (Figure 1d).

2.1.2. Chemical Purity of Al_2O_3-EDS/Mapping

The elemental composition of the generated alumina particles was evaluated using EDS analysis on the FE-SEM images (Figure 2). Based on the EDS analysis, there is an even distribution of Al and O atoms in the particles, confirming the successful synthesis of alumina particles from sol–gel. The EDS mapping did not show other elements such as chlorine that existed in precursors, so the synthesis method can be considered as appropriate for this purpose.

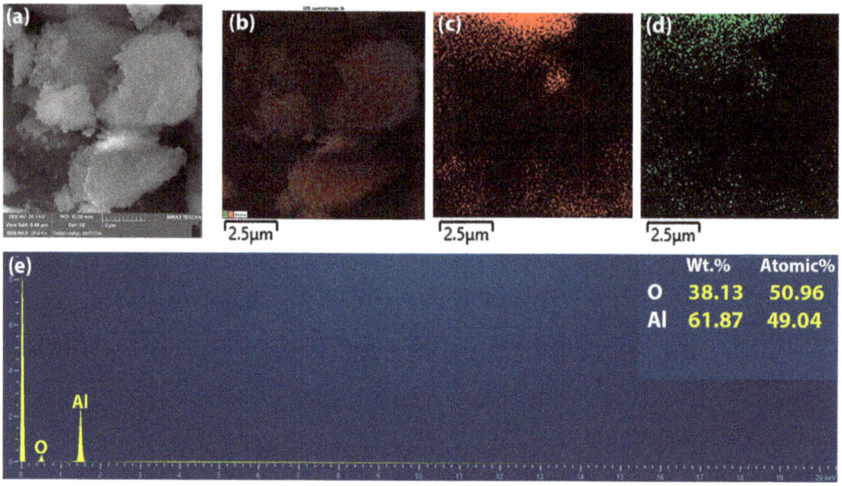

Figure 2. (**a**) SEM images of alumina particles, (**b**) elemental mapping, (**c**) Al, (**d**) O, and (**e**) EDS spectra.

2.1.3. XRD Analysis of Al_2O_3

An XRD spectrum of the sol–gel-derived nanoparticles of alumina powder is shown in Figure 3. All peaks were assigned by the standards No. 01-076-7776 (for α-Al_2O_3) [23].

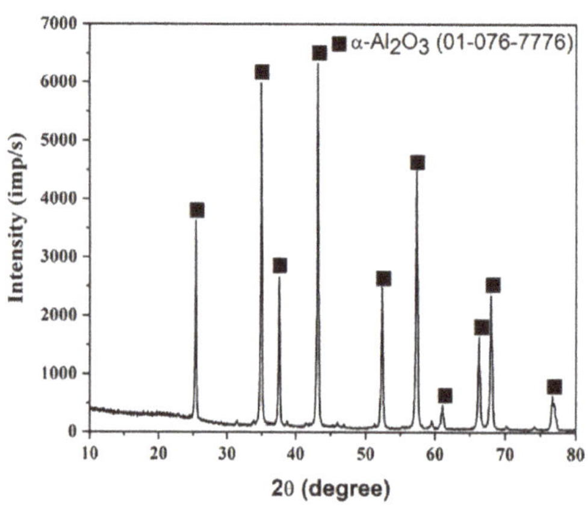

Figure 3. X-ray diffractogram of alumina powder obtained by sol–gel method.

The alumina with a rhombohedral structure that is most stable is called α-Alumina and is proven to be preset in the crystal structure. At 2θ values of 34.8°, 43.1°, and 57.3°, respectively, three significant distinctive peaks have been detected from the planes (104), (113), and (116) [23].

XRD patterns confirmed that most intensive reflections came from Al_2O_3 applying the pseudo-Voigt profile. Refined structural parameters include the lattice constant as well as the average crystallite size values obtained by applying the pseudo-Voigt profile. The values obtained for the lattice parameters were $a_h = b_h = 4.7495$ Å, $c_h = 12.9718$ Å, and $V = 253.41$ Å. The average crystallite size (D) was calculated based on the full width at the half-maximum intensity (FWHM) of the main reflections by applying Scherrer's formula (see Supplementary Material A). The size of the alumina crystallites is 32.7 nm. The fractional coordinates for O and Al are given in Table 1, where x, y, and z are the fractional coordinates for O and Al.

Table 1. Fractional coordinates for O and Al obtained according to XRD analyses of alumina particles produced using sol–gel method.

Fractional Coordinates for O and Al	*x*	*y*	*z*
O	0.68998 (38)	0.00000 (0)	0.25000 (0)
Al	0.00000 (0)	0.00000 (0)	0.35273 (10)

2.2. Characterization of Substrate, Copper Films, and MMC of Cu-Al_2O_3 Films

An investigation of the effect of alumina nanoparticles as reinforcement on the morphology, roughness, microhardness, adhesion strength, and wettability of electrolytically produced Cu films was conducted through the examination of Cu films formed with varying deposition times and alumina concentrations in suspension. To better understand how this reinforcement affected the Cu film characteristics, they were compared with films formed from the same acidic basic sulphate electrolyte that did not contain alumina nanoparticles.

The first factor influencing the structural, mechanical, and topographical properties of Cu and Cu-Al_2O_3 films are the properties of the chosen metallic substrate and the possibility of modifying its topography in order to achieve the best adhesion of the films.

2.2.1. Characterization of Substrate

Topography of Brass Substrate after Mechanical and Chemical Modification

The brass substrate was mechanically polished using SiC paper and the topography is presented in Figure 4a. Based on the roughness analysis, the average surface roughness parameter (S_a) was found to be 31.30 nm. The next step included a chemically roughened sanded brass surface aimed to remove deep channels and ridges and obtain sharp peaks for a better contact surface for the electrodeposition process (Figure 4b). The obtained roughness parameter S_a is 80.43 nm. The red color indicates the maximum height of the grain, while the blue indicates the depression.

Any modification of the surface of the substrate leads to a change in the microstructure, and therefore, the mechanical properties of the substrate. Before applying the films to a substrate, it is necessary to evaluate the mechanical properties (microhardness) of the chemically modified substrate without the film.

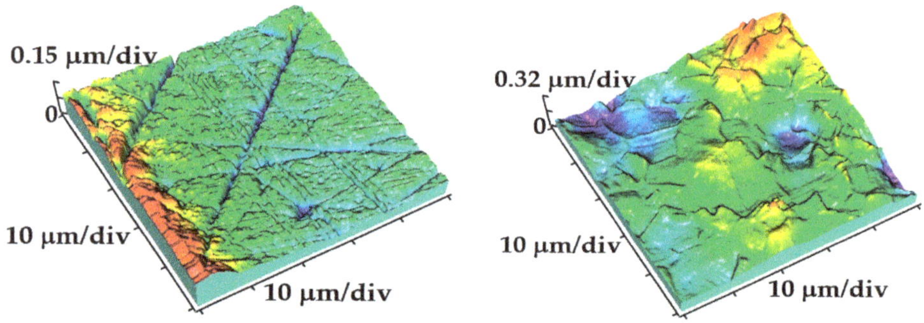

(a) (b)

Figure 4. The three-dimensional (3D) images of the brass substrate obtained on an AFM device: (**a**) brass surface after mechanical grinding and (**b**) after chemical etchings in a mixture of acids (HNO_3:H_3PO_4:CH_3COOH = 4:11:5 vol.%) at 10 s.

Microhardness of Brass Substrate after Chemical Modification

Figure 5a shows the relation between the applied load, P, and the measured values of the Vickers microhardness diagonal size, d, in the log–log scale. The size and character of the ISE effect (Indentation Size Effect) for brass substrates after chemical etching can be evaluated in different ways (Meyer's power law, Proportional Specimen Resistance (PSR), and Hays–Kendall methods) [24]. Meyer's power law and index n or "work-hardening index" for the prepared brass substrate was 2.01294. The value obtained for n (index $n > 2$) indicates the existence of a reverse ISE (RISE) [24]. The Proportional Specimen Resistance (PSR) method [25–28] was chosen to determine the absolute hardness of the substrate (Figure 5b). The calculated value of the microhardness of the chemically etched brass substrate was 1.55 GPa.

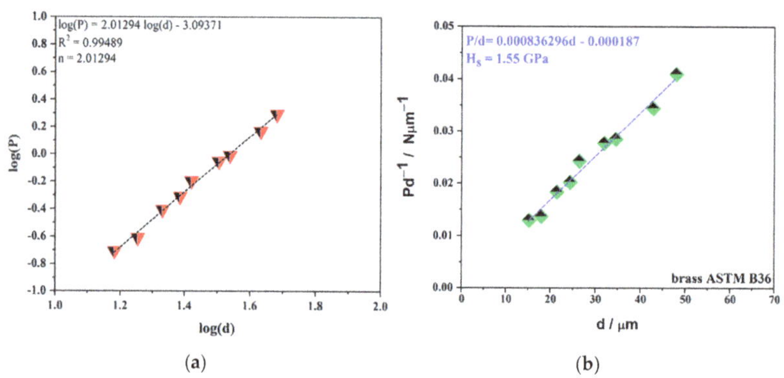

(a) (b)

Figure 5. Mechanical properties of brass substrate after chemical etching: (**a**) calculation of Meyer's index and evaluation ISE, and (**b**) calculation hardness of substrate with application PSR model.

2.2.2. Characterization of Copper Films and Cu-Al_2O_3 MMC Films

Microstructure of Free Cu and Cu-Al_2O_3 Composite Films-FE/SEM

Figure 6 shows the surface morphologies of the alumina-free copper films (Figure 6a,c,e) electrodeposited from ABSEs and Cu-Al_2O_3 MMC films electrodeposited from ABSE-Al-1% electrolytes (Figure 6b,d,f) with different film thicknesses (2, 22, and 52 μm). The electrodeposition mode was DC with a current density of 50 mA·cm^{-2}. Figure 6 also shows the corresponding grain size distribution histograms generated from image analysis of the FE-SEM micrographs.

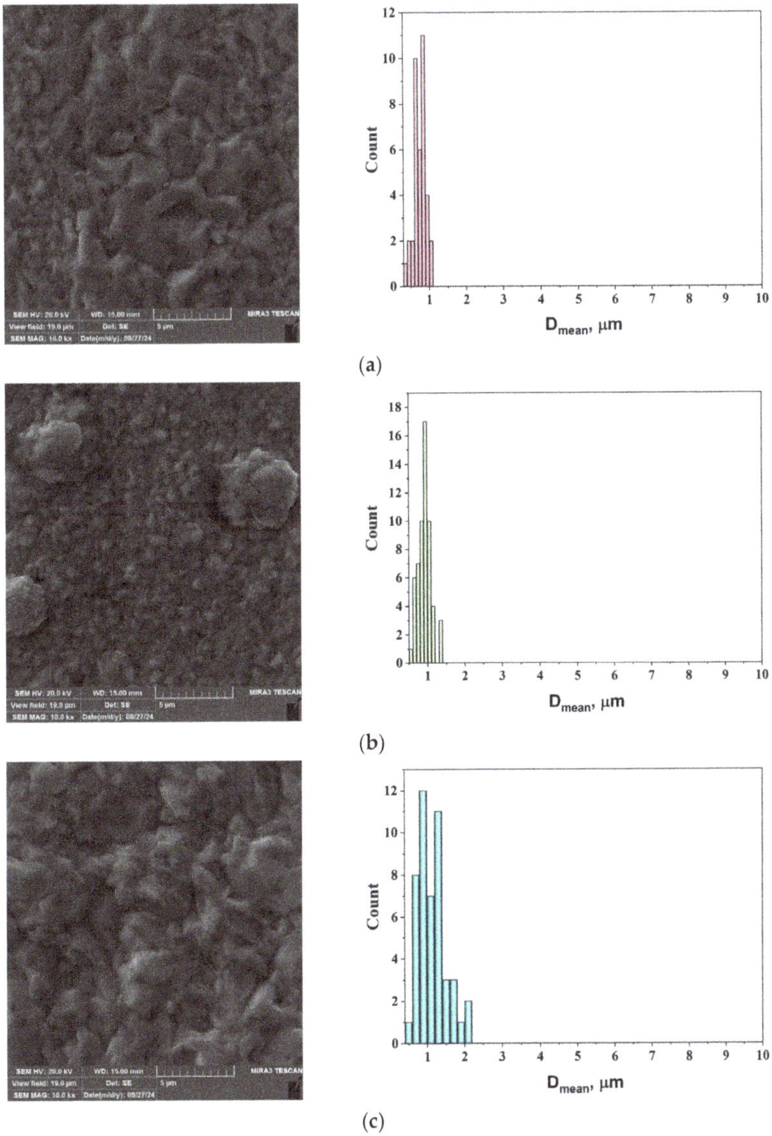

Figure 6. *Cont.*

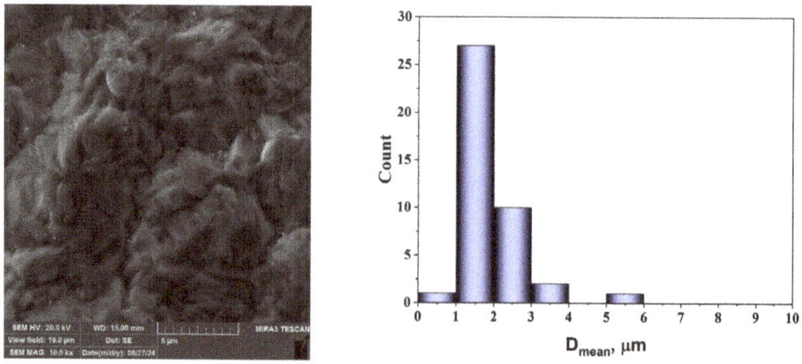

(d)

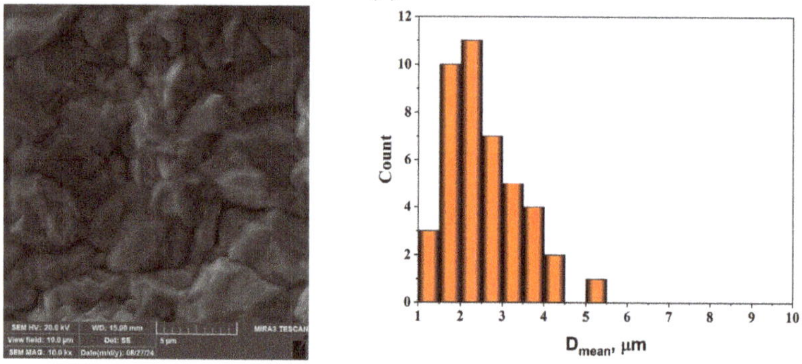

(e)

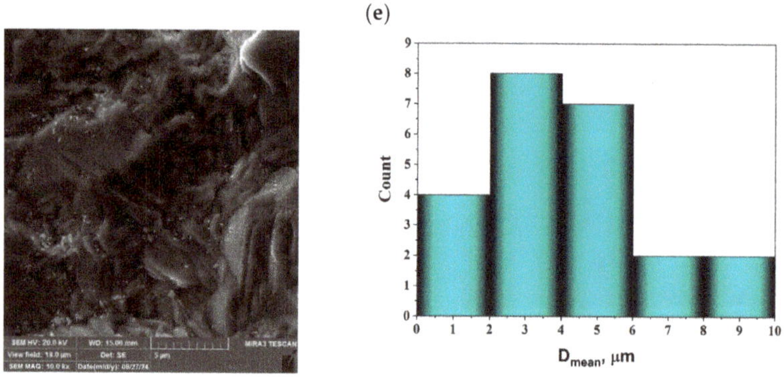

(f)

Figure 6. Surface morphology analyses of electrodeposited alumina-free Cu and MMC Cu-Al_2O_3 films electrodeposited/co-electrodeposited on the brass substrate with/without constant wt.% of alumina particles (1.0 wt.%) and with different thicknesses of the films: (**a**) 2 µm, Cu; (**b**) 2 µm, Cu-Al_2O_3; (**c**) 22 µm, Cu; (**d**) 22 µm, Cu-Al_2O_3; (**e**) 52 µm, Cu; and (**f**) 52 µm, Cu-Al_2O_3. The magnification was ×10,000 for all pictures.

The presented morphologies of all films are very coarse. The grain size and grain distribution are different for the Cu-Al_2O_3 and the pure Cu films. For the same thickness of the films with/without incorporated alumina particles, no discernible morphological difference in grain type was observed. The presence of relatively large crystals was observed in the DC mode in both cases, without/with alumina. This behaviour is expected since there is no change in the electrochemical deposition parameters. The only change in the microstructure was observed in the thinnest sample, which can be attributed to the inhibitory effect of alumina on the growth of copper crystals (Figure 6b). The structural change in co-electrodeposited metallic films is described by the electro-crystallization process, in particular, the combined process of nucleation and crystal growth [29]. Each Cu film is microcrystalline (mc) and compact. The presence of nano-alumina particles on the Cu surface is only observed in thicker films (Figure 6d, f). Bright, spherical, and small alumina nanoparticles were co-deposited on the surface, mostly at the grain boundaries. The mean grain size (D_{mean}) for each film produced was determined based on grain size distribution histograms, and Table 2 shows the clear grain boundary.

Table 2. The thickness and average grain size of Cu and Cu-Al_2O_3 films obtained from FE-SEM images.

Sample No.	Electrolyte	Experimental Thickness of Films, δ (μm)	Grain Size, D_{mean} (μm)
1.	ABSE	2	0.752
2.	ABSE-Al-1%	2	0.920
3.	ABSE	22	1.123
4.	ABSE-Al-1%	22	1.961
5.	ABSE	52	2.534
6.	ABSE-Al-1%	52	4.094
7.	ABSE-Al-3%	22	4.159
8.	ABSE-Al-5%	22	4.466

A 2 μm thick Cu film containing alumina nanoparticles has the smallest grain size (0.752 μm). As the deposition time increased, the grain size also increased, reaching a maximum film thickness of 4.094 μm for the 52 μm thick Cu-Al_2O_3 films (Table 2) when compared to samples co-electrodeposited from electrolyte ABSE-Al-1%. During the co-deposition of Cu and alumina particles, the increased concentration of alumina in the electrolyte interferes with the metal nucleation process by increasing the surface energy [30]. This leads to the growth of larger crystals and the appearance of a coarser deposit. The grain size of the Cu-Al_2O_3 film co-electrodeposited from electrolyte ABSE-Al-5% was 4.466 μm. This confirms that the dominant process is crystallization and that crystalline grain growth is favoured. The morphologies of the MMC films obtained from electrolytes ABSE-Al-3% and ABSE-Al-5% with the corresponding grain size distribution histograms are shown in the Supplementary Materials (Figure S1a,b).

Chemical Analysis of Metal Cu and MMC Films of Cu-Al_2O_3-EDS

Figure 7 shows the mapping analysis of alumina-free Cu, (Figure 7a) and Cu-Al_2O_3 films obtained at the same deposition times with 5.0 wt.% amounts of the alumina particles (Figure 7b).

Clear copper with dominant spectrum peaks was obtained from ABSEs. The atomic percentage of copper was 97.32%. The native oxide was present at 2.68% (Figure 7a). The surface area of the Cu-Al_2O_3 deposits shows the presence of elements from the alumina particles such as Al (0.32%) and O (1.35%) besides copper (98.32%) (Figure 7b). This suggests that the alumina particles were integrated or embedded in the copper film. To explain their integration into the deposit, the co-deposition of tiny inert particles floating in an electrolyte has generally been explained by the trapping process [31–33]. The EDS analysis of the films (thickness is 22 μm) with 1.0 wt.% and 3 wt.% alumina particles is shown in Figure S2a,b in the Supplementary Materials, respectively.

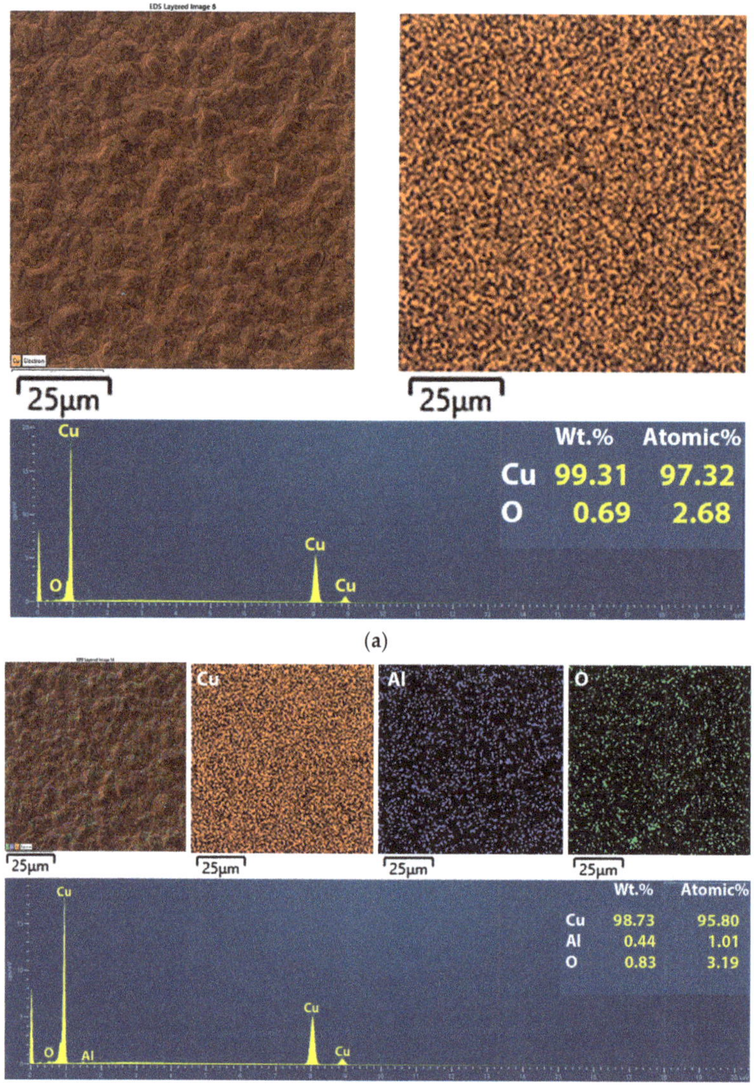

Figure 7. Element mapping of films: (**a**) alumina-free Cu film and (**b**) 5.0 wt.% of alumina particles. Film thickness was 22 µm.

The EDS results of the elemental compositions for Cu-Al_2O_3 films co-electrodeposited with varying concentrations of alumina powder in the ABSE for constant film thickness (22 µm) are given in atomic percentages in Table 3.

The amount of particles incorporated in the MMC films was determined by the wet or counting method [13], but the EDS method was suitable for evaluating the elemental amount on the film surface. As the concentration of alumina in the electrolyte increases, an increase in the concentration of Al on the surface of the Cu film is observed [30]. The oxygen content also increases, but there is also the possibility of a lower affinity of copper to form the native oxide of copper films with the incorporation of alumina particles.

Table 3. The EDS results (atomic %) for the Cu-Al_2O_3 films co-electrodeposited with constant thickness (22 μm) and varying concentrations of alumina powder in the electrolyte.

Sample No.	Electrolyte	c (wt.%)	Elements		
			Al	O	Cu
2.	ABSE	0	0	2.68	97.32
4.	ABSE-Al-1%	1.0	0.35	0.68	98.97
7.	ABSE-Al-3%	3.0	0.54	1.52	97.95
8.	ABSE-Al-5%	5.0	1.01	3.19	95.80

Figure 8 shows the elemental mapping analysis of the Cu-Al_2O_3 films co-electrodeposited with a constant concentration of alumina powder (1.0 wt.%) in the electrolyte for minimum and maximum values of film thickness. The mapping and spectra are shown for a thickness of 2 μm Cu-Al_2O_3 (Figure 8a) and 52 μm Cu-Al_2O_3 (Figure 8b). The content of each element is given in atomic percent in Table 4.

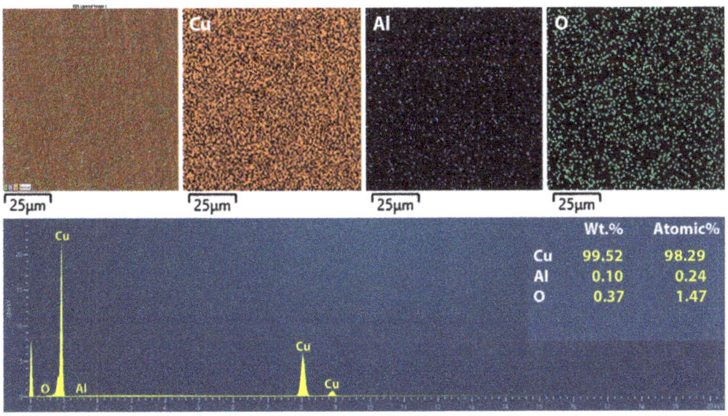

(a)

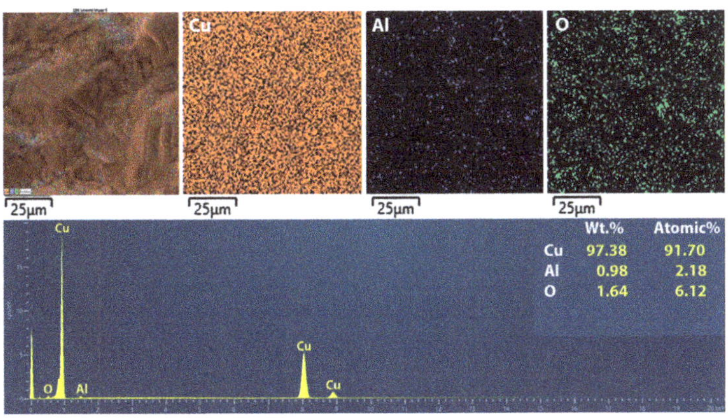

(b)

Figure 8. Element mapping of the Cu-Al_2O_3 films co-electrodeposited with a constant concentration of alumina powder (1.0 wt.%) in electrolytes with film thickness variation: (**a**) 2 μm and (**b**) 52 μm alumina powder.

Table 4. The EDS results for the $Cu-Al_2O_3$ films co-electrodeposited with a constant concentration of alumina powder (1.0 wt.%) and thickness variation.

Sample No.	Electrolyte	Experimental Thickness of Films, δ (μm)	Elements		
			Al	O	Cu
2.	ABSE-Al-1%	2	0.24	1.47	98.29
4.	ABSE-Al-1%	22	0.35	0.68	98.97
7.	ABSE-Al-1%	52	2.18	6.12	91.70

From the values shown in Table 4, it can be concluded that as the thickness of the film increases, the percentage of alumina in the film increases, reducing the matrix element (copper). The nano-alumina powder has a strong tendency to aggregate in the electrolyte due to its extremely high surface energy and small primary size, resulting in nanoparticle aggregates that are easily embedded in the composites, despite the continuous stirring of the electrolyte before and during the deposition process [34]. This effect was enhanced and intensified by increasing the current density and particle concentration in the electrolyte [30]. The variations in the content of nano-alumina particles in the electrolyte also had a considerable effect on the thickness growth [35].

2.2.3. Roughness Analyses of Free Cu and $Cu-Al_2O_3$ MMC Films—AFM Analyses

Figure 9 represents the 3D (three-dimensional) AFM images for the following films: the alumina-free Cu films with three different thicknesses (2, 22, and 52 μm) (Figure 9a,c,e) and the $Cu-Al_2O_3$ films with a variation of the same thickness: 2 μm (Figure 9b), 22 μm (Figure 9d), and 52 μm (Figure 9f). The variation in alumina concentration for $Cu-Al_2O_3$ films is shown in Figure 9g (for 3.0 wt.%) and in Figure 9h (for 5.0 wt.%). A light shade of purple corresponds to a maximum of grain peak, while a dark shade represent the base between grains.

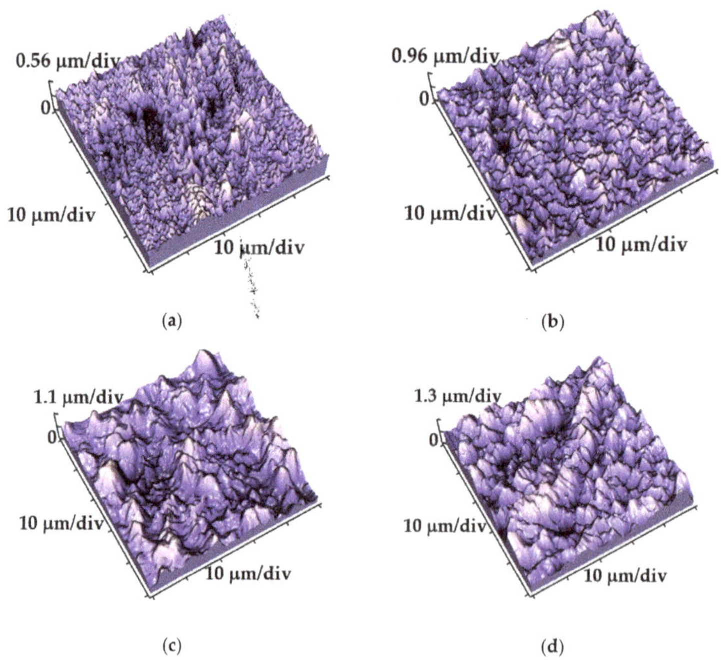

Figure 9. *Cont.*

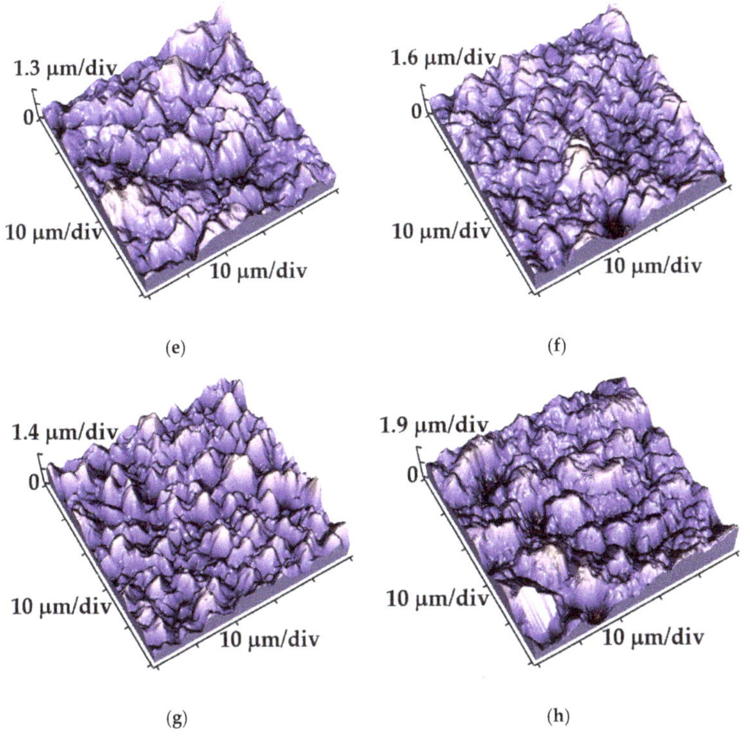

Figure 9. The 3D AFM images of all deposition Cu and Cu-Al_2O_3 films: (**a**) 2 μm Cu; (**b**) 2 μm Cu-Al_2O_3—1 wt.%; (**c**) 22 μm Cu; (**d**) 22 μm Cu-Al_2O_3—1 wt.%; (**e**) 52 μm Cu; (**f**) 52 μm Cu-Al_2O_3—1 wt.%; (**g**) 2 μm Cu-Al_2O_3—3%; and (**h**) 22 μm Cu-Al_2O_3—5 wt.%. The scan size was 50 × 50 $μm^2$.

The surface arithmetic means of the absolute roughness (S_a) values, and the root mean square roughness parameter (S_q) obtained using AFM software (SPMLab NT Ver. 6.0.2.) for the Cu films produced without/with alumina nanoparticles are given in Table 5.

Table 5. The values of roughness parameter S_a obtained by application of the AFM software from a 50 × 50 $μm^2$ scan area for the free-Cu and Cu-Al_2O_3 films with variations in the thicknesses of the films and concentrations of alumina powder in electrolytes.

δ (μm)	**2**	**2**	**22**	**22**	**52**	**52**	**22**	**22**
Electrolyte	ABSE	ABSE-Al-1%	ABSE	ABSE-Al-1%	ABSE	ABSE-Al-1%	ABSE-Al-3%	ABSE-Al-5%
S_a (nm)	96.5	312.2	153.8	354.7	366.6	584.7	420.7	849.0
S_q (nm)	126.7	387.1	195.8	436.7	452.4	695.8	519.6	1066.0

The increase in roughness of the Cu and Cu-Al_2O_3 films is a result of several parameters: the roughening of the brass surface with chemical etching, the cathode surface with increasing electrodeposition time, the incorporation of alumina nanoparticles in the films, and the change in the incorporation process with changing the concentration of alumina particles in the suspension [36,37]. The maximum roughness (1066 nm) has sample co-electrodeposited from electrolyte ABSE-Al-5% for maximum thickness. The minimum roughness (96.5 nm) is for the sample without alumina with the 2 μm thick Cu film. The general conclusion is that rougher films are obtained with increasing deposition time and alumina concentration in the electrolyte. When comparing films of the same thickness with and without alumina particles, all composite films were found to be rougher than the pure Cu films, which is not always expected with nanoscale reinforcements.

2.2.4. Microhardness Analyses of Free Cu and Cu-Al_2O_3 MMC Films—Vickers Indentation Method

Figure 10a, b present composite hardness data vs. indentation depth, h, together with fitting curves for all films (without/with alumina) and for three thicknesses (2, 22, and 52 μm), Figure 10a, and for all concentrations of alumina particles in the suspension, Figure 10b. The theoretical part related to the understanding of the composite hardness model is given in the Supplementary Materials. Figure 10c,d show histograms of the calculated and absolute microhardness, corrected for the influence of the brass substrate, for both cases of thickness and concentration variations. The phenomenological differences between the term's microhardness, composite hardness, absolute hardness of the film, and absolute hardness of the substrate are defined in the Supplementary Materials.

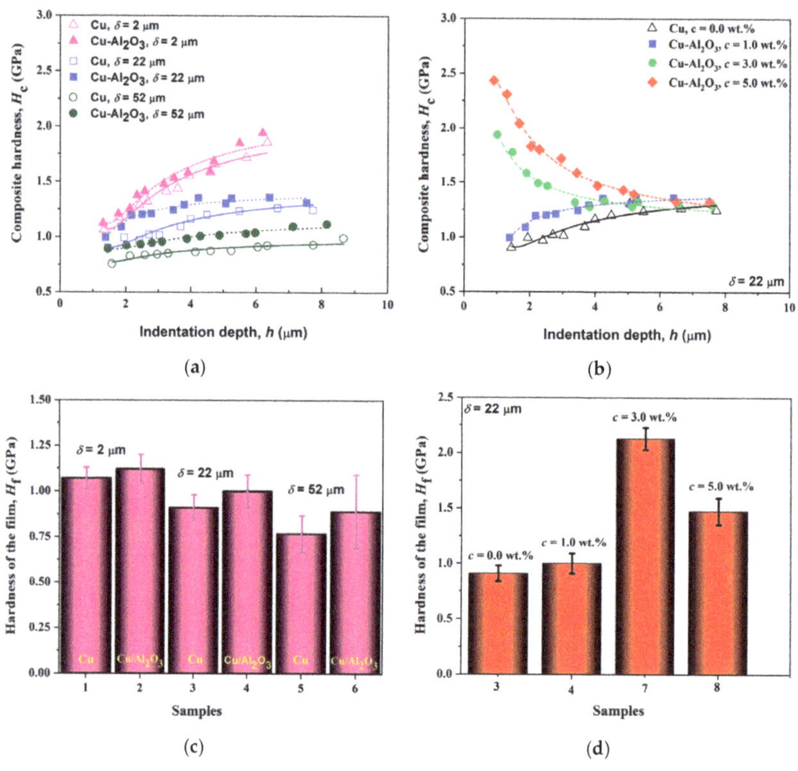

Figure 10. The microhardness analysis of Cu (normal line) and Cu-Al_2O_3 (dash line) MMC films applied to the C-G model: (**a**) with variations in the thickness of the films, (**b**) variations in the concentration of alumina particles in the suspension, and histograms with calculated values of an absolute hardness of the films (**c**) with different thicknesses, and (**d**) with 22 μm thickness and different concentrations of alumina particles.

For this purpose, the Chen–Gao composite hardness model was selected and applied [38]. The maximum hardness has the thinnest film reinforced with alumina nanoparticles. As the film thickness increases, the microhardness decreases. This can be explained by the transition from thin films to bulk material, which leads to an increase in the porosity and roughness of the film for free copper films. On the other hand, increasing the amount of alumina particles in the composite films seems to decrease film porosity [35]. However, there is also an increase in the standard error of measurement with increasing thickness, which can be explained by the appearance of agglomerates and the non-uniform distribution of the alumina particles on the surface of the film [39]. All Cu films and Cu-Al_2O_3 composite films with 1.0 wt.% alumina particles belong to the "soft film on hard substrate" composite system type, and all curves have a rising character in Figure 10a. The situation changes when it comes to Cu-Al_2O_3 composite films synthesized with 3.0 and 5.0 wt.% of alumina particles in the electrolyte. There is a change in the type of composite system (red and green lines in Figure 10b). The addition of Al_2O_3 nanoparticles during co-electrodeposition synergistically increased the hardness by about 4.73% for 2 μm thick films, 9.96% for 22 μm thick films, and 15.9% for 52 μm thick films, compared to pure Cu and Cu-Al_2O_3 films (Figure 10c).

The increasing Al_2O_3 nanoparticle concentrations during co-electrodeposition increased the hardness by about 9.96% for 1 wt.%, 134.1% for 3.0 wt.%, and 61.9% for 5.0 wt.%, compared to pure Cu films (Figure 10d). The consequences of this change can be explained through the change in film adhesion and the dispersion hardening effect. Thus, the higher density of grain boundaries in the nanocrystalline materials, which slows down the dislocation motion, can be linked to the hardening. The relatively soft copper matrix's plastic flow will be inhibited by the nanoparticles. However, an increase in hardness does not always correlate with grain size reduction or an increase in the content of incorporated particles within the film. Several additional factors play crucial roles, including surface roughness, grain shape, grain distribution, film thickness, porosity, the presence or absence of embedded agglomerates of particles, and film–substrate interfacial adhesion. Alumina, a brittle ceramic particle, especially in nanoscale form, tends to agglomerate within the Cu matrix due to its large specific surface area and the influence of van der Waals forces [40]. The second issue is the poor wettability between the alumina reinforcement phase and the copper matrix, leading to weak interfacial bonding. However, the presence of oxides on the surface of the composite films, derived from both alumina particles and natural copper oxide, can effectively enhance hardness while preserving plasticity.

A comparative study of the change in the microhardness of the copper matrix with the incorporation of alumina particles is shown in Table 6, where data are compared to other similar studies. For pure copper, the film had a microhardness of 80 HV (0.78 GPa), while composite films had a microhardness of 310 HV (3.04 GPa) [41]. According to published research, the hardness of Cu films from copper sulphate baths ranges from 50 to 105 HV (0.49–1.47 GPa) [42], depending on the electrolyte composition and direct current (DC) plating parameters. Values of nanohardness for pure Cu films, compared to DC plating and PC plating, have produced copper deposits that are five times larger [43]. The correlation between the hardness of metal films and composite films and their microstructure in addition to the amount and dispersion of the reinforcing stage were given in our previous publications [16,44].

Table 6. Microhardness data of Cu and MMC Cu-Al_2O_3 films: ED—electrodeposition; *c*—concentration; CED—co-electrodeposition; DC—direct current mode; PC—pulse current mode; HV—Vickers hardness.

System Film/Substrate	*c*	Alumina Type	Method of Synthesis	Parameter of Synthesis	Method of Characterization	Hardness	Ref.
Cu/Cu	0	/	ED/DC	11 A/dm^2, 20 min.	HV	81	[30]
Cu/Al_2O_3/Cu	30 g/L	Commercial/287 nm	CED/DC	11 A/dm^2, 20 min.	HV	247	[30]
Cu/Cu	0	/	ED/DC	10 mA/cm^2, 10 min.	HV	85	[35]

Table 6. *Cont.*

System Film/Substrate	*c*	Alumina Type	Method of Synthesis	Parameter of Synthesis	Method of Characterization	Hardness	Ref.
$Cu/Al_2O_3/Cu$	15.2 wt.%	Commercial/50 nm	CED/PC	$10\ mA/cm^2$	HV	149	[35]
$Cu/Al_2O_3/Cu$	17.2 wt.%	Commercial/50 nm	CED/PC	$10\ mA/cm^2$	HV	150	[35]
$Cu/Al_2O_3/Cu$	17.8 wt.%	Commercial/50 nm	CED/PC	$10\ mA/cm^2$	HV	156	[35]
Cu/Si	0	/	ED/DC	$200\ A/m^2$	Nanoindentation/GPa	1.29 ± 0.18	[39]
$Cu/Al_2O_3/Si$	5 g/l	Commercial/50 nm	CED/DC	$200\ A/m^2$	Nanoindentation/GPa	1.71 ± 0.14	[39]
Cu/Cu	0	/	ED/DC	$10\ mA/cm^2$	HV	100.4	[41]
Cu/Cu	0	/	ED/DC	$75\ mA/cm^2$	HV	138.2	[41]
$Cu/Al_2O_3/Cu$	8.2 wt.%	50 nm γ-Al_2O_3	CED/DC	$10\ mA/cm^2$	HV	182.9	[41]
$Cu/Al_2O_3/Cu$	3.2 wt.%	50 nm γ-Al_2O_3	CED/DC	$75\ mA/cm^2$	HV	241.7	[41]
Cu/Cu	0	/	ED/DC	$4 \pm 0.1\ A/dm^2$	Rockwell	≈65	[45]
$Cu/Al_2O_3/Cu$	50 g/l	Commercial/50 nm	CED/DC	$4 \pm 0.1\ A/dm^2$	Rockwell	≈50	[45]
Cu/Al_2O_3	1.0 wt.% 3.0 wt.% 5.0 wt.%	Synthesis from sol–gel	CED/DC	$50\ mA{\cdot}cm^{-2}$	HV	101.9 217.0 150.1	This work

2.2.5. Adhesion Analyses of Free Cu and Cu-Al_2O_3 MMC Films—Vickers Indentation Method

The adhesion of thin films is a very important feature that often determines the physical existence of a film on a substrate. Films that do not have good adhesion are delaminated from the substrate, and such films have no use value when it comes to microelectronic applications and electrical packaging. Testing the adhesion properties of the film can be achieved through the Vickers microindentation test, with theoretical approximations and the application of mathematical models. The basic theoretical equations of the model and the meanings of the variables are given in the Supplementary Materials.

Figure 11 shows the dependency of hardness difference, ΔH, on a ratio, δ/d, for the Cu films made from the electrolyte ABSE, and for the composite films Cu/Al_2O_3 made from electrolyte ABSE-Al-1%. A comparison of an adhesion parameter, *b*, for all three thicknesses, was also given. Table 7 provides the calculated values for the adhesion parameter, *b*, which was determined based on the slope of the linear dependencies depicted in Figure 11. The greater the adhesion parameter *b* value, the more the films adhere to the substrate [18].

Table 7. Fitting results of adhesion parameter *b* for the alumina-free Cu film and those produced with different film thicknesses and concentrations in the electrolyte.

Concentration of Alumina, *c*/wt.%	Thickness, δ/μm	Slope (*k*)	Intercept (*n*)	*b*	R^2
0	2	2.74958 ± 0.47254	-0.2128 ± 0.06956	1.4344	0.86793
1.0	2	2.6125 ± 0.29533	-0.12388 ± 0.0425	1.7210	0.93920
0	22	0.20341 ± 0.01504	0.02246 ± 0.0183	25.872	0.94787
1.0	22	0.23341 ± 0.0388	0.13497 ± 0.0444	26.774	0.77871
0	52	0.0495 ± 0.00782	0.37686 ± 0.0197	130.56	0.78104
1.0	52	0.04716 ± 0.00644	0.50972 ± 0.0152	165.25	0.82692
3.0	22	-0.48371 ± 0.0421	0.43142 ± 0.0532	15.330	0.98821
5.0	22	-0.21415 ± 0.0537	0.38714 ± 0.0572	4.6713	0.84983

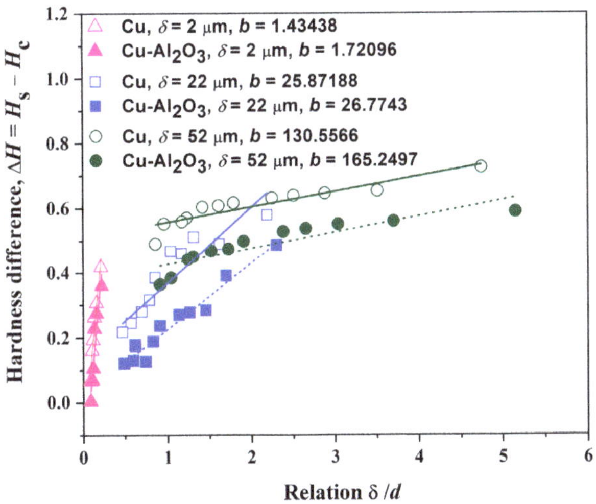

Figure 11. Adhesion properties of the films without/with reinforced alumina. Values of the adhesion parameter, called "critical reduced depth" for Cu (normal line) and Cu-Al_2O_3 (dashed line) are also given.

Table 7 illustrates that Cu-Al_2O_3 films formed from electrolyte ABSE-Al-1% demonstrated superior adherence compared to those obtained without alumina particles (electrolyte ABSE). Concurrently, the Cu films and Cu-Al_2O_3 films generated at a shorter deposition time (thinner films) from both electrolytes had the poorest adherence with the cathode surface. The value of composite microhardness data is influenced by the adhesion of the film to the cathode: stronger adhesion corresponds to higher hardness of the composite [46]. For harder films, the deformation zone at the interface of film/substrate is more extended, and critical reduced depth has a higher value for the "soft film on a hard substrate" type. This rule applies to the minimum concentration of alumina (1.0 wt.%). However, with Cu-Al_2O_3 composite films co-electrodeposited from electrolytes with 3.0 and 5.0 wt.% alumina particles (electrolyte ABSE-Al-3% and ABSE-Al-5%), there is a change in the type of composite system. By incorporating alumina particles, the composite film becomes harder than the brass substrate. An obvious decrease in the adhesion parameter was noted with an increase in the concentration of alumina in the electrolyte. Based on Lawn's law and the "spherical field stress model" [47], the contribution of the elastic component for both materials (films and substrate), expressed through Jung's modulus of elasticity E, as well as the contribution of the plastic component for both materials, expressed through hardness H, must be considered. Ceramic particles generally reduce the elasticity of metallic films [48], so the plastic component increases. However, ceramic particles with a high specific elastic modulus can effectively increase Young's modulus of the Cu composites [49]. The radius of the deformation zone under the indenter increases, so, critical reduced depth (b) decreases, too. Precisely because of this, there is a decrease in adhesion at a higher alumina concentration in the film, although the hardness of the film shows a higher value than the copper matrix.

2.2.6. Wettability Analyses of Free Cu and MMC Cu-Al_2O_3 Films—Sessile Drop Method

It is known that surface roughness directly affects the surface's wettability [19,50]. Measuring the static contact angles using different liquids is useful to determine the films' and composite films' wettability and affinity to polar and non-polar components. A sessile drop technique has been used to measure the water contact angle. The water contact angle (θ_{WCA}) of the pure Cu films and Cu-Al_2O_3 films at various thicknesses of the films were measured, and the results are shown in Figure 12a.

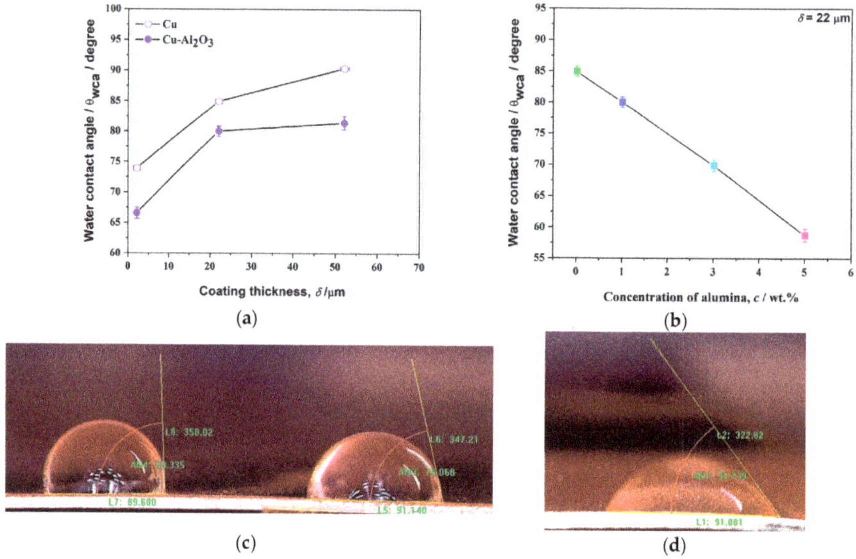

Figure 12. The values of static contact angles measured on the surface of Cu and Cu-Al_2O_3 composite films: (**a**) with varying thicknesses and 1 wt.% of alumina particles, (**b**) with variations in the concentration of alumina particles in ABSEs, (**c**) with the appearance of a water droplet on a particle-free copper film, and (**d**) with 5.0 wt.% of alumina particles.

The relationship between the concentration of alumina particles in the electrolyte and the impact on the water contact angle value of Cu and Cu-Al_2O_3 composite films is given in Figure 12b. Figure 12c,d show the method of measuring water contact angles on the surface of copper and MMC Cu-Al_2O_3 films with the maximum content of alumina particles.

The copper films without alumina particles electrodeposited on the brass substrate were hydrophilic and had a water contact angle value of less than 90° (θ_{WCA} = 84.94 ± 0.83). A more hydrophilic characteristic is exhibited by Cu-Al_2O_3 composite films that are produced by the co-deposition method. On the surface area of MMC Cu-Al_2O_3 films, the measured water contact angles ranged from 66.57 ± 0.94° to 81.42 ± 1.10° (Figure 12a). As demonstrated in our instance, rougher films, i.e., thicker films, are generally more hydrophobic than smooth and thick films [50]. Therefore, a change in the wettability of films results from an increase in film roughness with thickness variation. The lower contact angle of the Cu-Al_2O_3 films in comparison to the free Cu films may be due to the particle's incorporation on the surface of films and their relatively polar nature. On the other hand, it is known that ceramic particles increase the porosity of the composite [51], so this is an additional reason for better absorption of water. By altering the quantity of alumina particles in the electrolyte, the Cu-Al_2O_3 films' water contact angle changed (see Figure 12b). Although an increase in the roughness of composite films with the incorporation of alumina particles was recorded, here, in addition to the roughness, a more dominant factor on wetting is observed, namely the presence of polar alumina particles on the surface. With the increase in the concentration of alumina in the electrolyte to 5.0 wt.%, the value of the water contact angle drops from 84.94 ± 0.83° to 58.78 ± 1.03°. Numerous polar sites from unsaturated oxygen and aluminum atoms, which function as potent Lewis's acid and base sites, respectively, are present in alumina [52]. These surface aluminum atoms quickly create hydrophilic hydration structures by hydrogen bonding with interfacial water molecules [53]. Also, surface polarity is significantly increased by the substantial presence of unsaturated ions on the surface [54,55]. Hence, increased surface wettability is greatly influenced by the oxidation processes of copper films and the presence of oxide ceramic particles incorporated in the Cu matrix. Compared to our previous research on the values of the water contact angle, for the MMC Cu film reinforced with micro-sized pigment particles based on strontium aluminate, the water contact angles are very close to each other, in the range of 106.0–110.8°, but about 60% larger than for the pure Cu coating [16]. In this case, we had a transition from a hydrophilic free-copper film to a hydrophobic one. Identical wetting behaviour is shown for reinforced copper films with Fe/Al LDO (ferrite–aluminium-layered double oxide) nanoparticles [44]. The wettability of the Cu films changed from hydrophilic (for pure Cu) to hydrophobic (for Cu-Fe/Al LDO), and this change was dominated by an increase in the roughness parameter of the films.

In this case, the surface contact angles depend on the microstructure, roughness, presence of particles, type and size of particles, and film thickness. It is found that surface wettability increases as the roughness, thickness, and alumina concentration increases. As the parameters increase, the nature of the films becomes more hydrophilic. The dominant factor in this behaviour is the nature of the synthesized alumina particles and their presence on the top surface of the copper films [56].

3. Conclusions

Alumina particles were obtained from the sol–gel process from inorganic precursors. The particles had a corundum structure obtained at 1000 °C and a submicron size. Individual particles were small but tended to make agglomerates that could influence the process when longer deposition times and thicker films were made. Alumina particles were incorporated in copper films, and for thinner films, they increased the hardness of films, and the hardness for thicker films was decreased. From the point of view of hardness improvement, the optimal value for films is 22 μm. The incorporation of alumina also depends on the concentration of powder particles in the electrolyte, and the concentration of alumina particles in the film is dependent on this parameter. Adhesion of the film on the substrate depends both on the film thickness and alumina content. For the same film thickness, a pure copper film has a lower adhesion value compared to the composite film. The best adhesion for MMC films was obtained for the optimal alumina concentration in ABSEs, which is 1.0 wt.%. Roughness increases both with the film thickness and the content of alumina particles incorporated in the film. The grain size of copper films has the same trend as the roughness, showing an increasing character with increasing film thickness and with the incorporation of alumina particles. The wettability of the material is better when alumina particles are incorporated into the film, and this can be useful for some potential applications of composite films on brass substrates.

This sort of composite film coating, with improved hardness and wettability, is very suitable for microfluidic devices used for the trybocatalytic removal of pollutants in corrosive media. The industrial application of this type of composite film could also be useful in the construction of microheaters, which are used for the production of reactors, where it is necessary for the material to be resistant to corrosion and have improved resistance to the environment and at the same time assure the improved heat transfer conditions due to improved surface characteristics.

4. Materials and Methods

4.1. Materials and Method—Synthesis of Al_2O_3 Nanoparticles

The Clariant firm provided aluminum hydroxide chloride (Locron L; $Al_2Cl(OH)_5 \times 2.5\,H_2O$) in its crystallized form for purchase as an alumina precursor. Aluminum oxide Al_2O_3 nanoparticles were synthesized using the sol–gel technique. $Al_2Cl(OH)_5 \times 2.5\,H_2O$ was dissolved in demineralized water with a magnetic stirrer. They formed a sol that was transformed into a gel. The gel was heat treated at 1000 °C to obtain a suitable crystal structure for matrix reinforcement and to obtain fine small particles.

4.2. Materials and Method—Synthesis of Cu Films

Cold-rolled brass foil (ASTM B36 (70% Zn and 30% Cu)), K&S Engineering, Chicago, IL, USA), 250 μm thick, was used as a cathode, and copper foil, 1 mm thick (Alfa Aesar ThermoFisher GmbH Erlenbachweg, Kandel, Germany), was used as an anode.

The solution of a mixture of three acids (HNO_3:H_3PO_4:CH_3COOH = 4:11:5 vol.% at 10 s) was used for chemical polishing after mechanical polishing with SiC paper (#2000) for the preparation cathode. A solution of nitric acid (Merck KGaA, Darmstadt, Germany) in HNO_3:H_2O = 1:1 vol.% was used for the preparation anode. Copper (II)-sulphate pentahydrate, $CuSO_4{\cdot}5H_2O$, (Merck KGaA, Darmstadt, Germany), and sulphuric acid (98%) of p.a. quality were ordered from Zorka, Šabac, Serbia. The high-purity water, 18 MΩ·cm (Milipore, Burlington, MA, USA), was used for the preparation of an acidic basic sulphate electrolyte (ABSE). The cathodes with a total area of 5.0×1.0 cm^2, and an effective deposition surface area of 2.0×1.0 cm^2, were put in the middle of the electrolytic cell. The magnetic stirrer was used for the mixing of the electrolyte at 200 rpm. An ABSE composition of 240 g/L $CuSO_4{\cdot}5\,H_2O$ and 60 g/L H_2SO_4 was put in the 100 mL Pyrex glass. The pH values and temperature were 0.33 and 20 ± 0.5 °C, respectively. The constant current regime (galvanostatically) was turned off on the Keithley 2200 DC Power Supply Programmable device (Tektronix UK Ltd., Bracknell, UK). Based on Faraday's law, the deposition times were calculated to obtain the projected film's thicknesses of 2, 22, and 52 μm.

4.3. Materials and Method—Synthesis of Cu-Al_2O_3 Composite Films

A co-electrodeposition method was performed by the addition of Al_2O_3 nanoparticles (synthesized and described in Section 4.1) into the ABSE, and these electrolytes were noted as ABSE-Al electrolytes. Three concentrations of alumina particles in ABSEs were used (1.0, 3.0, and 5.0 wt.%) and electrolytes were marked as ABSE-Al-1%, ABSE-Al-3%, and ABSE-Al-5%. These electrolytes were stirred for 120 min prior to the co-electrodeposition process (CED). The current density value was identical for both conditions, without and with added nanoparticles. After achieving better dispersion and wetting of particles in the electrolyte, i.e., prevention of agglomerates, the mixing intensity was reduced to 200 rpm.

4.4. Characterization Methods

All experimental and measurement set data are given in the Supplementary Materials.

Supplementary Materials: The following supporting information can be downloaded at https://www.mdpi.com/article/10.3390/gels10100648/s1: Figure S1: Surface morphology and histogram analyses of electrodeposited MMC Cu-Al_2O_3 films co-electrodeposited on the brass substrate with a constant thickness of the films (22 μm) and with different concentrations of alumina particles in an ABSE: (a) 3.0 wt.% and (b) 5.0 wt.%. The magnification was × 10,000 for all pictures. The co-electrodeposition mode was DC with a current density of 50 mA·cm^{-2}; Figure S2: Element mapping of the MMC Cu-Al_2O_3 films co-electrodeposited on the brass substrate with a constant thickness of the films (22 μm) and various concentrations of alumina particles in an ABSE: (a) 1.0 wt.% and (b) 3.0 wt.%. The film thickness was 22 μm. Reference [38] is in the Supplementary Materials as number [12].

Author Contributions: Conceptualization, I.O.M., R.M.J.H. and M.M.V.; methodology, I.O.M. and S.S.M.M.; software, D.G.V.-R., I.O.M. and M.M.V.; validation, R.M.J.H., M.M.V. and A.D.M.; formal analysis, L.J.M. and S.S.M.M.; investigation, S.S.M.M. and I.O.M.; resources, R.M.J.H., A.D.M., M.M.V. and I.O.M.; data curation, S.S.M.M., L.J.M. and I.O.M.; writing—original draft preparation, S.S.M.M. and I.O.M.; writing—review and editing, all authors; visualization, M.M.V. and D.G.V.-R.; supervision, R.M.J.H. and A.D.M.; project administration, R.M.J.H., A.D.M., M.M.V. and I.O.M.; funding acquisition, R.M.J.H., M.M.V., A.D.M. and I.O.M. All authors have read and agreed to the published version of the manuscript.

Funding: This research was funded by the Ministry of Science, Technological Development, and Innovation of the Republic of Serbia (Contract Nos. 451-03-65/2024-03/200135, 451-03-66/2024-03/200026, and 451-03-66/2024-03/200017).

Institutional Review Board Statement: Not applicable.

Informed Consent Statement: Not applicable.

Data Availability Statement: The original contributions presented in the study are included in the article/supplementary material, further inquiries can be directed to the corresponding authors.

Acknowledgments: We acknowledge funding support from the Ministry of Science, Technological Development, and Innovation of the Republic of Serbia (Contract Nos. 451-03-65/2024-03/200135, 451-03-66/2024-03/200026, and 451-03-66/2024-03/200017).

Conflicts of Interest: The authors declare no conflicts of interest.

References

1. Kaunisto, K.; Lagerbom, J.; Honkanen, M.; Varis, T.; Lambai, A.; Mohanty, G.; Levänen, E.; Kivikytö-Reponen, P.; Frankberg, E. Evolution of Alumina Phase Structure in Thermal Plasma Processing. *Ceram. Int.* **2023**, *49*, 21346–21354. [CrossRef]
2. Abyzov, A.M. Aluminum Oxide and Alumina Ceramics (Review). Part 1. Properties of Al2O3 and Commercial Production of Dispersed Al2O3. *Refract. Ind. Ceram.* **2019**, *60*, 24–32. [CrossRef]
3. Boumaza, A.; Favaro, L.; Lédion, J.; Sattonnay, G.; Brubach, J.B.; Berthet, P.; Huntz, A.M.; Roy, P.; Tétot, R. Transition Alumina Phases Induced by Heat Treatment of Boehmite: An X-Ray Diffraction and Infrared Spectroscopy Study. *J. Solid State Chem.* **2009**, *182*, 1171–1176. [CrossRef]
4. Levin, I.; Brandon, D. Metastable Alumina Polymorphs: Crystal Structures and Transition Sequences. *J. Am. Ceram. Soc.* **1998**, *81*, 1995–2012. [CrossRef]
5. Drah, A.; Tomić, N.Z.; Kovačević, T.; Djokić, V.; Tomić, M.; Heinemann, R.J.; Marinković, A. Structurally and Surface-Modified Alumina Particles as a Reinforcement in Polyester-Based Composites with an Improved Toughness. *Mech. Compos. Mater.* **2020**, *56*, 249–260. [CrossRef]
6. Ashor, A.A.; Vuksanović, M.M.; Tomić, N.Z.; Petrović, M.; Dojčinović, M.; Husović, T.V.; Radojević, V.; Heinemann, R.J. Optimization of Modifier Deposition on the Alumina Surface to Enhance Mechanical Properties and Cavitation Resistance. *Polym. Bull.* **2019**, *77*, 3603–3620. [CrossRef]
7. Lazouzi, G.A.; Vuksanović, M.M.; Tomić, N.; Petrović, M.; Spasojević, P.; Radojević, V.; Jančić Heinemann, R. Dimethyl Itaconate Modified PMMA—Alumina Fillers Composites with Improved Mechanical Properties. *Polym. Compos.* **2019**, *40*, 1691–1701. [CrossRef]
8. Santos, J.S.; Márquez, V.; Buijnsters, J.G.; Praserthdam, S.; Praserthdam, P. Antimicrobial Properties Dependence on the Composition and Architecture of Copper-Alumina Coatings Prepared by Plasma Electrolytic Oxidation (PEO). *Appl. Surf. Sci.* **2023**, *607*, 155072. [CrossRef]
9. Mitra, D.; Kang, E.-T.; Neoh, K.G. Antimicrobial Copper-Based Materials and Coatings: Potential Multifaceted Biomedical Applications. *ACS Appl. Mater. Interfaces* **2020**, *12*, 21159–21182. [CrossRef]
10. Kumykov, V.K.; Sergeev, I.N.; Sozaev, V.A.; Gedgagova, M.V. Surface Tension of Copper in Solid Phase. *Bull. Russ. Acad. Sci. Phys.* **2017**, *81*, 357–359. [CrossRef]
11. Podlaha, E.J.; Landolt, D. Pulse-Reverse Plating of Nanocomposite Thin Films. *J. Electrochem. Soc.* **1997**, *144*, L200–L202. [CrossRef]
12. Thiemig, D.; Lange, R.; Bund, A. Influence of Pulse Plating Parameters on the Electrocodeposition of Matrix Metal Nanocomposites. *Electrochim. Acta* **2007**, *52*, 7362–7371. [CrossRef]
13. Stankovic, V.D.; Gojo, M. Electrodeposited Composite Coatings of Copper with Inert, Semiconductive and Conductive Particles. *Surf. Coatings Technol.* **1996**, *81*, 225–232. [CrossRef]
14. Wang, C.; Bai, L.; Xu, H.; Qin, S.; Li, Y.; Zhang, G. A Review of High-Temperature Aerogels: Composition, Mechanisms, and Properties. *Gels* **2024**, *10*, 286. [CrossRef]
15. Dascalu, I.; Hornoiu, C.; Calderon Moreno, J.M.; Osiceanu, P.; Somacescu, S. Layered Sol–Gel Deposition of a Sn, Ti, Zn, and Pr Mixed Oxide Thin Film with Electrical Properties for Gas Sensing. *Gels* **2023**, *9*, 638. [CrossRef]
16. Mladenović, I.O.; Vuksanović, M.M.; Dimitrijević, S.P.; Vasilić, R.; Radojević, V.J.; Vasiljević-Radović, D.G.; Nikolić, N.D. Mechanical Properties of Electrolytically Produced Copper Coatings Reinforced with Pigment Particles. *Metals* **2023**, *13*, 1979. [CrossRef]
17. Algellai, A.A.; Tomić, N.; Vuksanović, M.M.; Dojčinović, M.; Volkov-Husović, T.; Radojević, V.; Heinemann, R.J. Adhesion Testing of Composites Based on Bis-GMA/TEGDMA Monomers Reinforced with Alumina Based Fillers on Brass Substrate. *Compos. Part B Eng.* **2018**, *140*, 164–173. [CrossRef]
18. Mladenović, I.O.; Bošković, M.V.; Vuksanović, M.M.; Nikolić, N.D.; Lamovec, J.S.; Vasiljević-Radović, D.G.; Radojević, V.J. Structural, Mechanical and Electrical Characteristics of Copper Coatings Obtained by Various Electrodeposition Processes. *Electronics* **2022**, *11*, 443. [CrossRef]
19. Mogra, A.; Pandey, P.K.; Gupta, K.K.; Shivhare, S.; Bagal, V. Development and Characterization of Cu-Al2O3 Nanocomposite Coating Using Electrodeposition Process on Copper Substrate. *J. Inst. Eng. Ser. C* **2022**, *103*, 1103–1109. [CrossRef]
20. Souza, H.J.D.; D'Souza, N.; Ashith, V.K.; Nagappa Moger, S.; D'Silva, E.D. Effect of Deposition Time on Copper Incorporation of ZnS Thin Films by Low-Cost Technique. *Mater. Sci. Eng. B* **2023**, *294*, 116551. [CrossRef]

21. Mladenović, I.O.; Nikolić, N.D. Influence of Parameters and Regimes of the Electrodeposition on Hardness of Copper Coating *Metals* **2023**, *13*, 683. [CrossRef]
22. Alazreg, A.; Vuksanović, M.M.; Egelja, A.; Mladenović, I.O.; Radovanović, Ž; Petrović, M.; Marinković, A.; Jančić Heinemann, Mechanical properties of acrylate matrix reinforced with manganese-aluminum layered double hydroxide (MnAl-LDH). *Poly Compos.* **2023**, *44*, 6783–6792. Available online: https://Mediacy.Com/Image-Pro/ (accessed on 27 August 2024). [CrossRef]
23. Gafur, M.A.; Al-Amin, M.; Sarker, M.S.R.; Alam, M.Z. Structural and Mechanical Properties of Alumina-Zirconia (ZTA) Compo ites with Unstabilized Zirconia Modulation. *Mater. Sci. Appl.* **2021**, *12*, 542–560. [CrossRef]
24. Petrík, J.; Blaško, P.; Markulík, Š.; Šolc, M.; Palfy, P. The Indentation Size Effect (ISE) of Metals. *Crystals* **2022**, *12*, 795. [CrossRe
25. Gong, J.; Wu, J.; Guan, Z. Examination of the Indentation Size Effect in Low-Load Vickers Hardness Testing of Ceramics. *J. Eu Ceram. Soc.* **1999**, *19*, 2625–2631. [CrossRef]
26. Li, N.; Liu, L.; Zhang, M. The Role of Friction to the Indentation Size Effect in Amorphous and Crystallized Pd-Based Alloy. *Mater. Sci.* **2009**, *44*, 3072–3076. [CrossRef]
27. Li, H.; Bradt, R.C. The Microhardness Indentation Load/Size Effect in Rutile and Cassiterite Single Crystals. *J. Mater. Sci.* **1993**, 2 917–926. [CrossRef]
28. Chuah, H.G.; Ripin, Z.M. Quantifying the Surface Roughness Effect in Microindentation Using a Proportional Specimen Resistanc Model. *J. Mater. Sci.* **2013**, *48*, 6293–6306. [CrossRef]
29. LaGraff, J.R.; Gewirth, A.A. Nanometer-Scale Mechanism for the Constructive Modification of Cu Single Crystals and Alkanethic Passivated Au(111) with an Atomic Force Microscope. *J. Phys. Chem.* **1995**, *99*, 10009–10018. [CrossRef]
30. Maharana, H.S.; Ashok, A.; Pal, S.; Basu, A. Surface-Mechanical Properties of Electrodeposited Cu-Al2O3 Composite Coating an Effects of Processing Parameters. *Metall. Mater. Trans. A* **2016**, *47*, 388–399. [CrossRef]
31. Guglielmi, N. Kinetics of the Deposition of Inert Particles from Electrolytic Baths. *J. Electrochem. Soc.* **1972**, *119*, 1009. [CrossRef
32. Li, Y.J.; Zhang, X.Z.; Zhi, C.C. Kinetics of Ni/Nano-SiO2 Codeposition on the Sintered NdFeB Surface. *Strength Mater.* **2021**, 5 134–144. [CrossRef]
33. Dordsheikh Torkamani, A.; Velashjerdi, M.; Abbas, A.; Bolourchi, M.; Maji, P. Electrodeposition of Nickel Matrix Composit Coatings via Various Boride Particles: A Review. *J. Compos. Compd.* **2021**, *3*, 91–98. [CrossRef]
34. Kumar, N.; Kishore, K.; Yadav, S.; Sharma, P. Characterisation of Ni-Al2O3 Composite Coatings at Different Al2O3 Concentration *Mater. Today Proc.* 2024; *in press*. [CrossRef]
35. Allahkaram, S.R.; Golroh, S.; Mohammadalipour, M. Properties of Al2O3 Nano-Particle Reinforced Copper Matrix Composit Coatings Prepared by Pulse and Direct Current Electroplating. *Mater. Des.* **2011**, *32*, 4478–4484. [CrossRef]
36. Lim, J.D.; Susan, Y.S.Y.; Daniel, R.M.; Leong, K.C.; Wong, C.C. Surface Roughness Effect on Copper–Alumina Adhesior *Microelectron. Reliab.* **2013**, *53*, 1548–1552. [CrossRef]
37. Yang, J.; Huang, Y.; Xu, K. Effect of Substrate on Surface Morphology Evolution of Cu Thin Films Deposited by Magnetror Sputtering. *Surf. Coatings Technol.* **2007**, *201*, 5574–5577. [CrossRef]
38. Chen, M.; Gao, J. The Adhesion of Copper Films Coated on Silicon and Glass Substrates. *Mod. Phys. Lett. B* **2000**, *14*, 103–10 [CrossRef]
39. Kim, M.; Sun, F.; Lee, J.; Hyun, Y.K.; Lee, D. Influence of Ultrasonication on the Mechanical Properties of Cu/Al2O3 Nanocom posite Thin Films during Electrocodeposition. *Surf. Coatings Technol.* **2010**, *205*, 2362–2368. [CrossRef]
40. Yan, Y.-F.; Kou, S.-Q.; Yang, H.-Y.; Shu, S.-L.; Qiu, F.; Jiang, Q.-C.; Zhang, L.-C. Ceramic Particles Reinforced Copper Matri Composites Manufactured by Advanced Powder Metallurgy: Preparation, Performance, and Mechanisms. *Int. J. Extrem. Manu* **2023**, *5*, 032006. [CrossRef]
41. Thiemig, D.; Osborne, S.; Sweet, W.; Talbot, J. Electroplating of Copper-Alumina Nanocomposite Films with an Impinging Je Electrode. *ECS Trans.* **2008**, *11*, 35–44. [CrossRef]
42. Lamb, V.A.; Johnson, C.E.; Valentine, D.R. Physical and Mechanical Properties of Electrodeposited Copper. *J. Electrochem. Soc* **1970**, *117*, 381C. [CrossRef]
43. Tao, S.; Li, D.Y. Tribological, Mechanical and Electrochemical Properties of Nanocrystalline Copper Deposits Produced by Puls Electrodeposition. *Nanotechnology* **2006**, *17*, 65–78. [CrossRef]
44. Sasi Maoloud Mohamed, S.; Nikolić, N.D.; Vuksanović, M.M.; Vasilić, R.; Vasiljević-Radović, D.G.; Jančić Heinneman, R.M. Marinković, A.D.; Mladenović, I.O. Hardness and Wettability Characteristics of Electrolytically Produced Copper Composite Coatings Reinforced with Layered Double Oxide (Fe/Al LDO) Nanoparticles. *Coatings* **2024**, *14*, 740. [CrossRef]
45. Tey, E.; Hashim, M.; Ismail, I. Characterization of Cu-Al2O3 and Ni-Al2O3 Nanocomposites Electrodeposited on Copper Substrate *Mater. Sci. Forum* **2016**, *846*, 471–478. [CrossRef]
46. Magagnin, L.; Maboudian, R.; Carraro, C. Adhesion Evaluation of Immersion Plating Copper Films on Silicon by Microindentatior Measurements. *Thin Solid Films* **2003**, *434*, 100–105. [CrossRef]
47. Lesage, J.; Chicot, D. Models for Hardness and Adhesion of Coatings. *Surf. Eng.* **1999**, *15*, 447–453. [CrossRef]
48. Hsieh, C.L.; Tuan, W.H. Elastic Properties of Ceramic–Metal Particulate Composites. *Mater. Sci. Eng. A* **2005**, *393*, 133–139 [CrossRef]
49. Hashin, Z.; Shtrikman, S. A Variational Approach to the Theory of the Elastic Behaviour of Multiphase Materials. *J. Mech. Phys Solids* **1963**, *11*, 127–140. [CrossRef]

50. Wang, S.; Feng, L.; Liu, H.; Sun, T.; Zhang, X.; Jiang, L.; Zhu, D. Manipulation of Surface Wettability between Superhydrophobicity and Superhydrophilicity on Copper Films. *ChemPhysChem* **2005**, *6*, 1475–1478. [CrossRef]
51. Samal, P.; Vundavilli, P.R.; Meher, A.; Mahapatra, M.M. Recent Progress in Aluminum Metal Matrix Composites: A Review on Processing, Mechanical and Wear Properties. *J. Manuf. Process.* **2020**, *59*, 131–152. [CrossRef]
52. Cardoso, J.T.; Garcia-Girón, A.; Romano, J.M.; Huerta-Murillo, D.; Jagdheesh, R.; Walker, M.; Dimov, S.S.; Ocaña, J.L. Influence of Ambient Conditions on the Evolution of Wettability Properties of an IR-, Ns-Laser Textured Aluminium Alloy. *RSC Adv.* **2017**, *7*, 39617–39627. [CrossRef]
53. Argyris, D.; Ashby, P.D.; Striolo, A. Structure and Orientation of Interfacial Water Determine Atomic Force Microscopy Results: Insights from Molecular Dynamics Simulations. *ACS Nano* **2011**, *5*, 2215–2223. [CrossRef] [PubMed]
54. Li, W.; Jin, Y.; Gu, J.; Zeng, Z.; Su, X.; Xu, J.; Guo, B. Critical Surface Characteristics for Coating Adhesion and Friction Behavior of Aluminum Alloys after Laser Cleaning. *J. Mater. Process. Technol.* **2024**, *332*, 118549. [CrossRef]
55. Tong, W.; Cui, L.; Qiu, R.; Yan, C.; Liu, Y.; Wang, N.; Xiong, D. Laser Textured Dimple-Patterns to Govern the Surface Wettability of Superhydrophobic Aluminum Plates. *J. Mater. Sci. Technol.* **2021**, *89*, 59–67. [CrossRef]
56. Gun'ko, V.M.; Yurchenko, G.R.; Turov, V.V.; Goncharuk, E.V.; Zarko, V.I.; Zabuga, A.G.; Matkovsky, A.K.; Oranska, O.I.; Leboda, R.; Skubiszewska-Zięba, J.; et al. Adsorption of Polar and Nonpolar Compounds onto Complex Nanooxides with Silica, Alumina, and Titania. *J. Colloid Interface Sci.* **2010**, *348*, 546–558. [CrossRef]

MDPI AG
Grosspeteranlage 5
4052 Basel
Switzerland
Tel.: +41 61 683 77 34

Gels Editorial Office
E-mail: gels@mdpi.com
www.mdpi.com/journal/gels

www.ingramcontent.com/pod-product-compliance
Lightning Source LLC
Chambersburg PA
CBHW061833210926
51850CB00006B/878
* 9 7 8 3 7 2 5 8 5 4 2 9 5 *